MAKING KNOWLEDGE IN EARLY MODERN EUROPE

MAKING KNOWLEDGE IN EARLY MODERN EUROPE

Practices, Objects, and Texts, 1400–1800

Edited by
PAMELA H. SMITH
and BENJAMIN SCHMIDT

THE UNIVERSITY OF CHICAGO PRESS
CHICAGO AND LONDON

The University of Chicago Press, Chicago 60637
The University of Chicago Press, Ltd., London

Printed in the United States of America

25 24 23 22 21 20 19 18   3 4 5 6

ISBN-13: 978-0-226-76328-6 (cloth)
ISBN-13: 978-0-226-76329-3 (paper)
ISBN-10: 0-226-76328-5 (cloth)
ISBN-10: 0-226-76329-3 (paper)

LIBRARY OF CONGRESS
CATALOGING-IN-PUBLICATION DATA
Making knowledge in early modern Europe: practices, objects, and texts, 1400–1800 / edited by Pamela H. Smith and Benjamin Schmidt.
p. cm.
Includes bibliographical references and index.
ISBN-13: 978-0-226-76328-6 (cloth : alk. paper)
ISBN-13: 978-0-226-76329-3 (pbk. : alk. paper)
ISBN-10: 0-226-76328-5 (cloth : alk. paper)
ISBN-10: 0-226-76329-3 (pbk. : alk. paper)
1. Knowledge, Theory of. I. Smith, Pamela H., 1957– II. Schmidt, Benjamin.
BD161.M153 2007
001.2094—dc22

2007034835

♾ The paper used in this publication meets the minimum requirements of the American National Standard for Information Sciences—Permanence of Paper for Printed Library Materials, ANSI Z39.48-1992.

CONTENTS

ILLUSTRATIONS

PLATES (APPEARING AFTER PAGE 60)

FIGURES

ACKNOWLEDGMENTS

In processes of making knowledge, a productive combination of resources, intellectual *and* material, is always necessary. In the making of this volume, the intellectual resources are readily evident, we believe, in the individual essays, and we are immensely grateful to all the contributors for sharing their ideas and enthusiasms with us, as well as to those who took part in the original workshop, "Knowledge and Its Making." We would also like to thank the three anonymous readers for the University of Chicago Press, who offered rigorous and highly productive suggestions for revisions; and we are pleased to thank Susan Bielstein and Anthony Burton of the Press for so expertly ushering this book into production. The material resources, less evident in this volume but no less real, were generously provided by the Waldemar and Mabel Westergaard Fund of Pomona College. We are grateful also to the Department of History of Pomona College and to Martina Ebert of the European Union Center of the Claremont Colleges for supporting the workshop and this volume.

INTRODUCTION

Knowledge and Its Making in Early Modern Europe

PAMELA H. SMITH AND BENJAMIN SCHMIDT

We live in an age of exceptional knowledge making. Our postmodern world produces a surfeit of available knowledge, as we are regularly reminded, and this has caused a crisis of "information overload," in the argot of the day. Making knowledge, in many ways, has never been easier, and the quantity of accessible knowledge has never been greater. And if there are some who carp over their restricted access to certain categories of knowledge, the more common complaint describes a world awash with superabundant, perhaps even superfluous, floods of information. Inhabitants of the twenty-first century, especially those living in the postindustrialized West, possess ample means to know things, more so perhaps than at any other time in history. Knowledge is created and distributed in ever-expanding and far-reaching ways: information "superhighways" clogged with data; burgeoning "new media" spawned by ever-productive technological breakthroughs; mountains of cheaply produced books, struggling to be contained by the groaning shelves of commercial bookstores or in the dazzling, electronically sophisticated libraries that have lately proliferated. All of these developments have had a profound effect on knowledge making. It is a state of affairs that has even caused a putative crisis. To produce and consume knowledge has never been simpler; to access and circulate knowledge has never been more efficient; to digest and process knowledge, thus, has never been more challenging. In short, we fear we know too much.

Or so, at least, runs the cliché du jour. It is bracing in this regard to hear complaints of another age and place, in this case early modern Europe at the dawn of that great epoch of knowledge making, the Enlightenment. This earlier crisis is identified with hardly less urgency, the *cri de coeur* sounded by an otherwise sober scholar and a cleric of wide repute, Adrien

Baillet. Addressing the problem of what might be called "data management" at the twilight of the seventeenth century, Baillet discerns the same abundance of information and voices a strikingly similar fear of drowning in the tides of knowledge—for Baillet's peers, the torrential inundation of books in particular: "We have reason to fear that the multitude of books which grows every day in a prodigious fashion will make the following centuries fall into a state as barbarous as that of the centuries that followed the fall of the Roman Empire. Unless we try to prevent this danger by separating those books which we must throw or leave in oblivion from those which one should save and, within the latter, between the parts that are useful and those which are not."[1] It is the medium of print that this hardworking scholar fingers as the culprit—a concern that was addressed almost exactly a century earlier, it might be added, by the imaginative and ingenious "book wheel," which was proposed in Agostino Ramelli's *Le diverse et artificiose machine del Capitano Agostino Ramelli* and was meant to manage the ample printed data of the day (ca. 1588, or just over a century after the earliest proliferation of printed materials in Europe; see fig. I.1). It is worth noting, furthermore, that Baillet fears not for his contemporaries' sake but for pursuers of knowledge in "the following centuries" (and, thus, for our sake), and in doing so, he has keenly anticipated the debates of our lately branded "crisis" of knowledge.

There are significant distinctions of technology and hermeneutical habit that separate early modern and late modern producers and consumers of knowledge, yet in an important sense, the questions of the former period can speak directly to the latter. Our current crisis of knowing illustrates numerous telling patterns, which suggest a critical need for historical perspective. Along with Baillet, we tend to focus overwhelmingly on shifts from old to new conduits of knowledge. We struggle mightily to construct proper metaphors even to describe the matrices these agents of data traverse: super-"highways," inter-"nets," global "webs," personal "logs." In all cases, the idea remains that the old (read "print") medium has been supplanted by the new, yet this assumption fails to account for the many rich ways in which knowledge has traditionally been made—and, for that matter, the myriad manners in which it is still produced today. Debates on making and consuming information have tended to privilege certain instruments of knowledge—books and bookishness, for example, in the earlier period and scientific instruments in a slightly later period—and this has narrowed our perspective on the enormous range of ways in which making knowledge has taken place and continues apace. More generally, studies of knowledge and its making assume an expansion of

Fig I.1. Book wheel. From Agostino Ramelli, *Le diverse et artificiose machine del Capitano Agostino Ramelli* (Paris, 1588).

consumption—an explosion of available materials, which induces debates over crises of information management—yet without adequately studying the expansions of production that make the new age of information possible. We are in the habit of identifying the end products of knowledge making—texts and books, data and ideas—rather than the manner and means of their production.

Attention to the active production of knowledge, however—knowledge and its making, as we have termed it—can be enormously rewarding. It can point to a whole other roster of historical themes and questions, which move beyond the surfeit of objects of knowledge. Knowledge is made abundantly, yet it is also made in abundant ways. "Just how, then, are things known today?" asks Daniel Brewer in a provocative essay that explores the shifts in epistemology that took place in the literary Enlightenment.[2] To this "massive question," as Brewer frames it, we would add another, one that addresses how things are *made* known and that highlights the fascinating history of both of these queries. It is our purpose in this volume to explore the many ways knowledge was produced and consumed in early modern Europe and to do so in innovative ways. Indeed, a common misstep taken in the approach to early modern knowledge and its generation is the instinctive convergence on print media to the exclusion of other sources. This volume, which includes contributions from a wide span of disciplines, follows Nick Jardine and Marina Frasca-Spada in interrogating the primacy given to book learning and texts in the process of knowledge making.[3] Shifting away from a wholly (or even primarily) bibliographic approach makes double sense. First, we now have a much better sense of how books function as media of knowledge and just how "unstable" (to borrow from Adrian Johns) printed texts can be; this diminishes their face value.[4] And, second, we have sought to move well beyond the traditional, exclusive focus on textual knowledge and on to a keener appreciation of the myriad ways that Europeans of the early modern period gained, disseminated, and practiced knowledge. This volume, accordingly, seeks to expand the range of sources and methods that are implicated in early modern knowledge making. *Making Knowledge in Early Modern Europe* focuses above all on active processes of knowledge production. These are commonly social, and they involve the agency not only of historical actors but also of objects and material culture. This does not mean that we neglect such products of knowledge making as books and ideas. Indeed, it is this volume's attention to the active nature of making knowledge *and* to the ways this can revise our views of the products of knowledge that we believe to be its distinctive contribution to the field.[5]

In our conception of knowledge and its making, our nets have been cast widely. Consider, for example, the following instructions for metalworking and stone engraving, which derive from the pseudonymous author of *De diversis artibus,* Theophilus (in all probability, the monk Roger of Helmarshausen):

> If you want to carve a piece of rock crystal, take a two- or three-year-old goat and bind its feet together and cut a hole between its breast and stomach, in the place where the heart is, and put the crystal in there, so that it lies in its blood until it is hot. At once take it out and engrave whatever you want on it, while this heat lasts. When it begins to cool and become hard, put it back in the goat's blood, take it out again when it is hot, and engrave it. Keep on doing so until you finish the carving. Finally, heat it again, take it out and rub it with a woolen cloth so that you may render it brilliant with the same blood.[6]

This recipe for carving rock crystal was penned in the twelfth century. Four hundred years later the anonymous German author of a freshly printed treatise on metalworking, *The True Use of Alchemy,* offered a strikingly similar formula for cutting and engraving crystal and precious stones:

> Take goose and goat blood and dry it until it is hard. Take crystal or any stone. Pound the blood to powder, pour ashes on it. Let it mix well in a container. Mix in strong vinegar, then lay the stone in it, warm it a little. And the stone allows you to cut or form it as you want. Throw it in cold water and it will become hard again in an hour.[7]

The fact that these recipes remained constant over four hundred years and perhaps longer, despite their unlikely efficacy for softening stones, points to a number of interesting issues and important questions on the methods of knowledge making in the premodern period. We note, first, the persistence of certain knowledge in the face of obvious technological change: the relevance of an older recipe in the newer medium of delivery, namely print. It is also worth highlighting the status of this brand of knowledge, contained as it is in anonymous texts and the form of recipes. This prompts several further questions: To whom were these texts directed, and who may have read them? What relationship do these recipes have to actual processes of carving stones? Are they meant for the workers of real objects, or do they have a less material significance? What relationship does the knowledge contained in these books have to the discipline of natural philosophy as actually practiced at the time? And what accounts for the persistence of such recipes? Perhaps the advent of print motivated the replication of recipes, or perhaps the endurance of these recipes points to an underlying philosophy of nature or a set of principles within which metalworking artisans organized their knowledge. All of which begs a final question: what was the ultimate function of this knowledge?

These questions, which could easily be multiplied and expanded, provide an example not only of the difficulties but also the suggestive potential of attempting to connect objects, practices, texts, and ideas. Whereas such recipes might once have been relegated by scholars to the margins (cited as proof of the irrationality of natural philosophy before the advent of modern science, or of the primitive level of artisanal expertise in preindustrial Europe), recent approaches to the study of material culture, to the history of the book, and to analysis of "local" knowledge, as well as new methodologies in the history of science that direct attention to creative practices and indigenous knowledge, have dictated a closer look at phenomena previously dismissed. These new approaches suggest, in short, that we seek out new places to explore and fresh ways to understand knowledge-making practices of the past.

Our aim in this volume is to investigate the production and consumption of knowledge in early modern Europe in light of these new scholarly practices. The remarkable colonial expansions, commercial innovations, technological advances, and religious reformations that took place in Europe from about 1400 to 1800 brought about striking changes in material culture, religious practice, and political organization. We wish to explore how processes of making and using knowledge were changed by these events. The questions that motivate our inquiry are deceptively simple: How did people go about obtaining knowledge during the critical period of intellectual and cultural transformation that took place in early modern Europe? How did they demonstrate or prove that knowledge once obtained, and how did they convince others of the truthfulness, or validity, of their knowledge? How did they practice or perform their knowledge, and how, ultimately, did they disseminate it? It is the simple questions that are the hardest, of course; and while they may appear enticingly straightforward, these questions are anything but. They are vast in scope, and their disciplinary demands are certainly too broad to be covered in any single volume. Our goals, accordingly, are exploratory, and our collective approach places an emphasis on multidisciplinarity. In bringing together these essays, we aim less to provide definitive or comprehensive answers than to reframe questions. By collecting essays from diverse fields, we hope to bring together a number of new approaches to the study of knowledge and knowledge making in this critical moment in European history.

Our title, *Making Knowledge in Early Modern Europe,* points to our interest in action: in productive ways of knowing, not simply passive bodies of information, and in the active making of knowledge. We wish to gain a better understanding of how early modern "practitioners" acquired, authenticated, assimilated, disseminated, and represented knowledge in the period ranging from the voyages of European discovery to the eve of the Industrial Revolution. The first point of intersection among the essays, then, is their unified focus on practices of making and methods of producing knowledge. These range widely: from painting to printing, from reading to writing, and from building to collecting. The work of anthropologists on cultural practices and of scholars of religion on devotional habits is evident in many of the contributions to this volume, as are literary critics' interests in the mechanics of genre. Some of the essays add a further layer to the study of "practice" by relating habits of knowledge making to those objects enlisted in the production of knowledge. This marks a departure. Scholars have commonly studied material objects with a primary concern for the intellectual work that produced them. But over the last two decades art history, for example, has begun to move away from its earlier approach to visual materials, which was almost solely in terms of the concepts represented within them. Recent works of theorists and historians of material culture, in particular, have offered new models for the study of material objects, and revelatory technical analyses of paintings (most stunningly, by means of infrared reflectography) have caused wholesale revisions in the history of art.[8] Such approaches have begun to make their way into the mainstream and will do much to expand our horizons on what constitutes knowledge and how it is made.

Taking its cue from studies in distributed cognition, this volume also investigates the production of knowledge as a social process that includes many different communities of practitioners. One of our aims is to show how these different communities and their various forms of knowledge and practice intermingled in this period. Essays explore the interaction, for example, between natural philosophers and artisans in the making of scientific knowledge, and between natural philosophers and indigenous peoples in the production of botanical knowledge.[9] In bringing these essays together, we want to emphasize the shared and collective nature of knowledge making: the communications between different modes of cognition and between different strata of society. We have tried to do so in the terms of the early modern world instead of those of our modern world in the hope of breaking down the traditional disciplinary boundaries that have commonly occluded the practice of knowledge in modern scholarship. And to this end, we have incorporated in this volume the work of anthropologists, sociologists, historians of all kinds (of science, especially), and scholars of literature, art, and religion. This emphasis on the social dimensions of knowledge making leads to the final point of convergence among these essays: they point the way toward new narratives about the development of knowledge in the early modern period, histories that bring out the contingency and complexity of making and legitimating knowledge. There is no inevitable march of secularization or drama of scientific progress in these pages.

Making Knowledge from the Margins

These essays range over issues of objects and material culture, over practices of local and universal knowledge, and over the history of texts and books. While many essays speak to more than one of these themes, we have attempted to divide them in ways that highlight the new approaches they incorporate. In the first section, "Making Knowledge from the Margins," the authors find knowledge being generated in places where scholars have not previously looked, and they identify hitherto-unnoticed activity that is certainly worthy of the status of "knowledge." This section deals with producers and objects of knowledge that have conventionally been seen as marginal to the knowledge-making process, or even ignored altogether. It includes some of the most recent approaches to the analysis of local and indigenous knowledge.

In her essay on the construction of the Canal du Midi, one of the great technological marvels of seventeenth-century Europe, Chandra Mukerji examines a specific, and plainly overlooked, manifestation of material culture and indigenous knowledge. Her research has recovered the previously unnoticed role played by a contingent of women who worked on the canal. Native inhabitants of the region through which the canal would run, these women offered local knowledge in the most literal sense by filling in where the research and measurements of formal engineering were not precise enough to produce the complicated system for supplying the canal with water. She concludes that the indigenous knowledge of these Pyrenean women—the expertise acquired by building and repairing local irrigation ditches, by constructing public laundries in streams, and by other, commonly domestic activities—was critical to the canal project. The local knowledge that went into the canal is still visible today in the physical evidence of the locks and irrigation ditches, as well as in the washing stones and public laundries of the region. Mukerji's focus on local knowledge draws on recent studies in the field, but it pushes our notion of indigenous knowledge even further by discovering it embedded in the very objects—stones, streams, patterns of masonry—of the locality.[10] Her study demonstrates, moreover, the way in which a large-scale technical undertaking—in this case, the building of Louis XIV's Canal du Midi—nearly always reveals the workings of distributed cognition; it is not simply the product of its designers or its chief architects.[11]

Linda Seidel also asks us to reconsider the making of a masterpiece, in this case, the work of the incomparable Netherlandish painter Jan van Eyck—an experimenter and innovator of Leonardo-like imagination. Seidel quite inventively compares the master's *Ghent Altarpiece* to ideas in Cennini's handbook for painters and concludes that van Eyck regarded his celebrated technique as a means to investigate the natural and human world. Painting was for him a way of knowing the world. Seidel's approach—viewing works of art and craft techniques as epistemological activities that both articulate and demonstrate a practitioner's knowledge of nature and the cosmos—reflects some of the most innovative studies lately conducted in the history of art and technology.[12] It also directs scholars toward the materiality of visual culture and toward ways of doing and knowing in the artist's atelier.

By focusing on the relation between local knowledge and universal science (to use that word anachronistically), the two other essays in this section each probe the circumstances by which knowledge comes to be regarded as "scientific."[13] Both contributions challenge the traditional story of the rise of modern science, which asserts the growth of scientific knowledge as an inevitable uncovering of truth and argues for an overly strict separation between artisans and practitioners, on the one hand, and so-called scientists, on the other. Simon Werrett surveys the interaction among European fireworks makers, gunners, and savants in a period spanning the seventeenth and early nineteenth centuries. His essay demonstrates the contingency of the alliance between pyrotechny and chemistry. Such a joining of fields emerged only in the eighteenth century and not in the manner one would be led to expect by standard histories of chemistry. The union of chemistry and pyrotechny was not the result of the inexorable progress of scientific method or the superiority of chemical theory in making fireworks but rather the outcome of local, social controversy caused by the clash of two different orders of pyrotechnic knowledge and production. One group of pyrotechnists took up chemistry as part of their struggle to maintain distinction in the competitive world of fireworks making. Werrett's essay points to the diverse communities and multiple ways of knowing that went into the production of pyrotechnic knowledge and into its final alliance with chemistry. Londa Schiebinger's contribution also deals with the contingency of creating what came to be seen as universal scientific knowledge, in this case, Linnaeus's system of nomenclature. Schiebinger illustrates how the practice of imperial natural history shaped botanical knowledge in the eighteenth century.[14] Her essay makes clear, moreover, the manner in which the language of botanical description that became standard at this time inscribed a somewhat inaccurate vision of science; Linnaean nomenclature, that is, credited European individuals (mostly men) with botanical discoveries while ignoring the contributions of local knowledge and indigenous knowledge systems. Both Werrett's and Schiebinger's essays illustrate the ways in which the mechanics of knowledge production, almost always a collaborative project, must be viewed as a part of larger social dynamics.

Practices of Reading and Writing

The second section of the volume, "Practices of Reading and Writing," brings together essays that emphasize literary practices while focusing also on books as objects. Recent research on the history of the book has forever changed our naïve assumptions about how reading took place in the past. Books were handled and digested differently in the early modern period, and "texts" hardly influenced readers in the straightforward ways critics would have us believe. Research on books as objects can tell us much about material culture—about the production and consumption of books as commodities and about the acquisition and circulation of texts—but it still does not give much insight, as Arianne Baggerman's essay reminds us, into how people read or what they derived from their reading.[15] All five essays in this section deal with practices of reading and writing and with the production and consumption of textual knowledge. All contend with the basic question of how individuals (or communities of readers) made knowledge from books.

Herman Pleij's sweeping essay takes up nearly all of the themes in this volume—through the medium of print. He explores the manner in which the growing cities of the Low Countries in the late fifteenth and sixteenth centuries ushered in new ways of knowing and being. Urban centers, especially in the thickly populated provinces of Flanders and Brabant, were sites of innovation and manufacturers of novelties; and print was the newfangled medium—the next big thing—that circulated so many of the novelties. Inhabitants of the intensely citified Netherlands, Pleij argues, learned to read from local book publishers, who were all too eager to repackage old knowledge in the new, salable form of print. This fostered certain attitudes among readers toward individual agency, *curiositas,* and novelty itself. The vernacular literature written and read (aloud) by the *rederijkers,* or members of the chambers of rhetoricians, gave particular voice to such attitudes. Pleij demonstrates how urban life generated new audiences for the commodified knowledge of printers, which in itself gave rise to new practices of reading and writing. These new practices, in turn, had the effect of articulating and inculcating ever more deeply the new alignment of social and cultural attitudes.

In Rudolf Dekker's essay on the simultaneous rise of autobiographical writing and timekeeping, individual agency is once again instrumental. Dekker's contribution explores innovative practices of record-keeping in the seventeenth century: recording the self and recording time. It also identifies a "chronometric" sensibility that flourished in northern Europe around this time, which might be linked, Dekker suggests, to the new scientific experimentation taking place in the Netherlands and England. Autobiographical writing and other novel forms of record-keeping arose partly out of a new urban sociability in the early modern period, partly out of fresh religious practices born of the Reformation, and partly out of novel methods of legitimating information from overseas (data pertaining, in other words, to lands unknown to the classical *auctores*).[16] The self is central to Dekker's inquiry, as it is to Arianne Baggerman's essay, which examines pedagogic practices in the late eighteenth century. Concentrating on printed children's literature and certain relevant diary materials, Baggerman demonstrates how powerful autobiographical documents can prove for historians of reading practices. Such documents shed light on, for example, readers' responses to J. F. Martinet's *Katechismus der natuur* (Catechism of the Natural World), an eighteenth-century Dutch teaching volume that exhorted its audience to forsake the printed text—Martinet's catechism—and read the book of nature in its stead. The enthusiastic response of ordinary readers to this call for "active" reading—to make knowledge by investigating nature in situ—gives important insight into attitudes toward science and natural knowledge, attitudes that, if first developed in the early modern period, would continue to resonate into the nineteenth and twentieth centuries.[17]

The following two essays also treat practices of reading and writing, yet in a somewhat different fashion. Both concern the production of knowledge through practices of literature and highlight ways in which form or method can function not only to convey knowledge but also to constitute knowledge itself. Lori Anne Ferrell shows in her essay how the late-sixteenth-century scheme for shorthand developed by Timothy Bright exhibits many of the same aims as contemporary Calvinist pedagogy. The cumbersome and seemingly unworkable shorthand cultivated "judgment, memory, and dexterity of hand," as well as offering a universal language; and it was thus in line with both secular and religious pedagogical schemes of a more explicit sort. Ferrell shows how cognitive aims could go hand in hand with the development of manual dexterity and suggests that Puritan plain style involved a rather complex mix of visual and kinetic argumentation. In this sense, Bright's shorthand constitutes, not a method, but the message itself, analogous to logical schemes, such as Peter Ramus's "flow charts," that were supposed not only to organize knowledge but also to create knowing.[18]

Scott Black's essay reminds us that reading begins with writing and that, however active readers may become, authors endeavor to shape the parameters of readers' activity.[19] Black follows the evolution of the essay as a genre that moves "from a practice of humanist moral philosophy to an instrument of experimental natural philosophy." Using as his case study Robert Boyle's "Proemial Essay" and Boyle's adoption of the essay form in particular as a way to write about natural philosophy, Black argues that choices of genre matter greatly in the making of knowledge. Boyle's use of the generally brief, purposefully private, and highly flexible form of the essay was neither accidental nor unimportant; the essay served as the perfect medium for the occasional thoughts or leisurely reflections that Boyle wished to convey.[20] The essay was the form best suited to discuss the particular and unfinished; it was a generic option that perfectly conveyed ongoing knowledge making rather than previously completed and conceptualized systems of knowledge. Boyle selected the essay for his natural philosophical efforts because of assumptions that he and his early modern readers shared about this genre's particular efficacy to constitute "ways of knowing."

The Reform of Knowledge

Reform pervaded virtually all corners of society and all aspects of culture in the early modern period. Along with the church, myriad other institutions, from universities to guilds (especially, though not exclusively, that of physicians), were swept up in the effort to effect world-changing renovations. Practices of devotion, methods of philosophizing, and patterns of pedagogy were also drawn into the surge of reformation, as the very foundations of Europe's religious houses were restructured. Such reforming instincts drew their energies in part from eschatological worldviews and from the very palpable expectation, shared by many in early modern Europe, that the Last Days were imminent. And although such chiliasm faded in the final years of the seventeenth century, this occurred only as individuals turned inward toward habits of personal reform. The study of nature was often implicated in these personal reformations, as, for example, in seeing nature as giving access to God's original revelation in the book of nature or by locating in the natural world certain ancient wisdoms of divine significance. The study of nature was seen to provide new models of philosophizing and to offer a common ground, or even a transcendent reality, for intellectuals who had been torn apart by confessional divisions. Natural knowledge also held the potential of productive activity and of commodity goods, not to mention the distraction of wondrous *naturalia* and the experiments that nature's marvels enabled. The movement for reform was pervasive. It encompassed so many areas of life in this period that the final section of this volume, "The Reform of Knowledge," can only scratch the surface of this phenomenon, and the five essays that conclude this collection have been selected to highlight the different realms in which the reform of knowledge manifested itself.

One of the most important components of reform in the early modern period was a new valorization of sensory-derived knowledge. The senses and personal experience came to form the bedrock of the so-called new philosophy and of the "new method of philosophizing" in the sixteenth and seventeenth centuries. This prompted, in turn, an intensified concern with the reliability of the senses and the experiential knowledge that was founded upon them.[21] This final section of the volume examines some of these new valuations of the senses. How precisely do the senses convey knowledge? How does the mind know the world at all through the senses? Aristotle, too, had struggled with this question, positing a *sensus communis* in which the five commonly accepted senses flowed together and were judged by inner senses. The five senses were only the external ones, to be contrasted with the internal senses, which included fantasy, common sense, the faculties of estimation and cogitation, and memory.

Claudia Swan and Ole Peter Grell focus on two different ways in which knowledge was legitimated by appeal to the senses in the seventeenth century. Swan's essay investigates the incorporation of *sinnelickheden* (roughly, "sensualities") into the collections of three leading medical practitioners of seventeenth-century Holland. Her essay brings to light the sheer variety of objects contained in these assemblages and the multitude of meanings these objects could possess.[22] The artifacts gathered in these collections were associated, on the one hand, with sensuality, the body, and a sense of luxury; on the other hand, they were affiliated with sensory observation and the scholarly collection of information. The distinction between sensuality and sensory observation could be a fine one at this time, as studies of Dutch still lifes and the body have begun to show.[23] Swan goes on to demonstrate the mechanics of knowledge making by means of object collecting and concludes by showing how an appreciation of the multivalent significance of natural and artificial wonders is necessary to understand the meaning and function of early modern medical collections.

The theme of collecting is also taken up in the contribution of Ole Peter Grell on the great Danish physician and polymath Ole Worm. Worm was among the outstanding assemblers of objects and knowledge in his day (the first half of the seventeenth century), and he was also a passionate proponent of the new philosophy. The evidence of objects made knowledge tangible and hence reliable, he believed. He perceived the collecting and dispersing of knowledge (he was also a wide-reaching correspondent, as Grell demonstrates) as a vehicle of reform. This is seen both in Worm's fascination with the Rosicrucian Brethren and in his initial rejection (and eventual acceptance) of William Harvey's theory of the circulation of the blood on the grounds that it did not tally with the evidence of the senses. Worm's response to Harvey demonstrates how crucial a commitment to visual observation had become in the assessment of new knowledge by the early seventeenth century. Grell's essay provides a view into the mechanics of making knowledge during the Scientific Revolution by charting the breadth of Worm's correspondence and reviewing his extensive journals. Taken together, these sources provide superb insight into processes by which new knowledge was gathered, disseminated, and evaluated at the height of this Baconian moment of experimentation, and they indicate what was at stake for individuals interested in new natural knowledge and its objects.

Both Carina L. Johnson and Jonathan Sheehan highlight the place of religion and theology in the making of early modern knowledge. Johnson's essay underscores the vital role played by religious debates in the process of reception and interpretation of the New World. The religious context for making knowledge about the New World is an area that has received relatively minimal scholarly attention, and Johnson's essay strikes out in new directions by examining the way in which objects from the New World collected by Habsburg princes became implicated in Counter-Reformation debates about idolatry. The interaction between texts and these objects encouraged Counter-Reformation scholars to redefine their own culture with respect to that of the New World. The reigning wisdom about the similarity of humans in the Old and New Worlds was discarded in the last half of the sixteenth century as debates on idolatry ushered in a new paradigm of cultural difference.

Scholarly endeavor and theological inquiry also merged in the work of those Protestant antiquarians examined by Sheehan. The clerics under review in this essay, which focuses on early modern knowledge of the ancient Jewish Temple, looked not just to the earliest texts but to the earliest devotional practices described in the Hebrew Bible. In the process, they drew all of religion under the microscope, for their gaze shifted from local and confessionally defined understandings of the sacred to more universal ones, which pointed the way to an anthropology of religion. As Sheehan observes, "they helped to make religion itself into a category of scientific study." Sheehan thus shows that increased secularism had it roots, not in the growth of science at this time, but in the "pressures of theological controversy." The "anthropology of religion was produced, not in spite of confessional commitments," he argues, "but precisely because of them."

Not only was the urge to reform felt in the religious sphere, but it also permeated academic exchange. The reform of philosophy occupied scholars of all stripes, and the university, naturally, became a critical site of knowledge reformation. As André Wakefield shows in the final contribution to this collection, in the eighteenth century the new philosophy was employed by the reforming University of Göttingen as a means to increase the territorial fisc. The Hanoverian ministers saw in enlightened education a resource for the treasury; that is, they saw academic knowledge as a kind of industry for drawing money into the flagging region around Göttingen. There was income to be gained from offering a new scientific curriculum at this German territorial university. Part of the university's mission, then—part of the goal of making knowledge in this quintessential institution of learning—was to make money, "to enlist academic knowledge in the generation of state revenue," as Wakefield puts it. Similarly, Saxon officials established an entirely novel institution of higher learning, a mining academy in Freiberg, which was to be harnessed to serve the fiscal needs of the state. Producing knowledge was intended to be productive in more ways than one.

Wakefield's essay, like the others in this volume, highlights the complex processes by which knowledge was made and cultural change wrought in early modern Europe. Moneymaking, canal building, fire working, timekeeping, idol collecting, plant naming: these and much more, along with such standbys as writing, reading, and painting in their multiple forms, were all part of the varied means of knowing in early modern Europe. Many of these topics have long been overlooked in the study of early modern culture and intellectual history. Many of these processes—the active ways of producing knowledge that we highlight in these essays—have long been portrayed as the result of inevitable intellectual progress and have traditionally been attributed to the spread of science, the secularization of Europe, and the arrival of the Enlightenment. The essays in this volume show them, by contrast, to have been the result of much more contingent, circuitous, and particular forces. The essays also demonstrate how intricate the process of knowledge making could be, and how far and wide scholars need to look in order to locate the many methods of producing knowledge. Early modern Europe, we agree, deserves its reputation as one of the most intellectually explosive and culturally innovative places in Western history. This volume hopes to encourage an even richer and more expansive view of the information overflow of that earlier age and, more broadly, to offer an instructive lesson in the diverse ways that knowledge could be produced and consumed in a particular place and time.

PART I

Making Knowledge from the Margins

CHAPTER ONE

Women Engineers and the Culture of the Pyrenees: Indigenous Knowledge and Engineering in Seventeenth-Century France

CHANDRA MUKERJI

The history of the Canal du Midi is usually told as a story of misunderstood genius. The entrepreneur for the project, Pierre-Paul Riquet, is celebrated for standing up against the vigorous opposition of local nobles and the ridicule of powerful men at Versailles in order to realize a piece of engineering described by some period commentators as a wonder of the world.[1] But the project actually depended on problem-solving capacities that surpassed those of any individual or even any group of experts from the period. It required, not individual genius, but the collective application of diverse forms of tacit, as well as formal, knowledge. Women laborers were among those who brought such expertise on site. How they got there, whom they worked with, and what they contributed help reveal the importance of collaboration to the canal's engineering.

The Canal du Midi itself was a navigational canal built between the 1660s and 1690s in southwestern France. It stretched roughly 150 miles from Toulouse to the Mediterranean north of the Pyrenees and required over a hundred locks to carry the waterway over hills. It was made in a period when elevations were still difficult to measure precisely, particularly over long distances and in varied terrain. Locks were a new invention, too, and never used before in such large numbers and with such depths. Hydraulics—although a subject of interest to academics and discussed in classical texts circulating at this time—did not provide the necessary tools for calculating how much water was needed to keep the highest locks filled or how to maintain the inclines to manage the flow through the canal. In this historical moment, canal engineering was not just the application of contemporary knowledge to a new problem but a matter of developing new techniques to address difficult but crucial problems. The effort was too much for a single genius.

Fig. 1.1. The Canal du Midi in the area of the "second enterprise."

Pierre-Paul Riquet was an unlikely man to manage such a complex intellectual undertaking. He was a salt tax farmer and a wealthy, provincial financier—not an engineer. Because his ambitions ran above his station, his powers were deeply resented by those of better pedigree; many struggled to make his grand dream fail. Still, his success in organizing the construction of the Canal du Midi was unquestionable in the end. So in retrospect, he became a regional hero and man of natural genius.

There is archival justification for representing Riquet this way. He wrote that he had found unexpected intellectual powers in himself as he worked on the project. He wrote Jean Baptiste Colbert, the minister of the treasury, on 27 March 1670: "I am beginning to realize that I am the one who knows more than others. . . . I am at the moment a kind of experimenter, and [designing] new implements is easy beyond my dreams. . . . God has inspired me with thoughts that I consider sacred, so that with them my works will be advanced, and I will not fail."[2] Read out of context, Riquet appears in this letter as either a stunning egoist or a misunderstood man of talent. But he was making a different point when he wrote. In the preceding year, Colbert had put the contract for completing the canal up for bid and was seriously considering awarding the second half of the project to another entrepreneur. The specifications for the locks, drains, and other technical components for this "second enterprise" were prototypes developed under Riquet. Now someone else might use them and cut him out. It was in this context that he claimed authorship of the canal; he was the only one who could finish the job. No other entrepreneur, no matter how socially connected or well educated, had the experience to complete the canal. Colbert worried that Riquet was a dangerous dreamer. Later historians made him out to be the gifted martyr. But he was more precisely a well-placed local man who could assemble regional experts to do the job; he was a gifted intellectual impresario or organizer of social intelligence. His weakness was to confuse collective abilities with his own.

Riquet might have publicly founded his authority to build the Canal du Midi on divine inspiration, but he knowingly took advantage of the abundant local talent among the people of Languedoc.[3] The problem was that Colbert did not trust his ability to judge such things and sent experts from Paris to oversee the work, including an academic engineer, Pons de la Feuille. The latter was asked by the minister to be his eyes and ears for the second enterprise.[4] In this capacity, he did the most surprising thing. He advocated using more women laborers, extolling their abilities as well as lower cost.[5] Exactly what he admired about their work he did not say. But other evidence suggests that they were indigenous canal builders who brought practical experience in measuring and managing elevations and water flow. Their practical methods of routing canals and terracing land could be checked against the mathematical techniques of formal engineers like Pons de la Feuille. Their collaboration helped make the second enterprise possible.

Historians have long recognized that women laborers were used on the canal because Riquet said so explicitly in letters to Colbert. Some, such as L. T. C. Rolt and Michel Adgé, have minimized their numbers but still mentioned their contributions.[6] André Maistre, who has focused more explicitly on the labor process and economic history of the project, has been more generous about the numbers and significance of the women.[7] Still, none of these authors has seriously considered their contributions to the enterprise. Investigating the women's work more closely is a particularly good technique for revealing the social dynamics of collective intelligence used to engineer the Canal du Midi.

Women were deployed disproportionately where the canal or its feeder system cut through uneven terrain. It is likely they came from the Pyrenees, where women traditionally built canals for irrigation, domestic use, and public water supplies. Some laborers were explicitly recruited from Bigorre and Perpignan, towns rich with Pyrenean canals. If these workers included women (and many women arrived on site during major recruitments), they

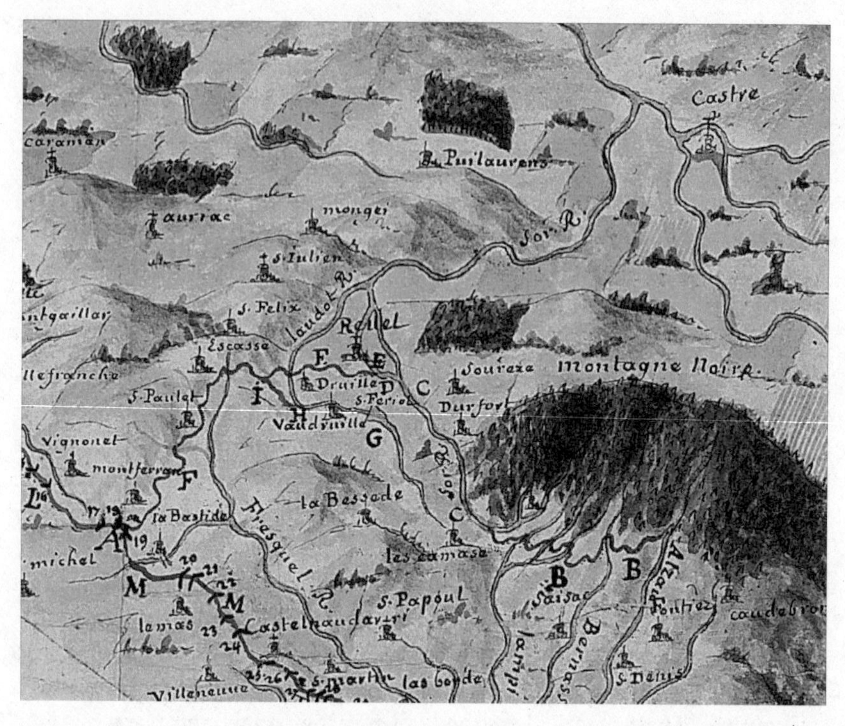

Fig. 1.2. Water system in the Montagne Noire, ca. 1666. Bibliothèque Historique de la Marine, Château de Vincennes, Paris.

would have brought to the Canal du Midi practical experience in building similar waterways in the mountains.[8] Pyrenean women knew how to use contours to direct water to where it was needed. They also diverted river water for their town supplies, using complicated systems of sluices, weirs, settling ponds, and drains to control the quality of the water and reduce the dangers from floods. In addition, they knew how to improve existing streambeds and use them in their water systems. These were all techniques employed along the Canal du Midi and in its water supply.

But how could Pyrenean women (or even peasant men, for that matter) begin to speak about technical matters with educated gentlemen like Pons de la Feuille? There was a powerful cultural, as well as social, barrier to breach. The local language was Occitan (langue d'oc) rather than French. And peasants from Languedoc were notably hostile to the French state and its tax collectors.[9] But when the tax farmer paid peasants by the basket to haul dirt for a dam in the region called the Montagne Noire, women joined the workforce in large numbers. They would have learned there that Riquet spoke Occitan as his native tongue and that he hired locals for their knowledge as well as their physical abilities to dig and haul dirt. To develop his proposal for the canal, Riquet hired not only a trained engineer named Andreossy but also a local *fontanier* (lay hydraulic engineer), Pierre Campmas, who had tacit knowledge of building water supplies.[10]

One way to analyze *how* different groups worked together on the Canal du Midi is to determine *where* they were deployed and with whom. Differentiating the engineering strategies used to construct separate parts of the canal, one can figure out what workers and supervisors (carriers of informal and formal knowledge) were doing at the time. Women were used primarily to finish the alimentation system in the Montagne Noire and to help with the "second enterprise," in which the canal reached the mountains around Bézier.[11] In other words, they worked where the canal had to be threaded through rough terrain, following contours with complicated twists and turns. Educated gentlemen and peasant women were an unlikely pairing for this work, but their collaboration accomplished the job.

Carriers of indigenous knowledge would not normally have been trusted with the weighty matter of building a navigational canal to glorify the reign of and please the king, but the tacit knowledge of Pyrenean

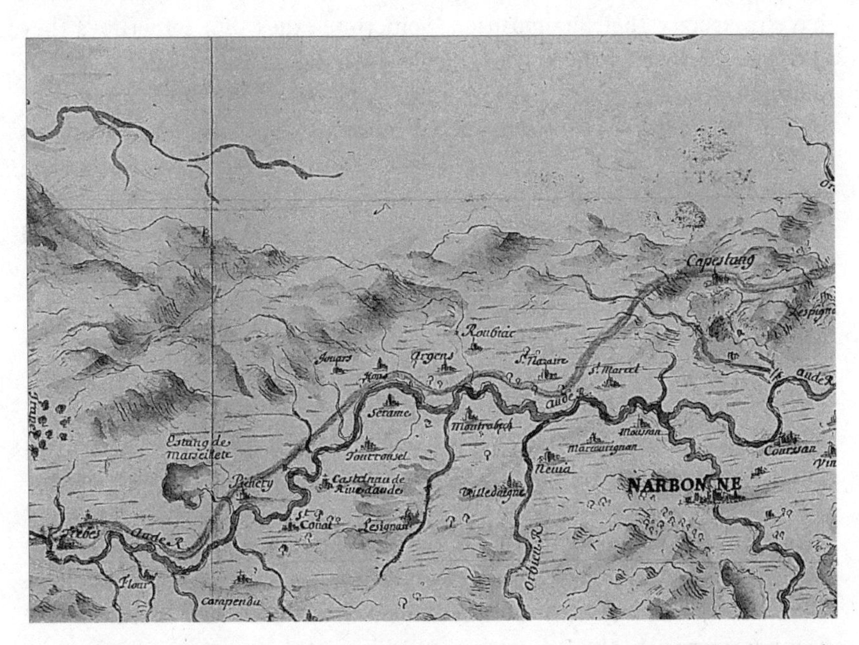

Fig. 1.3. Map of the second enterprise. Bibliothèque Historique de la Marine, Château de Vincennes, Paris.

peasants was easy to hide. Their contributions vanished inside account books and behind official verifications and Riquet's claims of authorship. In the historical record, their work became a product of genius and learned authority, but it was in fact one source of the collective expertise needed to build the canal.

The Written Record

The number of women who worked on the Canal du Midi is hard to determine from the archives because of inconsistencies and obscurities in the accounting techniques.[12] There were at least three thousand women clearly mentioned in letters and public documents, and roughly double that number suggested by the accounts. Of course, some of the women paid in the Montagne Noire may have also worked near Bézier and been counted again there. They were not listed by name in the records, so it is impossible to know if they kept working or went home after a season of labor. On the other hand, there were likely some women lost in the records, since brigades of workers could contain men, women, or both. There was a formula for paying women less. Three women counted as two men, so a brigade could contain forty men or sixty women, but the records did not have to specify their composition. Sometimes they did; sometimes they did not. Chillingly, the account books that did specify gender generally distinguished between "*maneouvriers*" and "*femelles.*" The latter word was usually reserved for animals, not humans, suggesting the low status of these women in the eyes of those who paid them. No wonder they were easy to ignore in the documents.

During the first enterprise, members of brigades were frequently listed by gender, although individual supervisors kept different kinds of records. During the second enterprise, however, the accounting system changed, and gender mostly disappeared. Workers were listed mainly by name or by task. Women's names are not apparent in the lists. Where "*femelles*" are specifically mentioned, they are often the majority of workers. Since official documents indicate that women were being recruited in greater numbers for the second enterprise, their relative invisibility in the books is not evidence that men were doing all the work. It simply indicates that many women workers from the second enterprise were not documented in these sources, and so we may never know how many there were.

Luckily, where gender was mentioned in the account records for the second enterprise, the documents provide a clearer picture of what the women were doing. We know where they were and (vaguely) what tasks they had been assigned. For example, in the area of Somail on 14 May 1678, the books show that Estiene Valletter supervised 185 workers, 125 of them "*femalle* [*sic*]"[13] and Jean Sabarié employed 400 "*femelles*" and 21 "*maneouvriers.*"[14] In these and other cases, women made up the majority of workers. Female laborers were generally hired for terracing and "*controlle des sables*" (reengineering sections of the canal that had filled with silt or been damaged by floods). Their terracing skills probably included controlling inclines and following contours in topographically complex areas. For *controlle des sables,* they probably were building settling ponds, skimmers, and silt-retaining barriers where streams entered the canal. These were common techniques that were used in the mountains for reducing the input of suspended materials and that were also employed on the Canal du Midi.

Women often worked for the same supervisors. Some managers who hired women to work on the alimentation system in the Montagne Noire also used women in the region between Somail and Béziers. We do not know whether they were the same women or whether these men recruited new ones. Other subcontractors used women for single projects. One example was the eight-lock staircase at Fonseranes, where a majority of women laborers were hired to set out the steps and build the locks. (The work was called *controlle de sables* even though the activity was mainly terracing; perhaps the women were simply classified this way for financial reasons or to keep them in gender-appropriate work categories.)

If the documentary sources detailing the work of women laborers are both explicit and obscuring, the written sources about indigenous engineering are limited but rich. Hydraulic engineering in the mountains was first documented by Louis de Froidour, one of Colbert's forestry officials, in letters he wrote to his patrons.[15] He had come to the southwest of France to study the forests but was also asked to "verify" work on the water system for the Canal du Midi.[16] After he had studied the canal and written favorably about it, he returned to the Pyrenees. In the former Roman spa town of Bagnères de Bigorre, he developed a devastating illness and stayed there to take the waters and recover. He took careful note of his surroundings and wrote later about the wonderful indigenous hydraulic engineering he found there:

> The greatest advantage that people of the country here derive from these rivers is that they divert them everywhere they want and that, since their sources are at high elevations and come down steep inclines, they can divert them into canals even in the highest mountains and on

> high precipices to make meadows there. They also route them around towns to serve as fortifications, and they run them into the majority of private homes for the well-being of the inhabitants; they also disperse water in all parts of the countryside to improve it and to water gardens, fields, meadows, pastures, and to turn millwheels to grind grain, tan leather, cut timber, forge iron, work copper, full fabrics, make paper, and, in a word, for all sorts of commodities, to such an extent that one could *say that to see all the uses one could make of water, one should see what they do in the Bigorre valley.*[17]

Of the town of Bagnères itself he wrote:

> The river [is diverted] into thousands of different canals that the inhabitants of the town use for watering their gardens, their fields, and their lands. The town itself is surrounded, in a fashion used in many locations, with a double and triple fosse of running water, and inside the city all the roads are washed by a canal that passes through, and at the same time, there are under the houses some small canals that provide water for all domestic uses; so much so that, together with the quantity of fountains in this town of cold water and hot water, one can say in all truth that there is no other spot on earth where nature has been studied to such advantage to make visible these marvelous waterworks.[18]

The women of the Pyrenees were major carriers of the vernacular tradition of hydraulics because the men did the more economically consequential work of taking flocks to upland meadows and of making cheese in the summertime. Women and children were left in the mountain valleys to tend to agriculture and gardening and to adapt the canals to domestic needs.[19]

Women were socially powerful in the region. Families in the Midi-Pyrenees lived in clan-based communities in which the oldest sibling—male or female—was the head of the family (*l'ainesse absolut*). This gave women surprising property rights in the central Pyrenees, sometimes inheriting domestic landholdings through the female line. The Basque culture on the Atlantic side of the Pyrenees took *l'ainesse absolut* as a firm social rule, whereas other areas maintained the general principle of preferring men over women but still allowed inheritance through female descendants when birth order and social assent dictated this.

The peasants in high mountain valleys lived in a world of quasi republics made up of affiliated villages and valleys that practiced collective land ownership. Confederations of towns in contiguous valleys coordinated their uses of natural resources for mutual advantage. They herded animals through commonly held meadows that they improved and used in culturally prescribed manners; they dug ore as they wished from local mines; and (with some restrictions) they cut trees from and fished and hunted without payment in local forests. Most of the land outside the villages was not private but collectively held. The combination of community-shared property and political alliances over vast areas helped to sustain large-scale, cooperative land-management systems. Some areas practiced advanced methods of forest management and timbering. And towns in valleys along rivers developed systems of canals that were used cooperatively not only for irrigation but also for household purposes. Spring-fed waterworks were tied to their sources with community labor, and women and children used the water systems in the domestic economy.[20]

Women in the Pyrenees were thought to have significant spiritual powers that legitimated their social autonomy and knowledge of nature. The Basques retained a strong strain of goddess worship, venerating the sun, the word for which in the Basque language means "grandmother." In most other parts of Europe and in parts of the mountains after the Romans arrived, the sun was a male figure. He became Apollo (and the symbol of the Sun King). But in the Atlantic and Midi-Pyrenees belief in the grandmother-goddess-sun was retained. The goddess was also represented in Pyrenean culture as the Virgin, who made many appearances in the central part of the mountain chain. In some towns like Ustou, the local church had an image of the Virgin rather than Christ above the altar. There were also repeated apparitions of the Virgin in Catalonia as well as in the Basque areas of the Pyrenees, and even at Lourdes in the nineteenth century. Fairies and female wood creatures abounded in Catalan and Basque stories about the forests; water sources throughout the Pyrenees were traditionally associated with fertility and often said to be controlled by fairies. The centrality of female imagery to this region and its association with water may help explain why the mountain springs were tapped for community canals only after the Day of Our Lady, or Notre Dame.[21]

Some laborers from the mountains who worked on the Canal du Midi were itinerant men of this region who often migrated into the valleys of Spain and France for grape picking and harvests;[22] sometimes women joined them as seasonal laborers and probably also on the canal. Their recruitment would explain why so many women laborers worked in the mountainous regions of the canal, and why techniques used for domestic water supplies in the Pyrenees appeared on the Canal du Midi.[23]

Physical Evidence

Although Froidour wrote about Bagnères de Bigorre as the place in the Pyrenees where such elaborate canal systems were built, there were similar canal systems in the mountain chain from the Atlantic Ocean to the Mediterranean Sea. We know that most of the towns date from the seventeenth century or earlier because they were mentioned by Froidour or were recorded on the early-seventeenth-century map of the region by Sanson. There is reason to suspect, although no way to prove, that the canals also dated from the earlier period and that the indigenous hydraulics in the Pyrenees that Froidour described were more widespread then he knew.

Domestic water systems that remain in Gerde, a town next to Bagnères, might have been part of the system that Froidour described. Small waterways off a main channel connect to gardens, orchards, fields, and (above all) houses. One house, sitting on a slight hill in Gerde, has a high channel from the street going into the house, a slightly lower one meandering through a lawn and flower garden, a third going just below to another part of the garden, a fourth transecting an orchard, and the lowest one watering a meadow. Elevation differences were used to deliver water to different places. Strategically placed stairs and stones allowed people to walk through the complex of waterways. Multiple sluices controlled the movement of water through the channels.[24]

Cheust, a small village on a steep hillside one valley west of Bagnères, is in an area where Protestants were said to have fled from the region of Bigorre during the wars of religion. The town's water system had all the attributes of those in Gerde but was threatened by the accumulation of cobbles. These small stones came with the waters rushing down the mountain during spring thaws or winter storms. They could damage the canal walls, impede the millstream, and clog up domestic water intakes. Therefore, the town water system was organized around a central settling pond that could capture the cobbles. This pond was constructed by widening the stream that flowed naturally through town, damming it up, and allowing water to be diverted from this area only over slightly elevated barriers that could hold back the stones. Around this pond were a set of sluices that could be closed to keep floodwaters from entering the town water supply or be opened to channel "settled" water into gardens, mills,

Fig. 1.4. Garden step in the town of Gerde.

homes, and the ubiquitous public laundry. Finally, at one end of the dam, the barrier for holding the cobbles could be lowered to allow excess water to be ejected along with debris from the floor of the holding pond.[25]

Simpler waterworks are also visible in other valleys of the Pyrenees. Take, for example, the town of Ustou, which Froidour called Houstou. It is on the road to the old Roman bath town Aulus-les-Bains. The people of Ustou did not create a massive town water supply but did set out weirs in the river to divert water to two mills, one for grinding grain and one for tanning. Although this water system does not seem to have the domestic

Fig. 1.5. Garden levels in Gerde.

Fig. 1.6. Settling pond, overflows, and the sluice to the mill and laundry in the town of Cheust.

Fig. 1.7. Mill and laundry in Cheust.

character of the canals at Gerde and Cheust, a woman in Ustou explained to me when I was taking pictures there that the stones along the channel to the grain mill actually constituted a kind of public laundry. She showed me the stone that her mother had used for her washing. The flat rocks along the channel were all washing stones, one assigned to each family. In this case, too, diversionary canals served both domestic and economic purposes.

Gerde and Ustou, although obviously located in formerly Roman areas, are clearly not direct products of Roman engineering. Roman systems were organized around baths, not public laundries. Classical engineering as a formal system of knowledge and the culture of public bathing that canals had served vanished from the Pyrenees, but an indigenous form of water engineering based on classical techniques was elaborated in these mountains for new purposes.[26]

In spite of the centrality of Bigorre to Froidour's accounts of Pyrenean canals, one of the most developed water systems was built outside the Bigorre region in the town of Mazères de Neste. Here the water system was built in a relatively flat valley and used extremely subtle changes of elevation to manage the flows. A weir consisting of large rocks built diagonally

Fig. 1.8. Tanning and grain mills in the town of Ustou.

across the Neste River diverted water through a sluice gate (that could be closed in case of flooding) and along a set of channels into town. Water from this system ran in two directions through the fields and into the village. Along one canal, stone pillars were scored vertically to hold pieces of wood to close the channel. This water supplied irrigation and drainage through ditches running up and down adjacent fields. If this sluice gate from the fields were left open, these ditches could be used for drainage. If the waterway into town was blocked, water would back up into the fields and irrigate them. If a gate at the bottom of the field was closed once it had been irrigated, the water would be retained in the ditches to seep into the dirt, and the sluice on the main canal could be opened to resume the flow of water into town. Two branches of river water flowed into town. One was diverted into houses for freshwater, and the other (from the irrigation area) ran along the streets, perhaps serving as a sewer. The water system met in the central town square, where it supplied a large public laundry, and then was carried back to the river. This whole complex constituted an irrigation/drainage system for agriculture, a domestic water supply, a water-based sewage system, and a public laundry.[27]

Public laundries in central squares constituted the most ubiquitous feature of Pyrenean town water systems. They suggest, from both a practical and a symbolic perspective, that women were indeed involved in

Fig. 1.9. Weir on the Neste River near the town of Mazère.

Fig. 1.10. Sluice gate in Mazère.

Fig. 1.11. Laundry in Mazère.

their design. The town laundry was the center of women's public lives, the place they would meet and work together, particularly in the summertime, when the weather permitted outdoor activity and most of the men were away, reducing domestic pleasures and indoor responsibilities.

Culturally, laundry also had surprising importance. The fairies thought to inhabit the inner recesses of the high mountains were supposed to be obsessed with their nice white, clean laundry. The periodic spring at Fontesorbes was one of the natural features in the mountains said to be run by fairies. The periodicity of the spring was explained by reference to the fairies doing their laundry. They turned the water on to wash their clothes and turned it off when they were finished.

These female spirits represented the powers of nature and reflected the importance of women to the happiness of men and communities. The most common stories about fairies describe them as rich and claim that they hid gold inside the mountains. Men wishing to acquire this gold had to find the fairies while the fairies were doing their laundry. These miraculous creatures, being shy of men, would hide, and the men could steal the linens and laces hanging up to dry. They could then use the laundry as ransom for gold or hold onto it. Textiles from these quasi deities would bring good luck to the men's families. Some good wives were even thought to be fairies, a belief that expressed the implicit continuity between fairies and ordinary women. Legend had it that some men did not steal the fairies' laundry but fell in love with the beautiful creatures and caught the fairies themselves. If they could convince their captives to become their wives, the families would enjoy extraordinary well-being.

Viewed through the lens of this folklore, there was nothing mundane about public laundries in this culture. They were places where fundamental principles of gender and happiness were understood and enacted, and where the powers of women in nature and over water could be publicly displayed.[28] In this context, it made sense for women to use Roman water-engineering techniques to control the flows of springs and rivers and to make this control a matter of public regard. Water from the mountains was associated not only with fertility but also with the miraculous powers of women (as fairies) to bring happiness to life.

Women Laborers and the Canal du Midi

Many techniques of canal construction in these mountain villages resemble forms used for the alimentation system and second enterprise of the Canal du Midi. The canal's feeder system began in the high forests of the Montagne Noire, using weirs, sluices, and diversionary channels to capture flow from local rivers near their sources. Some of the water diverted from the high rivers was dumped into the Sor River and then redirected from this natural channel to an artificial channel (*rigole*) near Revel. But this did not provide enough stored water for the dry summers, so some water was captured in the valley of the Laudot River, which had been closed with the St. Fereol dam. Below the reservoir, water from the Laudot joined the alimentary channel, which then ran over long distances through fields, along roads, past orchards, to where it was stored and used. This feeder channel followed the contours of the long fingers of land that reach from the high mountains toward Carcassone. The system as a whole resembled an extended and elaborated mountain water supply that could have served a town in the Pyrenees.

The alimentation system for the Canal du Midi was very much a product of indigenous expertise. Riquet first employed a local man to develop the plan for the water supply: Pierre Campmas, from the small mountain town of Revel.[29] Campmas made his living diverting water from local streams and rivers to the towns and mills of the Montagne Noire, using tacit knowledge of the landscape and mountain streams. His original design was good enough that it needed only slight revision before it was built.[30]

Fig. 1.12. Rigole de la Montagne.

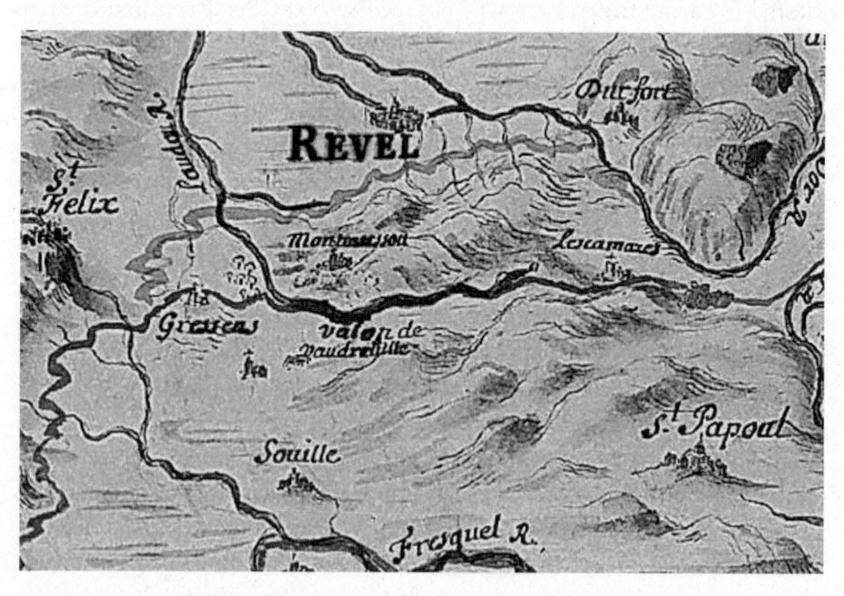

Fig. 1.13. Water supply system near Revel. Bibliothèque Historique de la Marine, Château de Vincennes, Paris.

Riquet and his collaborators recruited fifty-three men from the three closest towns to build the dam for the water supply, immediately turning again to indigenous laborers for help. When this small group did not make fast enough progress, more workers were recruited. By 1669, there were 7,000 men and 1,000 women at work in the Montagne Noire.[31] These numbers were cut by the 1670s, when the labor force consisted of 400–500 women and 200–300 men. The men were mainly stonecutters and masons who did the vaults and facing of the dam as well as some of the work for the supply channels. The women worked on the "terracing" for the waterway: they dug out the channels along the contours of the land and then brought the dirt up from the channels to the dam in baskets.[32]

Employing so many women laborers for hauling dirt up and down hills would make the most sense if they were mountain women, accustomed to such physical tasks and skilled at making waterworks in the Pyrenees. Unlike their male predecessors, these women were not inhabitants of local villages, so Riquet asked the locals to house and feed them.[33] Froidour described some Pyrenean people as like animals, the women being as bad as the men. This would also support the idea that women from the Pyrenees were the most likely source of the "*femelles*" who worked on the alimentation system.

For cutting the canal through the mountains to build the second enterprise, indigenous experts in canal construction again were important, including women. Pierre Campmas did not figure in the second enterprise because he continued to perfect the alimentation system; however, two of his collaborators from the Montagne Noire, Roux and Contigny, became supervisors of an area of the canal proper that ran through the mountains.[34]

Women were not listed in the crews working on the locks between Trèbes and the Orbiel in the second enterprise; men remained the surveyors, carpenters, and masons used for this kind of work. Women showed up in the accounts for the Somail and Béziers regions, precisely where the canal had to thread through the mountains and in just the area of the canal where this was accomplished using few locks.[35] Here they served both their immediate supervisors and two mathematicians: the sieur de la Feuille and his Jesuit assistant, Père de la Morgue (or Mourgues).[36] Academic mathematicians and indigenous engineers worked in tandem to thread the canal carefully around hills, over small ravines, and along scarps toward the sea.

In the mountains between Cesse and Béziers, where the canal had to be blasted out of rocky hillsides, inclines were particularly difficult to maintain.[37] Women did terracing for the canal near Cesse, and their work there has been called "a classic example of contour canal cutting, winding

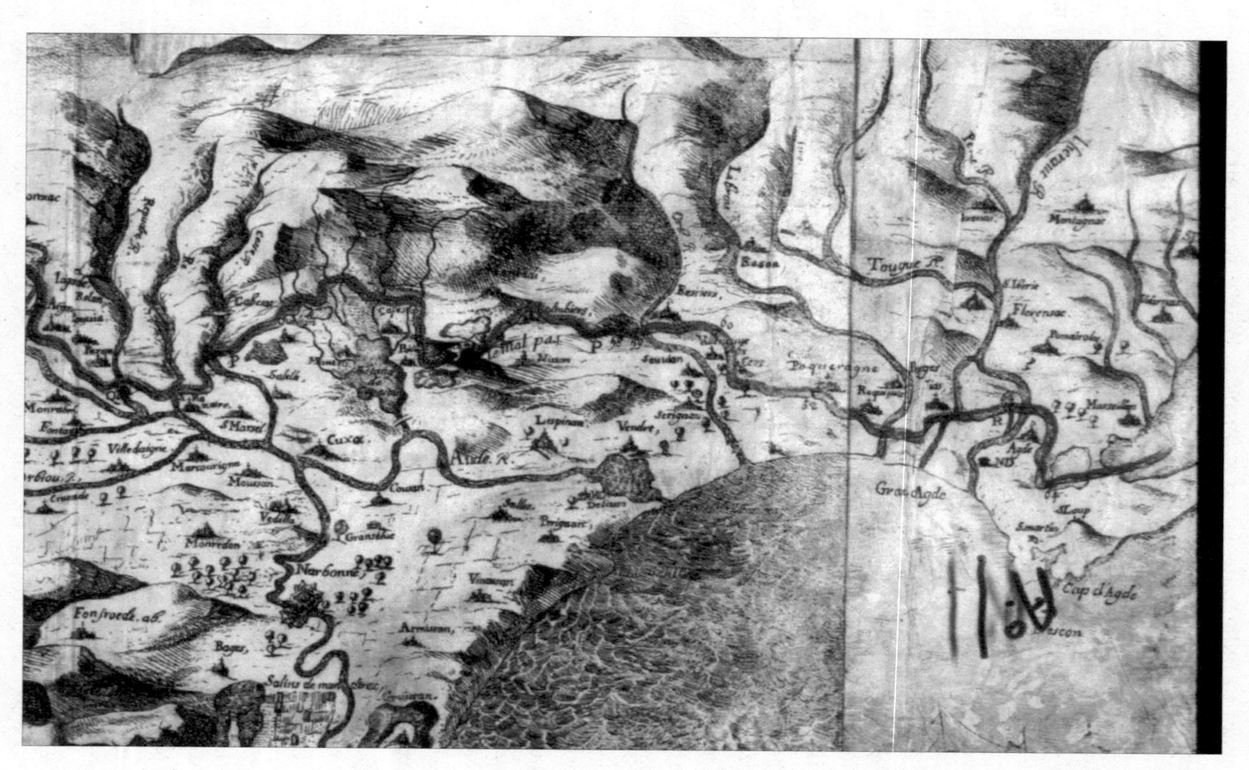

Fig. 1.14. Mountainous area of the second enterprise. Bibliothèque Historique de la Marine, Château de Vincennes, Paris.

this way and that.[38] Women who had proved themselves in the Montagne Noire may have been brought specifically to work in the Somail region by Contigny and Roux. The two men were familiar with what women could do.[39]

There were also silting problems in the Somail area that the women addressed. In 1678 and 1679, they were hired for *"controlle des sables,"* or redigging the canal where the silting problems were severe. The connections to some feeder streams were reengineered in the period, using a system of water and silt control well known in the Pyrenees. At the side of the canal where the rivulets entered, water was held in simple settling ponds, a widening of the streambed used to slow the movement of water before it entered the canal. In these "ponds," materials held in suspension by churning waters could fall to the floor of the streambed (like cobbles at Cheust). Where the water was allowed to enter the canal, a rock lip was placed at the bottom of the rivulets to hold back precipitated materials, and a stone skimmer was laid across the top of the streams to remove floating debris. This minor bit of engineering helped purify the water before it entered the Canal du Midi.[40]

Fig. 1.15. Water intake with silt barrier and rock for skimming floating debris.

Floodwaters were also a problem in the area of the second enterprise. The embankments across from stream entrances had to be reinforced, and drains installed for excess water. The drains were placed slightly downstream and across the canal from the water sources. Some overflow drains had sluice gates that could be lifted to expel water and silt from the floor of the canal. Most drains were simply low stone embankments that allowed excess water to overflow into lowlands.[41] These simple techniques were standard elements of indigenous hydraulics used in the mountains to reduce debris in the water supply and handle excess water.

The most dramatic site where women workers were recruited in large numbers was the eight-lock staircase at Fonseranes. Men appeared to do most of the other lock construction for the second enterprise, but the staircase at Fonseranes required complicated terracing. Women laborers were good at this. Each lock in the staircase had to contain the same volume of water, and the set of locks had to descend a hill that was not a straight slope. The depth of the locks had to be carefully controlled or the volume would be larger or smaller from one lock to the next. This would either create spills over the side of the locks or inadequate water for keeping boats afloat.

There had been repeated problems with double and triple locks built earlier. In some cases, the interior volumes of the series were distinctly out of proportion. For relatively simple structures, the locks could be given extra intakes and drains to manage the disproportions. These ameliorations worked on a small scale, but there was no comparable way to jury-rig an eight-lock staircase. Using surveyors to measure the elevations for the locks was no solution. This practice had been abandoned after 1673, when measurement inaccuracies had necessitated complete reconstruction of a series of locks. It was decided that the only way to verify lock volumes was

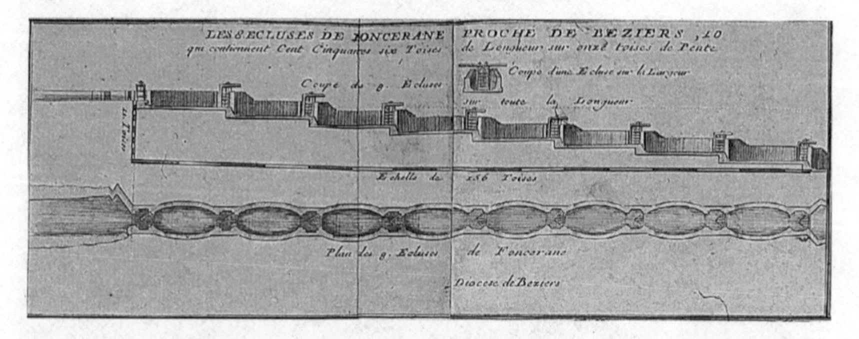

Fig. 1.16. The "staircase" at Fonseranes (post hoc formalization). Bibliothèque Historique de la Marine, Château de Vincennes, Paris.

Fig. 1.17. Interior of a lock at Fonseranes.

with water. But who could they trust to estimate elevations and volumes well enough? Apparently the answer reached at Fonseranes was women, where their work was supervised by Roux, among others.[42]

Even more surprising, the entrepreneurs who subcontracted for this part of the project were two brothers who were both illiterate. Michel and Pierre Medailhes could not sign the account books, so they wrote an X, which was witnessed and noted by others.[43] Clearly, many entrepreneurs had worked on the Canal du Midi. Many were more experienced in building locks and most were more educated than Michel and Pierre

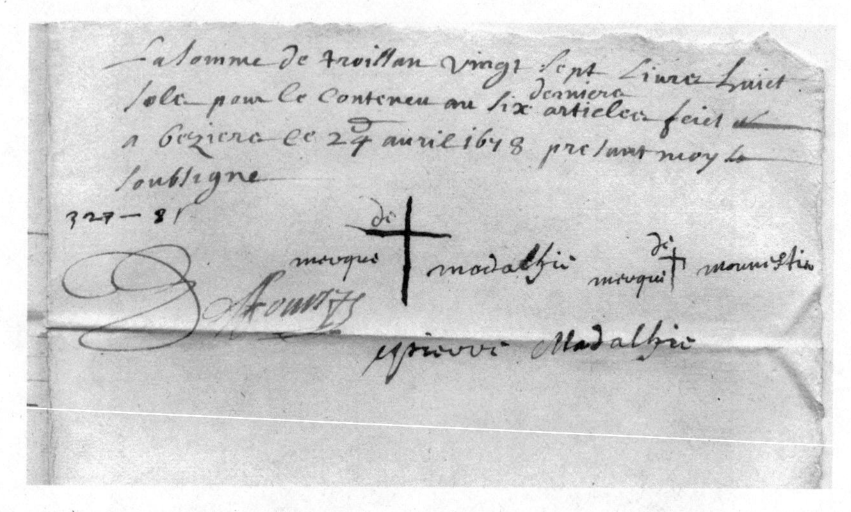

La somme de troi[s cent] vingt sept livres huict sols pour le contenu au six derniers articles faict a Beziers le 24 avril 1678 present moy le soubsigne

327 — 8 s

marque de Medalhe marque de Mounestier

Pierre Medalhe

Fig. 1.18. Accounts for Fonseranes.

Medailhes. Why were these brothers given the contract for such a visible and important piece of engineering?

Subcontracting engineering helped spread the risks of the project, and the work at Fonseranes was clearly risky. Indigenous experts were the best candidates if they were capable and efficient. The lock staircase was a great success. It was built properly, cheaply, and quickly. And, unlike many of the earlier locks, it did not have to be rebuilt in the 1680s. It was a tribute to the powers of indigenous knowledge and collaborations between illiterate women laborers and artisans. Nevertheless, what they accomplished was possible only because the gentlemen in charge of the project used survey skills to legitimate their work.

Conclusions

The story of Riquet as a lonely and misunderstood genius who had the vision and toughness to build the Canal du Midi against the odds has a pathos that has been too appealing to abandon easily. It has comforted a region with a long history of hostility from and against the northern monarchy. It is a source of pride within the complex, distinctive, and rich heritage of Languedoc. Riquet's numerous aristocratic enemies also helped to build his reputation among locals. It has been satisfying to prove these elites wrong. But the tales of Riquet's natural genius have obscured the other kinds of regional intelligence that made the canal possible, including the expertise of indigenous women engineers who came as laborers to the Canal du Midi.

The Canal du Midi was the product not of individual genius, but rather of a social intelligence that developed around the problems of threading a canal through a landscape that changed in topography, soil type, and habitation patterns. Riquet could never quite control the outcome of the process, not just because of Colbert's efforts to hold him in check, but because the emergent problem solving took many forms and required different talents. Engineering novelties were designed as solutions to particular local problems. But they were all realized through social action: common efforts, intense disputes, experiments, demonstrations, failures, repairs, insinuations, and collaborations that pushed all of those involved beyond their abilities. They produced together a kind of Durkheimian collective consciousness—what Durkheim suggested gave humans a sense of the transcendent.

The mysterious power of social learning was perhaps what made Riquet feel that God was allowing him to understand things previously far from his grasp. He sensed he was in the presence of a higher power, but really he was just in the presence of a conversation between the people

Fig. 1.19. Woman lock keeper in the area of the second enterprise.

of Languedoc and experts from Paris. Local knowledge based on making things was weighed against formal knowledge of engineering, allowing both indigenous laborers and educated gentlemen to see new possibilities for building the canal. The surprising intellectual fertility of this complex social environment gave Riquet a sense of well-being that was as powerful as his sense of despair about the criticisms of the project. In the Pyrenees, this kind of unanticipated and surprising pleasure would have been associated with fairies, which controlled not only the water of the mountains but human happiness as well.

CHAPTER TWO

Visual Representation as Instructional Text: Jan van Eyck and *The Ghent Altarpiece*

LINDA SEIDEL

In 1432, Jan van Eyck's painterly skills were put on public display when a massive, multipaneled altarpiece, on which he had worked with an older brother, was installed in a newly endowed chapel at the Church of St. John the Baptist in Ghent, now the Cathedral of St. Bavo (figs. 2.1–2.2). An extensive inscription, highly exceptional for the time, runs across the bottom frame of the exterior wings, beneath large-scale kneeling depictions of the work's donors, Jodocus Vijd and Elisabeth Borluut. This painted text praises the artists and situates the work as the earliest dated monumental piece in the Netherlandish canon.[1]

The polyptych could not have been an inconspicuous presence in the church, with its vast size, varied subjects, stunning realism, and what originally would have been an elaborately carved, mobile frame in accordance with the taste of the time. Because the inscription proclaims the work to have been the collaborative effort of Jan and Hubert (who died in 1426), the attribution of particular panels to one or the other of the brothers has long been the central preoccupation of art historians studying its many aspects. Throughout the twentieth century, Hubert was most often identified as the author of an initial, less complex overall conception and given credit for the execution of several of the interior panels, in particular those on the lower register with multitudes of small figures. One scholar boldly removed him entirely from the painted project by proposing that he was a sculptor, responsible only for the multistoried enclosure that originally had held the panels and enabled them to be pivoted. More recently, researchers have acknowledged the participation of multiple hands throughout the panels, indicating the presence of apprentices working under the supervision of a master painter.[2]

Fig. 2.1. Jan and Hubert van Eyck, *The Altarpiece of the Lamb,* 1432; Cathedral of St. Bavo, Ghent; exterior wings. Photo: Scala / Art Resource, NY.

Fig. 2.2. Jan and Hubert van Eyck, *The Altarpiece of the Lamb,* interior. Cathedral of St. Bavo, Ghent, Belgium. Photo: Scala / Art Resource, NY.

There has never been any doubt about Jan's responsibility for the most novel painted aspects of the work. These include, but are not limited to, tricks of spatial representation on the outer wings, astounding portrayals of precious minerals and fabrics on the central inner panels, lifelike figures of Adam and Eve on the upper register of the interior, and, throughout, luminous passages of chromatic transparency that display exceptional proficiency in the handling of oil glazes. Such features were surely among those most appreciated by Albrecht Dürer, who, during a visit to Ghent in April 1521, was taken by the dean and members of the Painters' Guild to the Church of St. John to see "something great." After viewing Jan's already-famous work, he wrote in his diary that it was "a most precious painting, full of thought."[3]

In this essay, I argue that Jan's vaunted technique of representation should be understood as something more than a supremely competent transcription of reality; it was, as Dürer's words intimated, a way of thinking.

Jan's meticulous renderings combined scrupulous visual examination of his physical surroundings with rigorous analysis of the material components of his craft. Together, these activities, integral to his art, constitute a systematic inquiry into the invisible, even mysterious, workings of the natural world. The paintings Jan produced provided a record of solutions to problems in visual representation; they set before his apprentices and followers a detailed, pictorial "how-to" manual of exempla for a variety of pictorial challenges. The panels of the Ghent altarpiece—with their virtuoso images of the draped and undraped human body, landscape elements, clouds, and shadows—stand as a visual counterpart to Cennino Cennini's celebrated *Craftsman's Handbook,* composed circa 1400, in which written instructions were provided for the accomplishment of similar tasks. Perhaps this is what the Nuremberg humanist Hieronymus Münzer implied when, after visiting Ghent in 1495, he remarked that the work displays such ingenuity and skill that it should be regarded not as "a painting" but rather as "the whole art of painting."[4]

Making Knowledge

Early written appreciation of Jan's work documents remarkable insight into its achievements. Bartolomeo Fazio, court historian to Alfonso V of Aragon, king of Naples, enumerated the significant features of Jan's art in a Latin treatise about famous men that was published in 1456. His rhetorical appreciation of *Johannes Gallicus* describes the painter's accomplishments in the areas of spatial representation, color, and liveliness of expression. The first two qualities distinguish Jan's unique talents; the third was the characteristic marker of early humanist criticism and the one that Fazio employed as well in his comments on the three other painters mentioned in his treatise.[5]

Fazio remarks that Jan "has been judged the leading painter of our time" and characterizes him as "not unlettered, particularly in geometry." He comments on Jan's sophisticated rendering of space in a triptych in the collection of his patron, praising the way in which Jan succeeded in rendering the illusion of the library in which the figure of St. Jerome is depicted: "for if you move away from it a little it seems that [the room] recedes inwards and that it has complete books laid open in it, while if you go near, there is only a summary of these."

Alfonso's triptych, Fazio notes, had depictions of the donor, Battista Lomellini, on one of its outer wings, with "the woman whom he loved (on the other). . . . Between them, as if through a chink in the wall, falls a ray of sun that you would take to be real sun light." His comment here draws attention to an effect of luminous transparency that the artist had succeeded in rendering, one that brings to mind the brilliant light that pierces a Gothic window behind the Virgin in the scene of the Annunciation on the outer wing of the Ghent altarpiece, casting its rays in the pattern of double lancets against the wall behind her. Additional effects of light, augmented by accomplishments in the rendering of texture and moisture, dominate Fazio's lengthy description of a third painting, a panel in the possession of a nephew of Federigo da Montefeltro. Like the other two, this one disappeared long ago and is only known through Fazio's account.[6]

In an often-quoted passage of remarkable detail, Fazio describes "women of uncommon beauty emerging from the bath, the more intimate parts of the body being with excellent modesty veiled in fine linen, and of one of them he has shown only the face and breast but has then represented the hind parts of her body in a mirror painted on the wall opposite, so that you may see her back as well as her breast. In the same picture, there is a lantern in the bath chamber, just like one lit, and an old woman seemingly sweating, a puppy lapping up water. . . . But almost nothing is more wonderful in this work than the mirror painted in the picture, in which you see whatever is represented as in a real mirror."

Fazio's descriptions go beyond acknowledgment of the artist's ability to simulate natural forms, celebrating instead Jan's handling of light's penetrating effects and stressing his capacity to produce unprecedented brilliance in painted form. Fazio's fascination with Jan's ability to render surfaces as though they were mirrors calls attention to the artist's marriage of lustrous pigments with the sheen of specific materials, a point the Dutch writer Karel van Mander later developed.[7] Fazio's admiration for painted passages involving water acknowledges Jan's accomplishments in the rendering of transparency, while praise for a gratuitous detail in which sunlight passes through a crevice celebrates a pictorial incident of exceptional brilliance and clarity. These elements that Fazio emphasizes as features of the artist's exceptional talent constitute the basis for our own astonishment at Jan's unprecedented accomplishments in his undisputed, extant works.

In the *Arnolfini Double Portrait,* Jan placed roundels that simulate reversed glass paintings on the frame enclosing the painting's famously reflective mirror (fig. 2.3). He incorporated into each minuscule representation the anticipated distortions that are produced when figures drawn on the undersurfaces of convex glass are viewed from above. By using pigments suspended in oil, together with oil glazes, Jan produced a (liter-

Fig. 2.3. Jan van Eyck, *The Arnolfini Double Portrait,* 1434; detail of the mirror. National Gallery, London. Photo: Erich Lessing / Art Resource, NY.

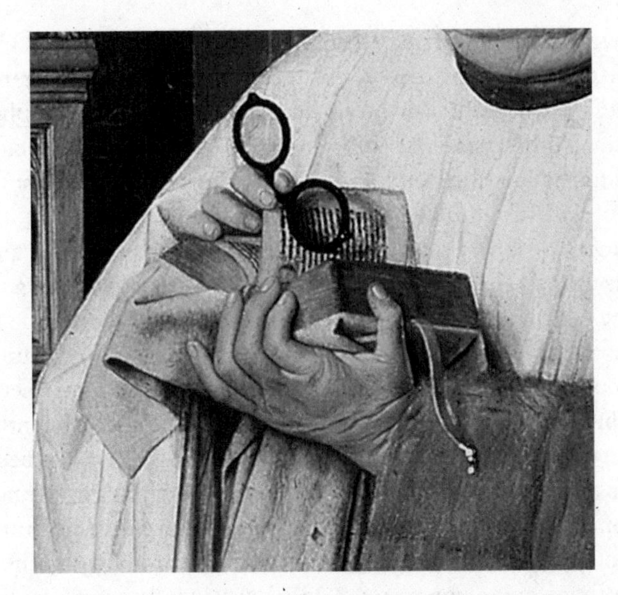

Fig. 2.4. Jan van Eyck, *The Virgin with Canon van der Paele and Saints,* 1436; Stedelijk Museum voor Schone Kunsten, Bruges; detail of the canon's hands. Photo: Erich Lessing / Art Resource, NY.

ally) layered image that mimics what it represents: the activity of looking through thick glass. This tiny passage, occupying no more than a quarter inch of painted surface, reveals the extent to which the mechanics of Jan's technique were integral to the effects he achieved.

Jan created a similar optical event in the altarpiece of *The Virgin with Canon van der Paele and Saints* in Bruges (fig. 2.4). Here he placed eyeglasses in the cleric's puffy hand, silhouetting one lens against the missal the canon holds and the other against the white vestment he wears. The prominent angle at which the spectacles are held emphasizes the lenses' powers of magnification and deformation; the glasses are positioned so that the outside viewer "sees" through their corrective surfaces into two miniature worlds of unfocused distortion, part text, part texture. Jan employs lightly tinted glazes to render the materiality of glass, drawing attention thereby to the process of looking through it. In these instances, and in the case of the leaded glass panels in the windows behind van der Paele, Jan exploits the translucent properties of his materials to render the movement of light through another kind of matter.

Jan's meticulous rendering of golden flecks in velvet pile, seen in abundance on this panel, has enabled textile historians to chart innovations in weaving techniques. A recent study of fabrics in Jan's paintings argues that viewers would have been able to identify the represented cloths, and establish their value, based on the precision of the artist's rendering of pattern and density and—above all—the inclusion of metallic fibers.[8] Such observations emphasize the ways in which Jan's painting technique inquires into the handling of a wide range of materials and is not simply a manner of faithfully imitating surface effects.

Instances of Jan's investigation into the composition and qualities of materials are particularly numerous, unusual, and systematic in the Ghent altarpiece. There the abundance of reflections and shadows, the rendering of jewels, the play of water, and the formation of clouds constitute as much a handbook on the structure of natural objects as the altarpiece's assemblage of biblical and historical figures forms a catalog of fifteenth-century Christian iconography.[9] In fact, this should not surprise us, because artists' shops had long been places where technical knowledge regarding the proper rendering of forms and themes circulated. Most of this information was transmitted directly, through apprenticeship, to judge by the few treatises that have come down to us; only occasionally were "trade secrets" written down for posterity.[10]

An early-twelfth-century German monk who called himself Theophilus produced a treatise, entitled *On Diverse Arts,* ostensibly to instruct a religious reader in the devotional component of his craft.[11] In it he treats the acquisition of raw materials as well as the sophisticated production of a variety of painted and metalwork objects. Some of his recipes report on techniques we would otherwise not have known were in use since material verification of these processes does not exist; the issue of bronze casting, which he discusses at some length, is most conspicuous here. The fact that there is a gap between what we read in Theophilus's book and what we can attest to with our own eyes reminds us of the existence, throughout history, of different kinds of evidence. Written texts provide information in a manner distinct from that in which visual media make knowledge known, even when they address the same subject. Their divergent instruments engage discrete interests and cultivate disparate audiences.

Cennino Cennini's *Craftsman's Handbook* is the most extensive work of this sort that has survived; it is closest in date as well to the years of Jan's productivity, having been written in Padua around 1400.[12] In terms of content, it overlaps aspects of Theophilus's earlier text, indicative of the relative stability of craft practices in western Europe and suggestive of the commonality of those traditions in the north, where Theophilus worked, and the south, where Cennino was active nearly three centuries later. Cennino's commentary on the topics and materials to which the artist needs to attend resonates in several of the visual novelties that appear in Jan's magisterial polyptych, in particular, the depiction of an array of materials and the use of glazes for capturing luminous effects. Juxtaposition of Cennino's written text (a primer for shop helpers) with Jan's painted panels (a summa of practices) nominates the latter as pictorial analogue to the former. A comparative reading of them underscores the experimental approach that early-fifteenth-century artist and artisan alike employed in the preparation of materials and the practical wisdom that governed their usage. It alerts us as well to early efforts by painters to elevate their craft's status, something that Dürer and Titian would work at more overtly and assiduously in the course of the sixteenth century.

Cennino invokes "all the saints of God" and dedicates his book to "the use and good profit of anyone who wants to enter this profession"; he then appropriates the main theme of the opening chapters of Genesis as introduction to his text. Theophilus had done something similar in his treatise. Both books begin by suggesting that Adam and Eve's error and the subsequent expulsion from Eden did not eradicate God's gift of knowledge to humankind; rather, this bequest of understanding persisted after the Fall, provoking Adam's awareness of the need to work for his living. Cennino notes that God's wisdom was passed down through the generations in the form of a series of labors.

This theme had already found visual expression in Florentine imagery several decades earlier. The mid-fourteenth-century reliefs of the creation of Adam and Eve on Florence's Campanile are juxtaposed with scenes of the first couple at work, followed by images of their distant offspring's labors. The latter are introduced at the end of Genesis 4 as fifth-generation descendants of Cain. Three sons of Lamech who are mentioned in the text by both name and trade figure in the Florentine imagery: Jabal, a herdsman; Jubal, the father of music; and Tubalcain, the first smithy. The Campanile reliefs, which omit scenes of the Temptation, Fall, and Condemnation, produce a novel message in which labor does not appear as punishment for sinful behavior but appears instead as the affirming continuation of God's purposeful actions in his creation of the first couple.[13]

Cennino includes the craft of painting as part of this process.[14] He defines it as an occupation that requires imagination, coupled with skill of hand "in order to discover things not seen, hiding themselves under the shadow of natural objects, and to fix them with the hand, presenting to plain sight what does not actually exist." Cennino identifies the sources of his own knowledge in apprenticeship: twelve years of training under Agnolo di Taddeo, who was taught by his father, Taddeo, who himself worked for twenty-four years with Giotto, who "brought the profession of painting up to date" and "had more finished craftsmanship than anyone has had since." Cennino ends his introduction by invoking for a second time "all the saints of paradise," adding to the list St. Luke the Evangelist, "the first Christian painter" (1–2). In him, the later Middle Ages saw the skills of the artist combined with those of the physician, a merging of two empirical practices, each of which made claims of efficacy in matters of healing.[15]

Just as Cennino began his book with God's creation of man and woman in his own image, Jan framed the upper portion of the Ghent altarpiece with the first couple; in the central panels, God the Father is shown seated between the Virgin and St. John the Baptist. Adam, at the left edge, shares his maker's high forehead, strong cheekbones, ruddy complexion, and trim, dark beard. In contrast, a full curly beard and mass of thick dark hair mask the features and contours of John's head, setting him apart from the heavenly Patriarch as well as the first man. On the panels below Adam and Eve's feet, a vast procession of holy figures—warriors, virgins, hermits, and pilgrims, in addition to apostles, popes, and cardinals—emerges from a landscape that stretches from one end of the lower register of the polyptych to the other. On one level, this multitude of figures constitutes a visual litany of "all the saints of Paradise," as mentioned by Cennino; on another, it plays a critical role in the complex sacramental scene that is enacted around the altar of the Lamb in the central panel.[16]

Like Cennino, Jan telescopes the story of the first couple within his representation of them, employing a variety of elements familiar from medieval manuscript illuminations to indicate the successive stages of their activity as described in scripture (fig. 2.5 and plate 1). The small fruit Eve holds between the fingers of her elevated right hand suggests the imminence of the Temptation, while the fig leaves with which she and Adam conceal their genitalia declare that they have already eaten of it. Eve's languid form, with its sinuous silhouette and tapered shape, resembles that of the tiny figures seen in Creation scenes like the one in the *Très riches heures*, a book of hours in the Musée Condé, Chantilly, made around 1415 by the Limbourg Brothers (fig. 2.6).[17] The bleached, pear-shaped figures that represent Eve in the miniature's depictions of her both before and after the Fall are usually identified as an expressive "mark of ideal feminine beauty," a euphemism that implies but does not explicitly acknowledge the female anatomy's broad-hipped potentiality for pregnancy.[18] In the *Ghent* Eve, this property is emphasized, markedly distinguishing her figure from the uniformly idealized ones on the manuscript page. The long left arm

Fig. 2.5. Adam; detail of fig. 2.2. Cathedral of St. Bavo, Ghent, Belgium. Photo: Erich Lessing / Art Resource, NY.

that falls diagonally across her abdomen emphasizes the girth of her lower body and points to a vertical line of darkened pigmentation that ascends toward her navel. This shadowy striation sets off the highlighted surface of her swollen belly.

The *Ghent* Adam also displays more anatomical detail than does his miniature counterpart, whose faintly modeled skin is rendered in pinkish flesh tones throughout. In contrast, Jan's figure displays dimples, bony bulges, and hair follicles, and his skin is rendered with variegated

Fig. 2.6. Limbourg Brothers, "The Garden of Eden." From *Tres riches heures* (1413–16); Musée Condé, Chantilly. Photo: Réunion des Musées Nationaux / Art Resource, NY.

coloration. The monumental Adam's angularly positioned arms crisscross his muscular torso, casting strong horizontal shadows on his lower chest and hip that dramatically emphasize his erect posture and draw attention to his robust physiognomy. The bold gesture further underscores a distinction between Adam's pale, luminous torso and the harsh redness of his hands—the tools with which he pursued work in his post-Edenic life.

Early viewers were so taken with the frank representation of the first couple that they used their names when discussing the whole polyptych, referring to it as the *Adam and Eve Retable.* Marcus van Vaernewick, a Flemish humanist writing in the middle of the sixteenth century, commented at length on the figure of Adam, remarking on the way in which light played on the body, arms, and hands and marveling at how real the painted flesh seemed, with its lifelike portrayal of veins and the depiction of tiny hairs growing on the skin. The treatment of Adam's right foot is so compelling, he observed, that no one can decide for sure whether or not it actually emerges in relief from the flat painted panel.[19] Van Vaernewick's master, the Ghent painter Lucas de Heere, in an ode to the altarpiece published in 1565, remarked on Adam's disturbingly lifelike pose, asking "who ever saw a body painted to resemble real flesh so closely?"[20]

Van Vaernewick has very little to say, however, about the handling of Eve's body, praising Jan instead for iconographic knowledge in his depiction of the fruit she holds. He is not alone among commentators in ignoring the subtle markings on her flesh. Modern scholars have also been silent in regard to the shading on her distended belly although they have taken note of Adam's reddened hands, attributing this exceptional feature to Jan's faithful transcription of reality. Both Otto Pächt and Erwin Panofsky understood this coloration as depicting heightened pigmentation following exposure to the elements. Pächt stressed this demonstration of observational fidelity on the artist's part, remarking that it "would have been inconceivable without the most intensive study of the living model." Addressing only the Adam, he noted that the figure offers "conclusive proof that it is painted from life: the flesh tone of the head and hands is markedly darker than the rest of the body. Jan's Adam is the portrait of a man who has laid aside his clothes and stands before the painter in the pose of our earliest ancestor and in his state of paradisal nudity. Head and hands, the only parts of the body exposed to the light in the normal life of western man, are tanned by comparison with the paler tone of the rest."[21]

Ever admiring of Jan's talent, Pächt invites the reader to follow him in appreciation of the painter's unparalleled novelty. As he observes, the audacious representation of sunburned extremities suggests toil in the fields, the punishment God meted out to Adam for his disobedience. The selective depiction of exposure to the sun provides dramatic indication of such activity, something that an overall darkened pigmentation would less readily suggest. With this unprecedented device, Jan revises understanding of the first man's status after the Fall. The discrete signs of sunburn replace gestures of shame and crudely fashioned garments, drawing attention to hands as enablers of manual work. They secure Adam at the origin of the chain of activities that began with his tilling of the soil, as Cennino, we recall, had similarly done in the opening passage of his treatise.

Jan's representation of Adam powerfully visualizes claims about the embodiment of the conditions to which God condemned future generations of mankind at the moment of the Expulsion; through the first man, the Lord's decree was to be passed down for all time. Adam's body markings intensify our awareness of his destiny; they are proleptic, avowing what is to be rather than confirming what has already occurred. Jan's representation of Eve adheres to this paradigm and removes her from the cycle of guilt and shame to which church writing subjected her. God mandated that she would suffer the anguish of pain in childbirth, a situation to which Jan's near-life-sized figure alludes with exceptional bluntness. The dark line that vertically bisects Eve's belly emphatically marks the potentiality for pregnancy already represented by her pear-shaped figure; indeed, it fully "realizes" it.[22]

The line corresponds physiologically to the juncture between abdominal muscle plates that is known as the *linea alba.* As these plates spread apart to accommodate the swelling belly, the line darkens, a phenomenon usually observed around the fourth month of pregnancy. The line's appearance has been most often noted by midwives on the bodies of darkly complexioned Caucasian women; it is commented upon in gynecological treatises from the late eighteenth century on as signaling the "quickening" of the fetus during the fourth month. Because of its pigmentation, it is referred to at this stage as the *linea nigra.* Description of this phenomenon is not known to me in any medical practitioner's treatise of Jan's time, but an image of a nude pregnant woman, shown squatting in a medical miniature of about 1400, suggests an emerging curiosity about the gravid female body at a time when firsthand descriptions remain, for the most part, infrequent and imprecise.[23] On the large panel in Ghent, visual representation trumps textual silence; Jan's painted depiction inscribes knowledge of this phenomenon of pregnancy into the historical record.

For Pächt, a representation like the one Jan made of Adam could result only from close scrutiny of a living model, something we do not readily recognize as common practice for the early fifteenth century even as we celebrate Jan for his advances toward naturalism.[24] That is the only way to arrive at the knowledge of the striation on Eve's abdomen as well. Jan could have become aware of localized changes in the pigmentation of a pregnant woman's abdominal skin during his wife's several gestations. Although we cannot know that for sure, he indisputably documents intimate knowledge of the process on the *Ghent* panel, making public the private experience of subtle and seemingly miraculous physical transformation.

The exceptional body markings on both Eve and Adam are the result of unobservable processes, the sun's rays and physiological displacement; one is external to the body, and the other, internal. By using pigmentation to depict the activity of unseen agents, Jan "present[s] to plain sight what does not actually exist," the work Cennino had said artists should do and one of the qualities that Jan's contemporaries took note of in their praise of his painting. Adam's unusual pigmentation additionally makes palpable Cennino's remark that God passed down knowledge through the first man as a form of labor. Cennino noted that the obligation to till the soil that was instituted at the moment of the Fall set off a chain of exertions that included the craft of painting. He did not in any way associate shame with the work of the hands or present these labors as sinful activity. Indeed, Cennino boasted that the sources of his own knowledge went back three generations to Giotto, who had trained Cennino's own immediate predecessors and had "brought the profession of painting up to date."

On the Ghent altarpiece, Jan reaches back further than Cennino to establish the distinguished antecedent for his art. His substitution of sunburned extremities and tense musculature for garments fashioned from animal skins makes vivid the actual words with which God assigned to Adam the burden of physical labor. The Lord ordered the first man and his descendants to work hard for all eternity, with their bodies serving both as tool and source of energy. By making the wear and tear on Adam's hands so explicit on the *Ghent* panel, Jan draws attention to them as the point of origin for all human endeavor, and with that exceptional detail, he makes a dramatic claim for both the ancestry of his craft and its scriptural authorization.

The erect, unashamed depiction of Adam in the Ghent altarpiece presents an avowal about the ultimate source of all labor, one that inscribes the artist into a biblical lineage. Jan's accomplishment, in its degree of technical precision, poetic imagination, and even hubris, remained unparalleled in artistic practice until the close of the fifteenth century.[25] Only Leonardo da Vinci pursued materials and natural structures with similar confidence and obsession, producing equally singular images replete with comparably transformative signification.[26] In the northern master's work, as in the Italian's several decades later, the artist's material means—pigments and glazes—enabled, even incited, pictorial achievement.

Sharing Knowledge

In the figure of Adam, Jan created sustained evidence of sunlight's luminous power. Elsewhere on the altarpiece, Jan analyzes other effects of light as it passes through lancet windows, glances off polished gems, inverts reflections in a carafe of water, and appears to cast shadows on the surfaces of his painted panels. The capacity of sunlight to cause change in coloration is something Jan surely had observed in daily practice; Cennino, his fellow craftsman, had remarked upon the working of such processes in his recipes for the manufacture of various artistic materials. In one passage, Cennino recommended exposing linseed oil in a copper or bronze pan to summer sunlight so that the oil changes and becomes "most perfect for painting" (59). Modern commentators on the treatise disagree over what is implied here since oil will not evaporate; possibly a "bleaching" is suggested or a reduction in volume through the formation of a surface skin, one that perhaps darkens the solution. The Italian also advised that linseed oil smeared on kidskin parchment made tracing paper; once the oil dried, he observed, the parchment is rendered transparent (13).

Jan and Cennino both address critical aspects of the painter's practice in their distinctive compositions. These relationships support my claim that the Ghent altarpiece, like Cennino's *Handbook,* demonstrated its author's knowledge and skills while serving as a primer of practices for students and contemporaries. Such functions, which are assumed for Cennino's written treatise, have not been proposed for elevated painted imagery like Jan's, even though the production of the "masterpiece," which was central to guild practices for a variety of craftsmen in the late Middle Ages, entailed similar demonstration of a worker's professional mastery.[27]

Cennino, like Jan, defines "the exact proportions of a man," noting that "the handsome man must be swarthy, and the woman fair" (48–49). He provides a recipe for producing the "apples" of the cheeks, observing that his "master used to put these 'apples' more toward the ear than toward the nose because they help to give relief to the face" (46). Jan, too, as we have seen, attended rigorously to features of proportion, relief, and skin color in his delineation of the first couple, visibly endorsing the observation of nature in his nude figures as well as in his study of all the components of landscape. Cennino limits his discussion of direct observation to a brief remark regarding the painting of a mountain: "If you want to acquire a good style for mountains, and to have them look natural, get some large stones, rugged, and not cleaned up; and copy them from nature" (57). Jan's panels proclaim the reverse, an approach that brings the artist *to* nature, immersing him in unmediated examination of geological, atmospheric, and botanical forms.[28]

The *Ghent* panels are stuffed with examples of every conceivable pictorial challenge, many of which Cennino mentions. The altarpiece, like Cennino's work, provides a representative selection, a sampler if you will, of solutions to the artistic tasks confronting an ambitious artist. In the lower panels, Jan demonstrates understanding of the habitats and structures of plants and grasses from which his pigments were drawn; Cennino comments at length on such matters in his *Handbook,* providing anecdotal information that indicates the importance he places on the painter's awareness of the sources for his materials. Of the six yellows Cennino de-

PLATE 1. Jan and Hubert van Eyck, *The Altarpiece of the Lamb*, interior (detail), Eve.

PLATE 2. Maria van Oosterwijck (1630–93), still life, 1668; Kunsthistorisches Museum, Vienna.

PLATE 3. "The simple Sand is a miracle in my eye." In J. F. Martinet, *Katechismus der natuur*, 5th ed., 4 vols. (Amsterdam: J. Allar 1782–89), vol. 1.

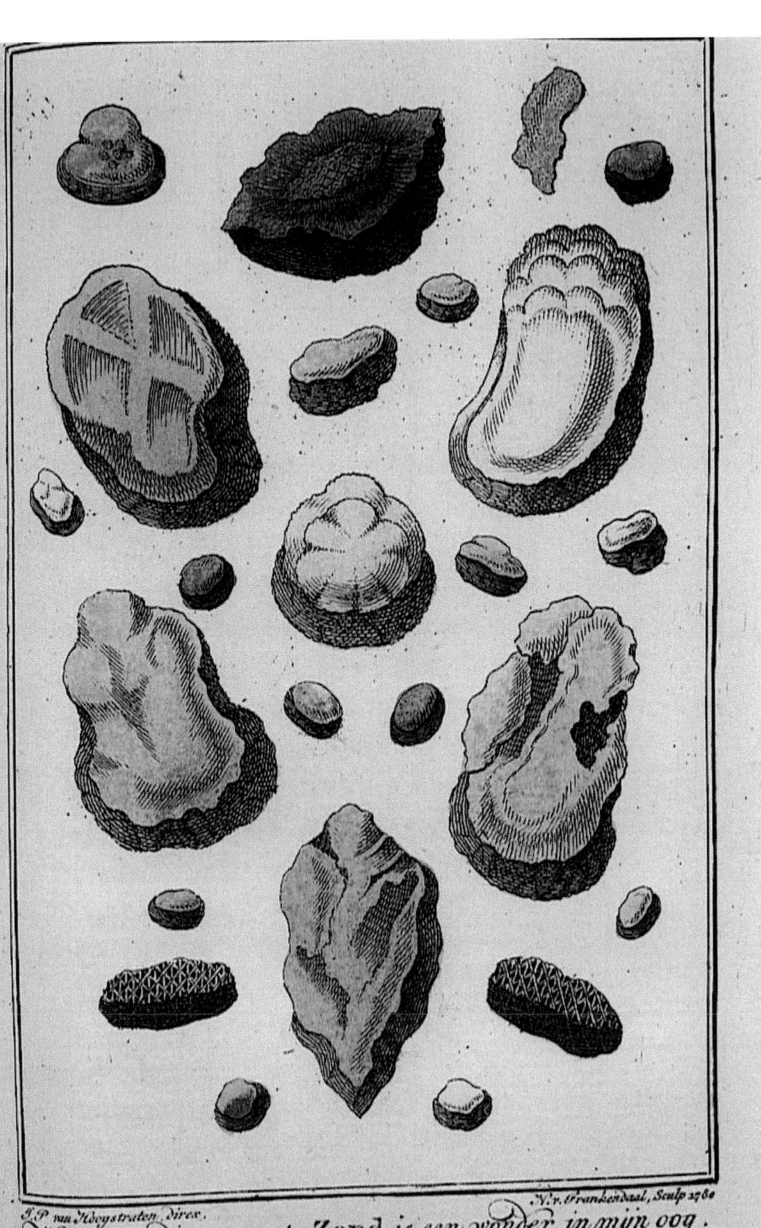

J. P. van Hoogstraten, direx. N. v. Frankendaal, Sculp 1780

Het eenvoudige veragte Zand is een wonder in myn oog. Gÿ zult het, met een Microscoop bezien, voor Keisteenen van allerleie koleuren houden, en onder dit uit den Rhijn, op den zoom van GELDERLAND gehaald, Stofgoud ontdekken.

I Druk, bladz. 195, 200. II, III Dr. bl. 195, 201. IV Dr. bl. 203, 209.

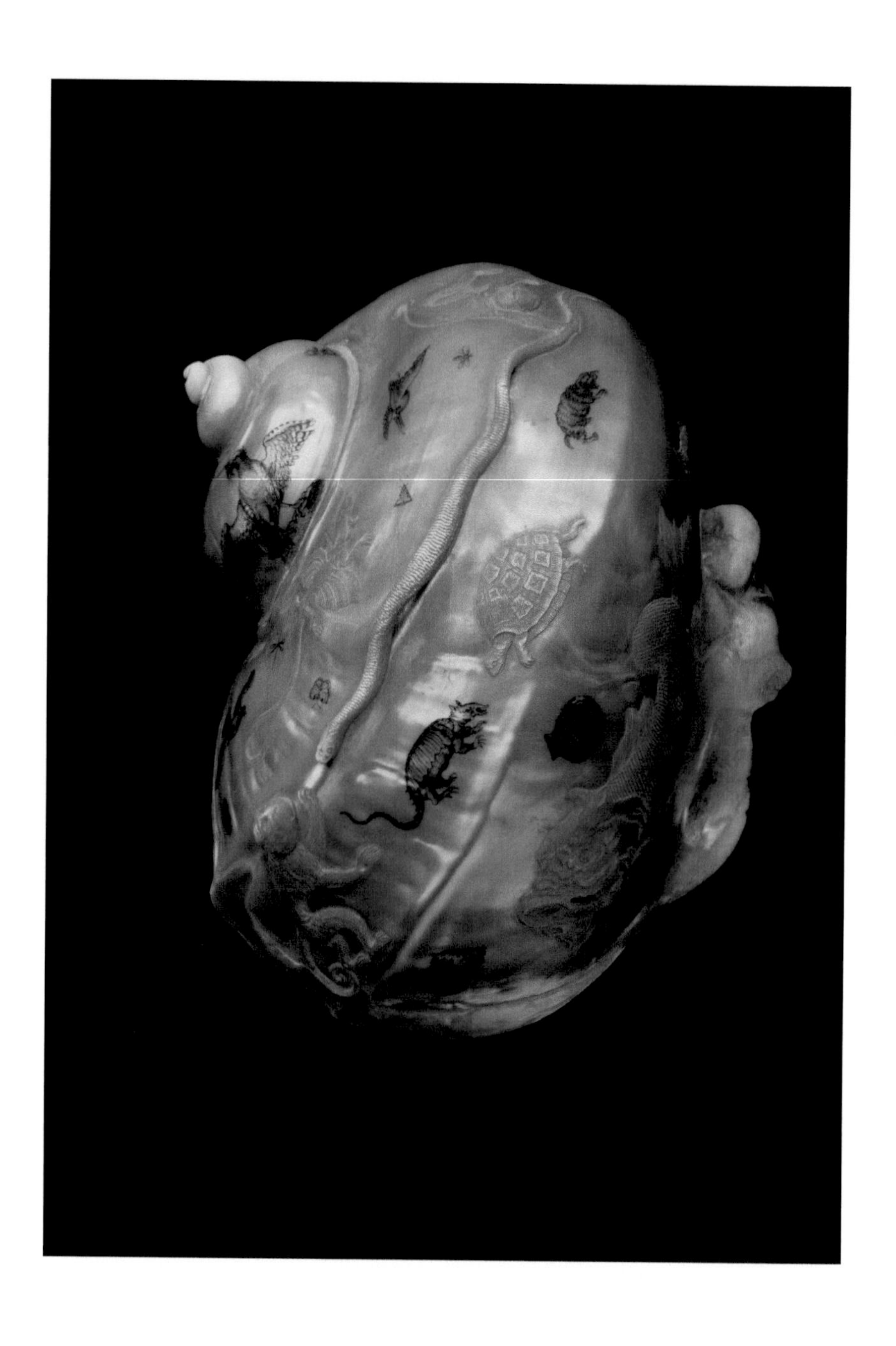

PLATE 4. Carved snail shell (*Turbo marmoratus*), seventeenth century, H: 9.5 cm, L: 16 cm; Herzog Anton Ulrich Museum, Braunschweig.

scribes, two come from minerals, one from a botanical, two are alchemically produced, and one is "manufactured." "The natural color known as ocher is yellow," he writes; it "is found in the earth in the mountains, where there are found certain seams resembling sulphur; and where these seams are, there is found sinoper, and terre-verte and other kinds of colors. I found this when I was guided one day by Andrea Cennini, my father, who led me through the territory of Colle de Val d'Esta. . . . Upon reaching a little valley, a very wild steep place . . . I beheld seams of many kinds of color: ocher, dark and light sinoper, blue and white; and this I held the greatest wonder in the world—that white could exist in a seam of earth" (27).

Cennino refers to the yellow color called giallorino as "manufactured," although it is not made by alchemy; he specifies that it "is actually mineral," since it originates in the neighborhood of great volcanoes" (28). However, orpiment, another yellow, is artificial and is "really poisonous." Although it has therapeutic value as a purgative for sparrow hawks, which are treated with it to prevent certain illnesses, he cautions his reader to "beware of soiling your mouth with it, lest you suffer personal injury," a word of warning very likely against moistening brushes or the tips of fingers with saliva during its use (28–29). Vermilion is so complicated to make that Cennino advises his reader to get it from the druggists "so as not to lose time (in its making) in the many variations of procedure" (24). Druggists are the source of other materials painters need as well: fish and leaf glue for making tinted and tracing paper (14); the sorts of sieves that are required for the preparation of the finest ground lapis (37); and pine resin for mixing into the powdered pigment (37).

Cennino advises his craftsman to resort to practices of the kitchen in the preparation of materials. One of the sources for the ground gesso coating that wood panels require before they can be worked comes from the "second joints and wings of fowl, or of a capon; and the older they are the better. Just as you find them under the dining-table, put them into the fire; and when you see that they have turned whiter than ashes, draw them out, and grind them well on the porphyry; and use it as I say" (5). Cennino recommends the use of bread crumbs for the effective removal of unwanted strokes in metal-point drawings (8) and likens the preparation of lapis lazuli to kneading, advising his reader to manipulate the pigment "just as you work over bread dough with your hand, in just the same way." The making of this precious color constitutes something of a secret, so the reader is advised to keep it to himself, "for it is an unusual ability to know how to make it properly." Finally, he remarks that young girls perform this work most successfully because "they are always at home, and reliable and they have more dainty hands" (38–39).

This anecdotal reportage, which, in several cases, provides alternative methods for producing a specific substance or accomplishing a particular task, demonstrates the utilitarian orientation of the painter's procedures and the degree to which, on occasion, they consisted of practices shared with workers in different, often more prosaic, sorts of pursuits. The painter's routine intersected with activities of the druggist and the kitchen maid, as well as the young girl, individuals whom we do not ordinarily think of associating with someone like Cennino, let alone an artist of Jan van Eyck's celebrity. Yet knowledge produced by the execution of everyday tasks was significant for the painter's enterprise, as was familiarity with skills that were learned through apprenticeship with a trained master. All workers, artists included, circulated in urban spaces, sharing forms of knowledge and professional secrets that a range of individuals labored to produce; they were linked as much, if not more, by technical needs than by a sense of group identity.[29]

Scattered throughout the references to materials and procedures that make up Cennino's handbook for artists are descriptions that appear in the far more frequently copied compilation of information known as the herbal.[30] This type of book, like Cennino's, preserves knowledge regularly handed down by word of mouth. It functioned primarily as a guide to the identification and use of botanical and mineral materials involved in the treatment of the sick. Herbals, which were produced in the vernacular and in several versions, contain information regarding the qualities and applications of indigenous, as well as foreign, plants, often presented in the way that Cennino offers his local lore, in anecdotal form. They note proper usage of botanicals as medicinal agents while also remarking on their utilization as dyes and fixatives and their importance in the kitchen. Herbals conclude, in some versions, with a discussion of the use of cosmetics: the art of makeup. As a genre, the herbal proliferated in the later Middle Ages and became one of the most popular early printed books in the latter part of the fifteenth century. Throughout this time, it remained a common, rather than elite, text, seldom rising to elevated levels of artistic refinement in its production.

The content of numerous illustrations in the herbal brings to mind information provided by some of the anecdotes in the *Craftsman's Handbook*. Indeed, the similarities are so striking in some instances that I cannot resist suggesting that Cennino's written vignettes were directly inspired by depictions in herbals. The ostensible purpose of the illustrations therein was to enable identification of particular plant specimens and to describe the places where they grew in order to forestall misidentification. Both books repeatedly caution their readers about the possibly harmful effects that might ensue from such error.

Alum, one of the first materials listed in the alphabetically organized herbal, is identified as a mordant for stabilizing fugitive colors and bringing out their hues. Cennino observes that it heightens ultramarine blue (38–39) and recommends its use to make a "very lovely pink" for application to parchment (103). He notes that alum is important for the manufacture of a red color called lac, an artificial pigment for which he provides several recipes; one of its sources is the gummy incrustation of certain trees. Cennino warns against easy efforts to extract lac from previously dyed cloth that contains alum, since that practice, he reports, results in unstable color (26).

Analysis of Jan's panels indicates that he pursued a variety of strategies to ensure the stability of his colors. Alum was one of the binding agents with which he experimented, presumably with an eye to enhancing the hues of his paintings. Pine resin, which raises the refractive index of pigments and contributes to the luminosity of colors, was another substance he employed to hold his powdered paint. And for reasons having to do most likely with the stability of color, Jan applied linseed oil to the polished gesso surfaces of his panels. This prevented pigments from sinking into the ground and becoming deformed through loss of clarity or brightness.[31]

Alum is also a household remedy for use in the treatment of illnesses; herbals recommend it both as an emetic to provoke regurgitation and as an astringent for tightening the skin. Artists' familiarity with the mineral salt connects them to suppliers and dispensers of the substance—pharmacists and physicians—with whom they would have crossed paths in daily life. The association of image makers with professionals involved in what we may loosely describe as health care is visually documented through a repeated motif. Jan frequently positions a carafe filled with clear liquid in the proximity of the Virgin in his paintings of her. In the Ghent polyptych, this vessel appears near the window in the depiction of the Annunciation on the outer wing, sunlight streaming through its glass. This sign of purity, understood to indicate Mary's chaste impregnation, depends for its signification upon a common practice in medical diagnosis. Contemporary manuscript illuminations record physicians, standing at bedside, studying similarly shaped, urine-filled vessels, frequently holding them up to the light: the clearer the contents, the healthier the patient.[32] St. Cosmos, one of the two physician saints in Rogier van der Weyden's *Madonna with Saints* (the "Medici" Madonna), in Frankfurt, conspicuously holds such an object as his attribute (fig. 2.7).[33]

Painters, who undoubtedly had personal exposure to this basic diagnostic procedure, were also involved in the production and dissemination of the claims to knowledge upon which it depended. A document reporting on the activities of Jan's distinguished Italian contemporary Gentile da Fabriano notes that the celebrated artist was engaged by a prominent doctor to create a colored urine chart for a treatise on the diagnosis of disease.[34] Such charts, many of which survive, present different hues corresponding to levels of clarity in urine; each hue is associated with a particular ailment. Physicians determined the nature of a patient's malady by matching the hue of the urine in the flask with one of the depicted colors. Unfortunately, Gentile da Fabriano was forced to leave town during an epidemic and was unable to finish his task. The existence of such charts, sometimes appended to herbals, provides us with evidence of the popularity of this diagnostic tool.

Cennino's treatise closes with a discussion of cosmetics.[35] He describes the way in which colors can be removed from flesh after a face has been painted and then launches into a paragraph on the perils of indulgence in the application of excessive makeup. He advises "certain young ladies of Tuscany" to make a practice of washing in spring, well, or river water, warning that if they use any artificial preparations their faces will wither and their teeth grow black (123). This passage, which recalls like-minded cautionary texts frequently found attached to herbals, contains a warning against the unhealthful effects of certain substances and colors and condemns the use of pigments for purposes of concealment. Artifice, Cennino implies, is the negative outcome of the abuse of art's proper materials.

Painterly practice was thus intimately intertwined with pursuits that had related needs for specialized substances and information. Jan's exploitation of materials and techniques to fix colors and capture the elusiveness of illumination expanded upon the know-how of pharmacists, while his scrupulous rendering of the structural details of a wide array of substances suggests familiarity with the analyses of botanical and mineral specimens that could be found in contemporary herbals and that were known to doctors. The knowledge on display in Jan's panels thus aligns him with diverse workers and professionals; the intersection of his life with theirs through the exchange of materials provided the route through which expertise was shared, even as individuals, like Jan, were independently involved in the advancement of their distinct labors. The success

Fig. 2.7. Rogier van der Weyden, *Madonnna with Saints,* 1450; Städel Museum, Frankfurt am Main, Germany.
Photo: Foto Marburg / Art Resource, NY.

of Jan's activities, like those of the doctor and the botanist, depended on public trust in evidence brought forth through observation and the deployment of visualizing skills; in the exercise of this talent, artists were one group among many.[36] Jan van Eyck was widely recognized in his own time as foremost among them.

Conclusion

Jan may not have been the "inventor of oil painting" in the way that earlier generations claimed and some art history books still assert, but his signature use of oils and his preoccupation with the depiction of the properties of light contributed to improvements in the handling of glazes by subsequent artists in Italy as well as the Netherlands.[37] His success in portraying the activity of the sun's rays as they produce reflections, transparencies, and transformations in pigmentation indicates intense study of the circumstances surrounding such phenomenological occurrences, not merely interest in their dazzling results. The simulation of so many of the enigmatic effects of light in details of the Ghent altarpiece imparts to this unprecedented work the quality of a pictorial instructional manual for contemporary artists, as well as a demonstration piece for the expertise of its master painter's shop. The written comments of those who saw the altarpiece confirm that Jan's "painted handbook" passed on knowledge of his achievements and served his successors as both compendium of learning and guide to practice; it contributed as well to the construction of a legend regarding his accomplishments in depicting miraculous luminosity in paint.

Although I have not read the polyptych as a theological program, as is usually done, in which the images are viewed through exegetical commentary, I have regarded it nonetheless as a kind of "text" and suggested that its closest analogue is the contemporaneous handbook of artist's recipes prepared by Cennino Cennini. Like the Florentine craftsman, the Flemish artist addressed an array of artistic challenges and introduced his viewer to a broad range of techniques, all the while documenting successful solutions to vexing problems. The response to Jan's panels in Ghent, both verbal and visual, confirms the special interest artists had for the work's unique properties. Van Mander likened the painters and art lovers who crowded the chapel to view the altarpiece to "bees and flies in summer swarming and hanging around baskets of figs or raisins." The Sevillian painter Francisco Pacheco, father-in-law and teacher of Velázquez, singled out the Ghent altarpiece as the van Eyck brothers' best work in the treatise on painting he wrote around 1625. He noted that the Spanish king had commissioned a copy of the highly praised altarpiece from the painter Miguel Coxie, and that Titian had been asked to provide the special blue pigment (lapis lazuli) necessary for its faithful execution (political activity in the eastern Mediterranean, where it is uniquely found, had made its acquisition extremely difficult). Van Mander subsequently included the anecdote about Coxie in his own work, noting as well Jan's investigation of numerous kinds of paints, an activity that he likened to the practice of alchemy.[38]

As the most publicly available of Jan's works, the Ghent altarpiece was the site where his exceptional innovations were made visually available to subsequent generations of artists throughout Europe. For centuries after the artist's death, the panels functioned as practicum and primer for a select, professional audience, putting on display, not just "the whole art of painting" as Münzer tersely proclaimed, but documenting, as Otto Pächt would so perspicaciously remark in our own time, Jan van Eyck's "cognitive breakthrough."[39]

CHAPTER THREE

Explosive Affinities: Pyrotechnic Knowledge in Early Modern Europe

SIMON WERRETT

In his 1949 *History of Fireworks,* the pyrotechnician Alan St. Hill Brock claimed that a revolution in pyrotechny had occurred at the beginning of the nineteenth century. Ever since their explosive entry into the spectacular festivals of European court culture in the fifteenth century, fireworks had always been the color of natural "white" fire. But with the revolution in chemistry of the late eighteenth century, a series of pioneers in France and America used metallic salts in combination with gunpowder to create brilliant and brightly colored flames, whose luminous colors still remain part of the experience of fireworks today.[1] Contemporaries also appreciated the powers of science to transform a humble art via the application of chemistry. James Cutbush, Philadelphia chemist and author of the vast *System of Pyrotechny* (1825), took it as self-evident that the progress of pyrotechny was due exclusively to the application of chemistry: since fireworks "cannot be understood without a knowledge of chemistry, it is obvious that science is a powerful aid to pyrotechny."[2] Brock argued the same and shared with other historians of fireworks the view that chemical expertise was the proper qualification for assessing the history of pyrotechnics. As did the chemist J. R. Partington in his celebrated *History of Greek Fire and Gunpowder* (1960), Brock evaluated historic fireworks recipes by comparing them to current pyrotechnic knowledge, identifying the pioneers who had contributed most toward advancing the chemistry of fireworks effects.[3]

Brock's and Partington's chemical assessments of historic pyrotechny reflect an ancient approach to imagining the arts that assumes their progress was necessarily achieved through the application of science, an assumption that dates back to Baconian visions of human arts improved by natural philosophy and the *Encyclopédie*'s assertions that practice could advance only via the addition of theory under the analytical gaze of the expert savant. Against an allegedly closed, secretive, and interested form of artisanal knowledge, savants posed an open, disinterested, and collaborative vision of scientific knowledge, which, by making explicit the tacit secrets of the arts, would render them amenable to enlightened management and scientific improvement. The history of arts then entailed locating moments when such enlightened application occasioned progress toward current techniques.

An alternative vision poses the relationship of arts and sciences as moving in the reverse direction. Historians have now long argued that the new sciences of the seventeenth and eighteenth centuries were themselves deeply informed by the practices and knowledge of artisans. Experimental sciences adapted artisanal interests, skills, and epistemology to the study of nature, formalizing tacit competencies into more systematic procedures for knowing the natural world. On this reading, enlightened management of the arts is often read less as an inevitable contribution to artisanal improvement than as a disciplinary tactic to gain control over artisanal communities and practices, rendering them open through systems of scientific surveillance.[4] And as boundaries between the arts and sciences are questioned, so crude oppositions of open scientific knowledge and artisanal secrecy are rejected in favor of a more complex view in which both practices entailed secretive and open formulations of knowledge.[5]

This essay adopts neither a progressive nor a revisionist approach to the relations of artisanal and scientific forms of knowledge but instead proposes a more social and reciprocal reading of their historical interactions. Brock's tale of progress through chemistry provides the starting point, but I identify an alternative set of events as leading to the transformation of fireworks into applied chemistry in the early nineteenth century. I take my cue from social studies of scientific controversy and offer a case study from the eighteenth century in which the relationship of the arts and sciences was repeatedly contested by a variety of groups of practitioners who sought to gain authority over the proper definition of pyrotechnic knowledge.[6] Visions of pyrotechnic history that begin with a chemical perspective miss the fact that in the eighteenth century there were many different forms of knowledge to which pyrotechny could be allied. Different communities of pyrotechnic practitioners and commentators posed different forms of knowledge as authoritative, including painting, poetry, and architecture, in addition to physics, mathematics, and chemistry. It was by no means self-evident that chemistry would eventually come to be closely associated with fireworks. That it did so, in the event of the transformation of fireworks from natural to colored fires, is here presented, not as the result of the inevitable convergence of science and art, but rather as the product of a series of complex social interactions and disputes which spanned the eighteenth century.

These disputes centered on two very different communities of pyrotechnicians. State gunners made and performed fireworks for the courts in northern Europe beginning in the fifteenth century and approached the making of fireworks following priorities and values emergent in European military practice. Fireworks were also made and performed by private families of skilled pyrotechnicians, but the form of their community and concomitant approach to pyrotechnic knowledge was quite different from that of the gunners. The first part of this essay sets out the different social forms and ways of managing pyrotechnic knowledge that each community practiced. Until the mid–eighteenth century, these communities operated within relative isolation, but in the 1740s, one of the most celebrated of pyrotechnic families, the Ruggieri of Bologna, traveled north from Italy to Paris, where their fireworks displays captivated royal and public audiences and established them as France's premier pyrotechnicians for many years. The second part of the essay follows the Ruggieri's fortunes and the reactions of different communities to their work in the midcentury. Now pyrotechny became a highly contested art, as disputes arose between the Ruggieri and the gunners traditionally in charge of French pyrotechny and as a variety of savants sought to gain authority over pyrotechnic knowledge. Travel thus brought together different pyrotechnic communities and their ways of knowing and prompted prolonged controversy. Participants used alliances with different forms of knowledge to promote their positions in this contested domain, and it is in this context, I suggest, that chemistry was adopted as a means to give distinction to the Ruggieri's practice. While the Ruggieri presented chemistry as a self-evident component of the progress of pyrotechnic art, it was never obvious that this should be the case, even for the Ruggieri. That chemistry and fireworks converged was rather the contingent outcome of the travel of pyrotechnicians and the clash of ways of knowing it produced. The proper relation of the arts and sciences was part of these contests over pyrotechnic authority and took variable and unpredictable forms.

The same was true of questions of openness and secrecy, which played an important role in the pyrotechnicians' debates. The fate of the Ruggieri in eighteenth-century Paris supports historians' criticisms of essentialized definitions of scientific and artisanal knowledge as open and secretive respectively, but also problematizes critiques of such essentialism that highlight the openness of art and the secrecy of science. Here I suggest that openness and secrecy were less elements of one form of knowledge or another than accusations or claims operating as tactics in efforts to forge alliances and gain distinction in the contested field of pyrotechnic knowledge. Different forms of pyrotechnic communities had different opinions about the degree to which pyrotechnic knowledge should be explicit, and when knowledge was not open, it was not necessarily secret. Accusations of secrecy could often serve to attack competing communities of practitioners. When gunners accused the Ruggieri family of secrecy, they did so only in contexts where such accusations might prove damaging. At other times, competitors' practices could be identified with openness. Approaches to openness and secrecy were thus, like the relations of pyrotechny to other arts and sciences, variable and complex. The same practitioners could adopt different positions at different times and did so depending on the social context of the current controversy over pyrotechnics.

The history of enlightened pyrotechnics thus offers a way to address and reconsider broader issues of the historical relations of the arts and sciences and their changing ways of making knowledge. This relationship, as the Ruggieri's experiences suggest here, was reciprocal, interactive, and frequently contested and brings to light the importance of considering a range of different communities and their ways of knowing in examining the history of science and art. This essay is also intended as a contribution to growing calls for a geography of knowledge, rather than a history of science, in early modern Europe, since understanding the interactions of art and science, and the emergence of different visions of their relationship, cannot be divorced from a mapping of the historical movements of artisans and other knowledge communities as they traveled around Europe in the early modern period. Understanding the "convergence" of artisanry and science, however this is manifested, demands tracing the physical convergence of communities and the contests and alliances which their coming together entailed.[7] That pyrotechny became allied to chemistry was ultimately a result of such movements.

Gunners and the Rise of Pyrotechnic Knowledge

In early modern Europe, executing fireworks could be a highly rewarding practice. Princely power depended on theatrical display, and fireworks formed central elements in the repertoire of court spectacle. Princes rewarded skilled pyrotechnicians with promotions, gifts, and official ranks, such as *artificier du roi* in France and Chief Firemaster in England. Such rewards prompted competition among an international array of pyrotechnicians, who constantly sought to perfect their techniques and provide novel or improved performances. Pyrotechnic knowledge therefore carried a high premium.

For many courts, the local state artillery provided pyrotechnic skills. Gunners in government employ spent much of their time making and testing explosives, cannonballs, and cartridges, but they turned to the production and display of fireworks when called upon. Early modern terminology reflected this martial emphasis. "Firework" referred to artillery for both war and "recreation," while "fireworker" in England, *Feuerwerker* in the Germanies, and *Feierverker* in Russia were artillery ranks, not an occupation. Treatises on "pyrotechny" typically discussed military fireworks, with smaller sections devoted to fireworks for recreation or triumphal displays. Until the eighteenth century, works dealing exclusively with festive fireworks were rare.

Gunners' knowledge was site specific. Beginning in the late seventeenth century, the chief site of fireworks production was the "laboratory," an explosives workshop typically located inside a city's fortress or the state arsenal.[8] Gunners manufactured ammunition and fireworks in wooden huts and courtyards, overseeing work by hired laborers on rockets, serpents, stars, and other pyrotechnics. Gunners also built and executed displays and in France and England often designed, or "invented," the (often allegorical) scenery, though by 1700 court architects and poets had taken over this task.

Gunners managed a complex array of skills, recipes, and technical knowledge. Pyrotechnic knowledge was composite, drawn from a variety of printed books, handwritten notes, records of technical procedures, and drawings, accumulated in laboratories by practitioners. From early on, efforts were made to systematize this knowledge; fireworks laboratories, like contemporary herbaria and academies, were "centers of calculation," compiling knowledge from distant localities through variegated networks and processing it for use in practice.[9]

Knowledge gathered for the laboratory was kept and used locally, reflecting the private, indeed fortified, space of the laboratory. Often recipes and techniques were compiled and recorded in illustrated manuscript books, which neatly laid out procedures otherwise too complex to commit to memory.[10] Like other artisanal "how-to" books, pyrotechnic manuscripts might also be prepared as gifts for officer-patrons, as gunners sought prestige for their art. Certainly, skilled artists, such as the Italian engraver Stefano della Bella, were employed to produce these works (see fig. 3.1).[11]

Fig. 3.1. Stefano della Bella's watercolor of a Parisian artificer lighting the fuse of a rocket before the Pont de Tuileries and Louvre. From *Traicté des feux artificielz de joye & de recreation* (Paris, ca. 1649), fol. 37. Reproduced by kind permission of the Bibliothèque Nationale de France.

There is little to suggest that gunners tried to keep pyrotechnic knowledge secret. No doubt certain recipes were restricted to laboratory personnel, but gunners were also responsible for publishing many pyrotechnic books in the seventeenth and eighteenth centuries, often explicitly attacking pyrotechnic secrecy. Casimir Siemienowicz, Lithuanian gunner and lieutenant general of the Polish Ordnance, included a long diatribe against pyrotechnic secrets in his *Artis Magnae Artilleriae* (*Great Art of Artillery*; Amsterdam, 1650), the most influential treatise on fireworks of the period. Siemienowicz compared "pyrobolists" who allegedly kept their knowledge secret "with the canting Alchymists of the Times past . . . who, tho' they dealed in nothing but Smoke, yet arrogantly took upon them to be Professors of so noble and excellent an Art as Chymistry." These were "sooty Adepts" with a "Science of Diabolical Extraction . . . immediately derived from Hell itself." Siemienowicz claimed to have seen numerous books of pyrotechnic secrets, which he had tried for himself only to find them "nothing but mere Smoke, the Effluvia of distemper'd Braines."[12]

Promoting openness was supposed to cultivate order. Authors argued that explicit practice might render artillery and pyrotechny more disciplined and so more effective: "Gunnes your Majesty hath, but want Gunners, because they want Respect and Encouragement; let Occasions be ruled with Reason, Wars managed with Discipline, Judgement and Pollicie."[13] Pyrotechnic manuals might eliminate sloth. Hence, Colonel Armstrong, surveyor general of the Ordnance, advocated translating Siemienowicz's *Artis Magnae Artilleriae* into English, so "that by naturalizing the most celebrated *Author* in this kind, a great Step would be taken towards recovering our *Pyrobolists* and *Fire-Workers*, from the Lethargy they seem to have been wrapped in for many Years past."[14]

To give order to their practices, gunners divided books into sections on fireworks for war and for recreation, each consisting of a series of short chapters devoted to the preparation of a particular type of material, manufacturing instrument, or firework and its variants, with a description of its components and dimensions and instructions for manufacture. Brief explanations of the firework's effect and instructions for firing it could also be added, and most chapters had illustrations of the finished piece. Implicit in this organization was a deference for geometrical reasoning. If nineteenth-century pyrotechnicians thought chemistry indispensable to their art, seventeenth-century gunners thought the same about mathematics. Historians have often noted the use of mathematics as a tool for disciplining early modern artillery, and the same was true for pyrotechny, often in the same books.[15] Verses dedicated to London gunner John Babington in the 1630s proposed that his book would

> deck the ayre
> With splendant stars, silver and golden showers;
> These are th'effects of *Mathematick* powers.[16]

Siemienowicz argued that pyrotechny must be based on a few clearly understood compositions, based on an experimental knowledge of their qualities. He had put, he said, "none into Practice till I had examined them by an exact Arithmetical Calculation, Geometrical Demonstration, and by solid Arguments drawn from Natural Philosophy."[17] Discussing the rocket molds on which rocket cases were made, Siemienowicz noted: "Workmen do not always observe the same Proportions in the Dimensions of them . . . and in this they verifie that Proverb, *As many Men, so many Minds*."[18] He then constructed a figure giving a series of heights for rocket molds in a steady proportion to their diameters, to produce rockets in a range of defined weights (see fig. 3.2).

Ken Alder has contrasted seventeenth-century artillery books with the highly mathematical knowledge of French engineers in the eighteenth century. Alder emphasizes how earlier artillerists relied on secret recipes and the tacit skills of the "gunner's eye" for accuracy rather than math-

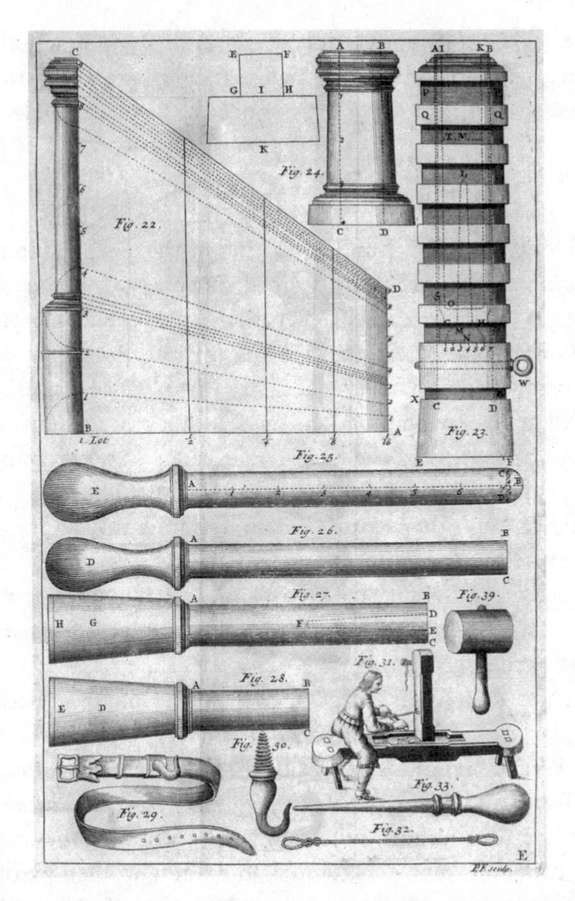

Fig. 3.2. In the engraving, figs. 22–24 show geometrically proportioned rocket molds. From Casimir Siemienowicz, *The Great Art of Artillery*, trans. George Shelvocke (London, 1729), opp. p. 144. Reproduced by kind permission of the New York Public Library.

ematics.[19] While the contrast is true to a degree, later engineering practice tended to intensify gunners' existing practice. If earlier gunners did not articulate tacit skills, this was perhaps because they wrote for other gunners, entailing much "assumed knowledge," which gunners took for granted, or "unrecognized knowledge," which gunners did not notice as significant to their practices.[20] While treatises gave exhaustive accounts of the manufacturing process, they rarely, at least in the seventeenth century, described the sites of pyrotechnic manufacture or the personnel and physical skills requisite to these places. This was not due to an epistemology fundamentally opposed to a later "expository" and mathematicized form of knowledge—gunners promoted this approach from early on. Rather, until the scale and number of persons involved in pyrotechnic production substantially increased, demanding knowledge that might travel from place to place, knowledge of tacit skills did not need to be objectified.[21]

Keeping It in the Family: The Ruggieri

The social form of pyrotechnic production in the arsenal and the military preference for explicit procedure helped shape a body of public pyrotechnic knowledge in the seventeenth century. Fireworks laboratories entailed a mixture of artisanal craft practice and military formality and acted as centers of calculation. In many places, this knowledge was allied with the discipline of geometry and mathematics, a model for disciplining unruly practice. But beginning in the 1740s, definitions of what counted as valid pyrotechnic knowledge and practice began to shift. This was largely due to the appearance in northern Europe of an Italian family of pyrotechnicians who dramatically altered the nature of courtly fireworks displays and presented a quite different tradition of more strictly "family" fireworks knowledge. This had a significant impact on the local gunners, and their reactions in turn affected the family's practice.

Besides the military, families of pyrotechnicians also performed fireworks for early modern Europe's courts. Fireworks skills often passed through several generations of a family dynasty, making names such as Ruggieri, Torré, and Brock synonymous with fireworks in the eighteenth century.[22] Typically, fathers, sons, and brothers headed the enterprise, but wives and daughters also contributed, and a few women pyrotechnicians became celebrated in their own right.[23] Many families hailed from Italy and traveled around Europe to display their talents. Their approach to pyrotechnic practice and knowledge differed from that of the gunners, however. In Italy, fireworks were closely connected to theatrical productions—indeed, festive fireworks originated in Italy as *intermezzi* to liturgical and later secular drama. Italian families' pyrotechnic knowledge and skills were thus frequently linked with a broader knowledge of stage scenery design and the creation of theatrical effects.[24]

Without the security provided by an institution like the artillery, pyrotechnic family fortunes sometimes depended on maintaining secrets as a source of independence and wealth.[25] Unlike the gunners, no members of the most famous families of pyrotechnicians published books on fireworks in the eighteenth century. However, the reason behind this lack of publication was not necessarily a deliberate attempt to withhold knowledge. Close-knit families did not need to make their methods explicit or public because this knowledge was passed down by apprenticeship and experience without the need for formal, published treatises. Furthermore, enduring patronage relationships could provide family pyrotechnicians with an income no less secure than that provided by more formal institutions, in which case secrecy was less pressing.[26] Even though family knowledge was not public, it was not necessarily secret, although, as Siemienowicz's attacks suggest, it might be construed as such by rivals.

While family techniques remained outside public knowledge, the only means to secure them was by hiring family members, and this strategy was extremely successful. By the mid–eighteenth century, one Italian family in particular supplied fireworks for numerous European courts. These were the Ruggieri of Bologna, whose fireworks became famous across Europe in the eighteenth century.[27] The Ruggieri made their name in France, first appearing in Paris around 1743, when five brothers, Gaetano, Petroni, Antonio, Francesco, and Pietro, the eldest and leading member, accompanied the Comédie Italienne to Paris to provide fireworks *intermezzi* for their dramas.[28] Their performances followed the Italian theatrical formula, taking place between acts. Known as *"spectacles pyriques,"* these consisted of displays of fixed and rotating fireworks, set on iron axles on the stage where the Comédie performed. They included "various transformations of different figures of fire . . . which succeed either by horizontal, vertical or inclined rotations, or by the interposition of figures varied by their arrangements or their shapes; such as suns, girandoles, pyramids, cradles, fountain jets or cascades, wheels, globes, crosses, polygons and pointed stars."[29]

Initially integrated into dramatic performances, the fireworks soon became dramas in themselves, advertised with titles such as "Magical Combat," "Gardens of Flowers," "The Forges of Vulcan," and "The Palace of Fairies."[30] The Ruggieri also innovated in technique, introducing an ingenious mechanism allowing fire to be communicated from fixed to moving pieces in a variety of complex combinations. While gunners utilized their artillery skills and fired fireworks one at a time by hand or in sequence using trains of gunpowder, the Ruggieri's method was peculiar to the recreative pyrotechnician, demonstrating the distance of Italian pyrotechnic skills from artillery practice. The Ruggieri's *spectacles pyriques* were a smash with the Parisian nobility and soon led to the Italians' appointment as *artificiers du Roi* by Louis XV. From the 1740s onward, the Ruggieri executed many grand fireworks for the king and City of Paris.

The Ruggieri family were typical of pyrotechnic clans, publishing no accounts of their works and depending on exclusive techniques to ensure their success. Knowledge traveled only with members of the family and not in books. It is thus easy to see how the spread of the Ruggieri's style of pyrotechny directly followed the movements of family members and their confidents. In 1749 Gaetano Ruggieri traveled to London with another Bolognese pyrotechnician, Giuseppe Sarti, to perform a fireworks display to celebrate the peace of Aix-la-Chapelle. Gaetano was rewarded with a position in London's fireworks laboratory, where he stayed for many years. Sarti moved to Russia, where he was given work in St. Petersburg's arsenal laboratory for his magnificent Italian fireworks. Back in France, Petroni Ruggieri's sons Michel-Marie and Claude-Fortuné, both born in Paris, continued the family tradition into the 1840s. Their sons in turn oversaw a new fireworks business until the 1890s, and the Ruggieri company continues to perform fireworks to this day.[31] The Ruggieri family thus came to control a substantial part of European pyrotechnic practice in the eighteenth century, with significant consequences for the forms of pyrotechnic knowledge.

An important difference between the Italian pyrotechnicians and the gunners was the former's emphasis on the theatrical and narrative components of displays. The Ruggieri's spectacles were "poems" and made reference to classical history and mythology, while the "machines" from which they fired fireworks were designed according to the rules of ancient Roman architecture. Although the Ruggieri did not publish books on fireworks practice, they did collect a substantial library of royal festival literature, prints, and poetry.[32] Besides serving as a functional collection of the kind discussed by Claudia Swan in this volume, the library evinces the importance given by the Ruggieri to allegorical "invention" in addition to practical "execution" in pyrotechny. Generally, Italian pyrotechnicians' knowledge joined artisanry with *artistry*, rather than *artillery*. This poetic style made Italian pyrotechny especially attractive to new classes of polite, educated society, who were much enamored of Italian artistic culture at this time.

One consequence was that enlightened savants increasingly sought to claim authority in the territory of pyrotechny. In fact, French fireworks had already begun a transition from gunners' displays to Italian theatricals before the Ruggieri arrived. Parisians first witnessed fireworks in the Italian style in 1649. The design of the spectacle had been given to an architect, much to the annoyance of the gunners, who traditionally produced French displays.[33] When the Ruggieri arrived, this trend intensified. Parisian men of letters, who placed a high value on classical culture and the Italian fine arts, took up the Ruggieri's style enthusiastically and recast pyrotechnic knowledge in a form amenable to the Italians but not to the gunners. Gunners, in turn, reacted angrily to the Italian incursion.

Fireworks in the Republic of Letters

The earliest response to the Ruggieri was enthusiastic. In 1747, encouraged by the popularity of Italian fireworks in Paris, the French architect, fortifications engineer, and celebrated voyager to South America Amédée-François Frézier reissued his *Traité des feux d'artifice pour le spectacle,* originally published in 1706.[34] Frézier was no gunner, having trained in architecture in Italy, and used his treatise, not to discipline practitioners, but as a means for social distinction. In the *Traité,* Frézier made his own skills authoritative, stressing the value of artistic and aesthetic knowledge above gunners' pyrotechnic skills as the most essential component in producing fireworks displays. Hence, he dispensed with the military part of pyrotechny to focus exclusively on fireworks for spectacle and denied that pyrotechnic artisanry alone could produce successful performances: "It is not enough to have made a pile of many fireworks in order to compose a pleasure fire if one does not know how to create effects by an ingenious arrangement in a specially made theater, decorated with those things that characterize the subject of the festivities."[35]

Such prescriptions well fitted the Ruggieri's new practice, and Frézier praised the family for moving beyond mere artifice to consider the arrangement of displays and particularly for their "genius" in manipulating fire in the *spectacles pyriques:* "as soon as one possesses the elements of artifice, all that remains is to have a genius for the mechanics of the disposition and communication of fires in order to foresee the effects. . . . it is necessary to be like the sieurs Ruggieri, entrusted with giving free rein to their imagination to divert the public."[36] Frézier held up the Ruggieri as exemplary and celebrated their knowledge of classical festivals, iconography, and theater, which he thought requisite to pyrotechny. French engineers of the early eighteenth century often embraced the idea of re-creating Rome in France. In the *Traité,* Frézier supplemented a traditional account of pyrotechnic manufacture with a description of the origins of fireworks and festivals in ancient Greece and Rome.[37] He insisted that the objective of fireworks must be to recapture ancient magnificence by centering the display on a "theater" of temporary, classical architecture, employing the requisite symbols and themes drawn from ancient history and mythology.

Again, this fitted the Ruggieri's practice, but it also placed importance on savant-architects like Frézier. Fireworks depended on "architects, painters, and sculptors, but particularly those men of letters who know how to present through agreeable ideas subjects which give occasion to rejoicings."[38] While Frézier praised the Ruggieri, he represented them as subservient to the authority of savants.

Frézier's perspective, however, was not typical. A few years after the publication of the *Traité,* new articles on fireworks appeared in the *Encyclopédie* of Diderot and d'Alembert. These emphasized the division of labor between savants and pyrotechnicians already posed by Frézier but left little room for praise of the Ruggieri. Instead, following the editors' enthusiasm for exposing the secrets of the trades, the *Encyclopédie* sought to make transparent the Ruggieri's practice to the advantage of French gunners.

The *Encyclopédie*'s article "Feu d'artifice" was not authored by a gunner, nor even a Ruggieri, but by a savant, the playwright Jean-Louis de Cahuzac. Cahuzac proposed a clear Cartesian division and hierarchy of labor in pyrotechny, further raising the savant above the pyrotechnician. Cahuzac sharply divided the artistic side of designing pyrotechnic displays from the labor of making and executing fireworks. Cahuzac deemed the latter to be the "physical" part of pyrotechny, and noted: "I do not believe I have to touch upon these objects. I have only sought to know them in so far as they seemed related to the great spectacles that the King, the cities, the provinces, etc. offer to the people on solemn occasions; there seemed to me in this case, a need to submit to the general laws, which have always been the rule of all the arts."[39]

Gunners' knowledge was thus subordinated to a separate article, "Feux d'artifice (artificier)," while Cahuzac examined these "general laws." The kind of specific knowledge of architecture, theatrical effects, and arrangement valued by Frézier and the Italians was itself subordinated to a higher, more metaphysical knowledge of aesthetics, belonging exclusively to the savant:

> It is in the nature of things that all spectacles represent something. Now, when we paint objects without action we represent nothing. The movement of the most brilliant rocket, if it does not have a fixed aim, displays nothing but a trail of fire that vanishes into thin air. . . . Those fireworks that only represent a kind of repetition through the play of different colors, movements and brilliant effects, and the decoration

> on which they are placed, no matter how cleverly designed, will never amount to anything more than the frivolous charms of paper cutouts. In all the Arts it is necessary to paint. In the one that we call *Spectacle,* it is necessary to paint with actions.[40]

Thus, the effects, decoration, and arrangement that Frézier in theory, and the Ruggieri in practice, had advocated as raising them above French gunners were here posed as insufficient for truly impressive fireworks. Cahuzac posed the appreciation of painting as a model for the assessment of fireworks, and in his view, the man of letters was not just a collaborator but a judge of pyrotechny, superior to all.

Meanwhile, the *Encyclopédie*'s article on "artifice" was authored by Parisian fireworker Jean Charles Perrinet d'Orval.[41] Perrinet d'Orval represented his art according to the tradition of gunners' treatises, with accounts of recipes accompanied by plates showing the techniques and tools for manufacturing pyrotechnics (see fig. 3.3). There was no sign here of Cahuzac's generalizing intellectualism, the emphasis being on safety and procedure rather than pretentious aesthetics.

Perrinet d'Orval's article corresponded well with the aims of Diderot and d'Alembert for the *Encyclopédie.* In many respects, the *Encyclopédie* followed the gunners' traditional concern to systematize and make explicit the procedures and skills implicit in craft and manufacture. Diderot, like the earlier gunners, celebrated the arts but had no time for what he viewed as the artisans' secrecy. Craft techniques were to be laid bare in transparent descriptions which stripped practices of their idiosyncrasies to present generalized procedures for each trade.[42] This would have posed no problem to the gunners responsible for manufacturing fireworks, though the savants' claim to authority over the process of explicating artisanal practice contrasted with the traditional authority of gunners themselves in describing their trade. Nevertheless, Perrinet d'Orval was evidently comfortable writing for the *Encyclopédie,* and more importantly, Diderot's interest in exposing secrets coincided with his own.

An issue of particular importance to Perrinet d'Orval was the Ruggieri's method of communicating fire from a fixed to a moving firework, the key to their success in France and Frézier's approval. In the *Encyclopédie,* Perrinet d'Orval took pride in exposing the technique: "The secret of this communication of fire was brought from Bologna to France in 1743 by the sieurs Ruggieri. . . . The author of this memoir, having found out this secret, had the pleasure to announce it to the public in his treatise

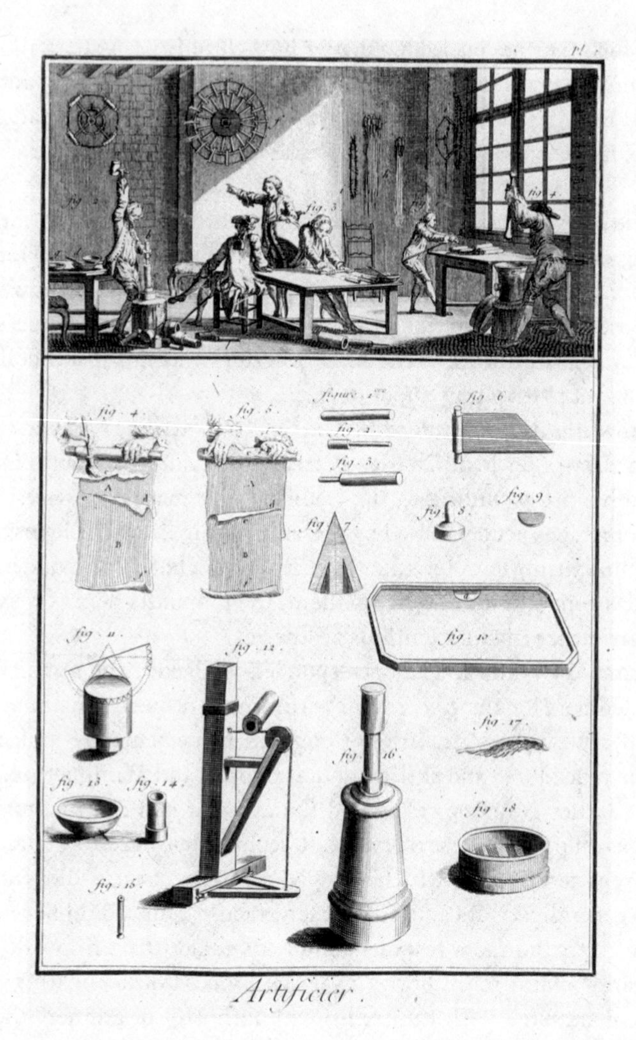

Fig. 3.3. Pyrotechnic workshop and tools. The first of seven plates under the title "Artificier," in *Receuil de planches, sur les sciences, les arts liberaux, et les arts méchaniques avec leur explication,* 11 vols. (Paris, 1762–72), vol. 1 (1762). The plate accompanied the article by J. C. Perrinet d'Orval, "Artifice," in *Encyclopédie, ou dictionnaire raisonné des science, des arts et des métiers,* ed. Denis Diderot and Jean d'Alembert, 17 vols. (Paris, 1751–65), vol. 1 (1752), 740–44. From the author's collection.

on artifice, published at Bern in 1750. It consists in a thing very simple, of bringing together two fuses [*étoupilles*] so close, without, however, touching, so that one cannot help but give fire to the other when it burns."[43]

The article went on to explain the technique. It might appear that the Ruggieri had kept their technique secret, but the issue was more complex. The treatise Perrinet d'Orval referred to was his *Essay sur les feux d'artifice,* first published in Paris in 1745 and with a second, identical edition published in Bern in 1750. Here Perrinet d'Orval offered an *imitation* of the Ruggieri's technique of communicating fire. He did not claim their method had been kept secret, only that neither he nor the public knew it: "The mechanism seemed so ingenious to me that I left it to myself to reproduce it, charmed by the idea of rendering it public if I should succeed."[44] Only in the *Encyclopédie* did Perrinet d'Orval explicitly call the Ruggieri's technique a "secret." In fact, the Ruggieri do not appear to have kept this method a secret at all. In another, abridged edition of Perrinet d'Orval's *Essay,* published the year after his *Encyclopédie* article, he wrote: "The secret of this communication of fire was brought from Bologna to France in 1743 by the sieurs Ruggieri. . . . This secret, which they have very obligingly communicated to amateurs of the art . . . consists of a thing so simple and easy in execution that the machine is at once made."[45]

As suggested above, family practitioners could keep methods to themselves not because they were secrets but because there was no need to make them explicit. Evidently, the Ruggieri explained their methods when asked. Moreover, as *artificiers du Roi,* the Ruggieri had precisely the kind of secure patronage relationship that could alleviate the need to keep techniques exclusive and therefore secret. Perrinet d'Orval was likely constructing the Ruggieri's technique as a secret just so he could expose it: "secret" was a derogative label, and the opportunity to gain inside knowledge would likely pique the interest of the *Encyclopédie*'s enlightened readers.[46]

The revelatory nature of the *Encyclopédie* was not in tension with the local knowledge tradition of gunners—indeed, it perhaps owed something to it. Claims to expose secrets provided Perrinet d'Orval with an opportunity to damage his Italian rivals and generated new public knowledge in the process. The price for this was savants' insistence on a social and epistemological hierarchy of pyrotechnic knowledge. In the *Encyclopédie,* the gunner was ranked below the generalizing savant, while the "secretive" Ruggieri were made lowest of all. Despite their fame in Paris, the family was noticeably absent from the list of authors on fireworks in the *Encyclopédie.*

Gunners' disdain for the Italian interlopers in print was matched by action at the sites of pyrotechnic performance. From the late 1740s to the early 1760s there were numerous reports of fights between French and Italian pyrotechnicians during displays, evincing a fierce nationalistic competition. At a display in 1749, for example, "there were 40 killed and nearly 300 wounded by a dispute between the French and Italians, who quarrelling for precedence in lighting the fires, both lighted at once and blew up the whole."[47] Similar reactions to the Ruggieri and their colleagues are evident in England and Russia and signaled the beginning of a resurgence of locally trained pyrotechnicians. During the 1760s the gunners' tradition of publishing pyrotechnic knowledge also revived, and numerous works by local gunners appeared in France, England, and Russia, as a response to the perceived incursion.[48]

Men of letters continued to seek authority over pyrotechny outside the confines of print. The republic of letters provided its own set of "family connections," channels of communication through which savants could obtain exclusive secrets from abroad, especially from that perceived center of pyrotechnic knowledge China. Savants such as Leibniz long coveted pyrotechnic secrets from the East, seeking especially the composition of "Chinese Fire," which, according to Jesuits and other travelers, gave a brilliant white flame.[49] In the 1750s Jesuit networks connected to the republic of letters made this discovery possible. The missionary Pierre Nicolas d'Incarville had taught humanities and rhetoric for the Jesuits in Quebec before being posted to China in 1740. D'Incarville now communicated the recipe of Chinese Fire (a composition of powdered iron sand) to savants in the Paris Academy of Sciences, to whom he had already sent much information on Peking flora and fauna. The recipe was published in the academy's prestigious journals in 1763 and was quickly utilized in fireworks displays across Europe.[50] Thus, it was d'Incarville, and not the Ruggieri, who became the most celebrated name in fireworks in this decade.

Fighting Back: The Ruggieri and Chemistry

The incursions of French gunners and savants left the Ruggieri at a low ebb by the late 1760s. To make matters worse, in May 1770 a firework by Petroni Ruggieri for the marriage of Marie-Antoinette and the dauphin in Paris went badly wrong. An explosion caused the crowd on the new Place Louis XV to panic, and scores were killed in the ensuing crush. In response, the City of Paris slashed the budget for fireworks from several hundred thousand livres to less than two thousand, effectively abolishing the Ruggieri's main source of income.[51]

The response to these encroachments and catastrophes came from the second generation of the Ruggieri family, headed by Petroni's sons Claude-Fortuné and Michel-Marie. Claude would become France's most prominent pyrotechnician in the early nineteenth century, but only after great efforts to restore the family's fortunes. To achieve this, Claude did not compete with the gunners or men of letters in the contested domains of artillery and aesthetics. Instead, his solution lay in chemistry. Chemistry offered a new, local means to create novel fireworks that would once again give prominence to the family and distinguish them from both gunners like Perrinet d'Orval and men of letters like Cahuzac, Diderot, and d'Incarville. In the process, Claude Ruggieri came to defer to his chemical colleagues. Although chemistry would provide a powerful boost to the Ruggieri's fortunes, the family did not contest the hierarchical classification of savants over practitioners promoted in the *Encyclopédie* and by the Academy of Sciences. This was the price for maintaining distinction.

Chemical practice had long been important in the production of fireworks, yet chemical knowledge was given relatively little attention in pyrotechnic accounts. Most treatises described methods for making gunpowder and procuring substances for firework compositions but were little concerned with the theoretical knowledge behind their actions. Gunners were more likely to pose geometry and mathematics as requisite to pyrotechnic knowledge, while savants highlighted the value of fine arts and classical learning. But in the first years of the nineteenth century, Claude-Fortuné Ruggieri began to argue explicitly that chemistry was a vital addition to these disciplines, setting a trend that has continued up to the present day.

Claude's interest lay especially in the new French chemistry that savants such as Lavoisier, Laplace, Fourcroy, and Chaptal were developing in the 1780s and 1790s. In 1801, Claude dedicated a new treatise on fireworks, the *Elémens de pyrotechnie,* to Jean-Antoine Chaptal, Napoleon's minister of the interior, former director of the Grenelle Saltpeter works, and author of the Lavoisian textbook *Elémens de chimie* in 1791, whose title evidently inspired Ruggieri.[52] Chaptal had much to say about the state of artisanry in France. Following the Revolution, he began putting into practice a long-standing vision of French prosperity founded on the application of the sciences, especially chemistry and mechanics, to manufactures. Against what Chaptal perceived as a debilitating divide between the arts and sciences in the Old Regime, he offered the image of a revolutionary "new man" who would combine theoretical (ideally chemical) knowledge and artisanal practice to accumulate private wealth and national prestige.[53]

Claude-Fortuné Ruggieri presented himself as exemplary of Chaptal's new man. A friend of aeronaut André-Jacques Garnerin, Ruggieri applied aerial philosophy and chemistry to fireworks throughout his career. He shot fireworks from air balloons, illuminated temples with burning oxygen and phosphorus, and in 1806 devised one of the first brilliantly colored fires, a green "palm fire" using metallic salts to intensify the color. Brock credited this colored fire with revolutionizing pyrotechny in the nineteenth century. Such exploits brought Claude fame and the prestigious rank of *artificier du roi.*[54]

Chaptal's egalitarianism likely appealed to Claude, whose family had experienced precisely the social distancing identified by the chemist. Moreover, the 1770s and 1780s had seen a change in the family practice that left it amenable to Chaptalian reform. With their dwindling financial security and reputation, the Ruggieri turned to the pleasure gardens of Paris to restore their fortunes. In 1766 an aging Pietro Ruggieri opened the Jardin Ruggieri on the rue Saint-Lazare, offering fireworks and attractions formerly reserved for the court to the more humble citizens of the capital.[55] In 1794, Claude and Michel-Marie took over the operation and performed fireworks in the pleasure garden for the next fifty years. By the time of the Revolution, the Ruggieri had transformed their practice into a diversified business, precisely the kind of enterprise that interested Chaptal.

Opening the Jardin Ruggieri prompted the family's involvement with the chemists. In 1766, Pietro Ruggieri asked a young Antoine Lavoisier to try producing colored fires to be used in the new pleasure garden. Lavoisier managed yellow and blue flames, but not green. For Lavoisier, pyrotechnic knowledge provided insights later recorded in his *Elémens de chimie.*[56] Lavoisier also advised when the family proposed making fireworks from inflammable airs in the mid-1780s.[57] Chemistry thus provided a local resource for adapting fireworks to the pleasure garden business, and Claude's enthusiasm for chemistry was a natural outgrowth of this collaboration.

Chemical knowledge also offered a means to compete with literary savants and hostile artillerists. In his *Elemèns de pyrotechnie,* Claude utilized the same tactics his adversaries had employed earlier in the century, accommodating rival forms of knowledge only to claim them as subservient to his own specialization. Print served as a means to achieve this. Publishing the *Elémens* brought the Ruggieri's practice into the realm of public knowledge. Publishing was itself a form of economic diversification, a way to increase revenues and compete with other pyrotechnicians. Print also created a forum for shaping knowledge. In the *Elémens,* Ruggieri presented pyrotechnic knowledge in a carefully ordered hierarchy. Like the men of letters, Ruggieri identified "artifice," the manufacture and execution of fireworks, as insufficient for true pyrotechnic expertise. Pyrotechnicians must also have a knowledge of architecture and art to warrant any higher status. But against the men of letters, he claimed that pyrotechnicians must additionally be skilled in physics and chemistry:

> Artifice, by itself, is nothing more than mechanical work, which demands no more than the measure and name of the materials which it employs; but to merit the name of an artist in this genre, it is also necessary to be a physicist [*physicien*], in order to foresee, without having to resort to trials, the effects of any operation; a mechanic, for the perfection of a piece that one has invented; an artist and architect, because the pyrotechnician must know how to make the effects of fire agree with all the rules of architecture. . . . Knowledge of chemistry is also of absolute necessity, to combine with certainty the materials that one employs and to make these compositions with the most economy possible.[58]

Claude thus posed his own skills as authoritative. Gunners' and savants' knowledge was necessary, but not sufficient, for the best pyrotechnic practice: "Pyrotechny . . . is a dark chaos which one cannot penetrate without the torch of chemistry."[59]

Claude then filled the *Elémens* with many discussions of chemistry and physics. His introduction detailed numerous new chemical substances for use in pyrotechny, while an appendix provided a *"vocabulaire"* of terminology, much of it derived from Lavoisier's new chemical nomenclature. Ruggieri also made sure to evoke the new chemistry in fireworks recipes, though he seems to have had his own ideas about the process of combustion in gunpowder:

> Saltpeter is . . . no other than the combination of oxygen gas and others, condensed and reduced to the smallest volume that it may occupy. It is not only the heat named caloric which suffices to dissipate the saltpeter and evaporate it into the state of a gas, but the fire itself, named phlogiston, in action, that is, in a state of causticity. Thus are all the parts of charcoal divided to infinity, which catch fire as fast as they are consumed. It is in this instant . . . that the active fire ignites the oxygen of nitre, evaporates it, and reduces it to the state of a gas.[60]

Evidently Ruggieri did not understand Lavoisier's dismissal of phlogiston in favor of caloric. Even so, the new chemistry, its language and application to pyrotechny, helped bolster profits in the Jardin Ruggieri and distinguish the family from threatening adversaries. The recipe was a success: Claude-Fortuné quickly rose to become the chief pyrotechnician in France, performing displays for Napoleon and Charles X. The *Elémens de pyrotechnie,* going through several editions, helped shape fireworks into a new form and established chemistry as a science deemed essential to pyrotechny, a view that continues to hold today.

Conclusion

The formation of a pyrotechny closely allied to the science of chemistry was not an inevitable outcome, as teleological accounts of the improvement of arts through the application of scientific knowledge might suggest. In the case of fireworks, such an outcome was the result of contingent disputes among various groups of pyrotechnic practitioners, set against the changing context of eighteenth-century disputes over the right forms of pyrotechnic knowledge. Knowledge was shaped in a constant and often-hostile exchange between competing groups of gunners, savants, and family pyrotechnicians. It took a diversity of forms in the process, stressing artillery, mathematics, and the skills of the gunner in the seventeenth century, theatricality, belles lettres, and architecture in the eighteenth, and finally chemistry in the early nineteenth century. Similar histories might be told of other practices surrounding pyrotechny: the art of engraving fireworks prints, the composition of the allegorical content of displays, even in the forms of music accompanying fireworks. Indeed, the social study of controversy might usefully be applied to rethink the history of any form of artisanal knowledge.

At the same time, this essay has questioned boundaries and characterizations of "artisanal" and "scientific" knowledge. Common accounts of artisanry and science as "secretive" and "open" essentialize what were often contingencies of local conditions of practice or claims within situations of dispute over the appropriate form that knowledge should take. In the case of pyrotechny, institutional affiliation or secure patronage relations permitted, or even encouraged, openness about technical knowledge, evinced in the publication of numerous treatises on fireworks throughout the early modern period and the Ruggieri's willingness to disclose their methods in the mid–eighteenth century. Alternatively, a lack of publication did not necessarily indicate secrecy, since social formations like the guild or family managed and reproduced knowledge without the need for its explicit articulation. However, competing groups could label such practices as "secretive" in order to advance their own interests in controlling practice. When gunners felt threatened by foreign incursions, it was useful to label the Ruggieri's techniques as "secrets" to be exposed to the public. Similarly, making knowledge explicit was not an exclusive attribute of savants' practice. Gunners encouraged openness and published their techniques to win patronage and bring order to their work. The value of exposition in the military arts may even have informed savants' desire to articulate craft knowledge. Then, when faced with difficult circumstances, the Ruggieri turned to publication and articulation to gain support from the French chemists and make manifest the importance of chemistry to fireworks. Ultimately, openness served as a tactic against the gunners.

CHAPTER FOUR

Naming and Knowing: The Global Politics of Eighteenth-Century Botanical Nomenclatures

LONDA SCHIEBINGER

> The names bestowed on plants by the ancient Greeks and Romans I commend, but I shudder at the sight of most of those given by modern authorities: for these are for the most part a mere chaos of confusion, whose mother was barbarity, whose father dogmatism, and whose nurse prejudice.
> —*Linnaeus, 1737*

Ordering nature's bounty was a central problem for Europeans in the eighteenth century. Voyaging outside Europe, naturalists, colonists, missionaries, and traders encountered all manner of new and strange fruits, vegetables, medicines, spices, and other economically valuable plants. Inside Europe, known plant species burgeoned in this period from the six thousand described by Gaspard Bauhin in 1623 to the fifty thousand recorded by Georges Cuvier in 1800. Botanists, doctors, pharmacists, gardeners, and collectors all had equal interest in establishing systems of classification capable of managing the growing number of known plants.

For Carl Linnaeus, the celebrated founder of modern systematics, botany had two foundations: classification and nomenclature. Linnaeus taught that nomenclature was a convention agreed upon by a community of practitioners. Like coins, he wrote, botanical names were used by "agreement of the commonwealth." Linnaeus, never a modest man, fancied himself a lawgiver bringing order to a botanical "commonwealth" threatened from without by the "invasion," as he called it, of "vast hordes" of foreign plants and their "barbarous" names.[1]

This essay explores the rise of botanical nomenclature through the linguistic history of plant names, stepping only lightly on issues concerning the well-trodden terrain of eighteenth-century taxonomy. Nomenclature interests us here for what it reveals about the social process of making knowledge. Historians have celebrated the rise of Linnaean systematics in the eighteenth century as the birth of modern, scientific botany. And, indeed, it was. Between Alphonse de Candolle's (1867) *Lois de la nomenclature botanique* and the first *International Code of Botanical Nomenclature* established by the 1905 Vienna Congress, botanists fought bitterly over whether 1737 or 1753 (both Linnaean) was to serve as the recognized "starting point" of botanical nomenclature. This story is well known and will not be recounted here.[2] One could, however, see the rise of Linnaean systematics also as a form of what some botanists have called "linguistic imperialism," a politics of naming that accompanied and promoted European global expansion and colonization.[3] Bruno Latour has emphasized that Europeans employed a number of "instruments" to mobilize natural objects so that information collected by voyagers could be re-created and exploited at key sites within Europe.[4] Botanical nomenclature, as it developed in the eighteenth century, became such an instrument.

The process of knowledge making discussed in this chapter comes from "the margins" only because our focus in this volume is on Europe. From a global point of view, the tropics stand at the center of the plant kingdom, as it is still called. The tropical regions of the world, occupying only 6 percent of the earth's surface, contain the greatest diversity of the world's flora. It is estimated that, of the 250,000 species of higher plants living on earth today, 20 percent (or 35,000–50,000) are found in the tropical rainforests of the Amazon alone.[5]

This chapter also comes from the margins because Europeans won the scientific naming game. Jorge Cañizares-Esguerra has recently pointed out that losses in struggles over naming, surveying, and remembering were as central to imperial control as territorial losses in war.[6] Cultures that grew and blossomed along with the rich vegetation of the tropics developed highly complex knowledge and naming systems; nevertheless, naming systems from genetically rich but today economically poor parts of the world did not prevail.

Cultural Practices of Naming

Naming—the way cultures come to refer to objects, whether animate or inanimate—is a deeply social process. It is also highly political, and botanical nomenclature should be considered in a larger context of the history of naming. In the twentieth century, for example, when professional women sought independence, they kept their original surnames and refused to take their husbands' names when marrying. Similarly, when Caribbean or African countries finally shook off the yoke of European rule, many renamed their republics, choosing names highlighting what they considered indigenous cultural traditions. Thus, when French rule collapsed in Saint Domingue in 1804, making it the second country in the Americas—after the United States—to win its independence from Europe, the former French part of Hispaniola selected the Arawak-derived name "Haiti," despite the fact that few indigenous people survived there.

One also sees the process of the loss and rebuilding of cultural identities that we will explore in the naming and renaming of plants in social practices surrounding West Indian slave names during the colonial period. When slaves were first transported from Africa to the Caribbean, many retained their African names and were entered on plantation lists with names such as Quashie (Sunday), Phibah (Friday), Mimba, Quamino, and so forth. By the 1730s, the proportion of African names diminished. Slaves in British territories often had more than one name: a name given them by their master or overseer (at time of purchase or at birth) and a name given by their parents or themselves over the course of their lifetimes (which masters might be aware of but rarely used). The names given by their masters were the ones recorded in plantation records and consisted primarily of first names (e.g., Dolly, John, Samuel, Betsy, Jenny), nicknames derived from some personal characteristic of the slave (such as Run-Away Mary or Big Tom), whimsical names (such as Time, Fate, Badluck, or Strumpet), or names taken from Greek or Roman classics (e.g., Venus, Caesar, Apollo—these grand names only pointing up more sharply the slaves' desperate situation). Masters assigned names arbitrarily, disregarding a slave's family ties or geographical origins. Masters, by contrast, had at least two names (first and last) and often three (first, middle, and last) acknowledging a set of family connections important for social status and property transfers. By the 1780s, the use of African names among slave populations had dropped to about 20 percent.[7]

With emancipation in the 1830s, former slaves in British territories began taking surnames for the first time. In choosing these names, they rarely drew from African or slave names but tended to model their names on those of their former owners, sometimes their white fathers, or from persons of European heritage whom they admired or whose status they hoped to acquire. (We should recall that, in the U.S. civil rights movement of the 1960s, a number of African Americans once again took African names as a source of "black pride.") The politics of naming can, however, cut different ways: in French territories free people of color were forbidden to use European names and after 1773 were required to take names of obvious African origin to emphasize their outsider status.[8] Names offer a sense of identity, cultural location, and history. It is within these broader contexts that I will discuss European botanical naming practices.[9]

This essay has three parts. In the first I explore the extent to which plants were uprooted from their native cultures and acclimatized to European naming systems. This history is epitomized, as we shall see, in the linguistic history of the flower sometimes (and in some places) called the *flos pavonis* (or peacock flower), in others the *tsjétti-mandáru,* and still others the *monarakudimbiia.* In the course of the eighteenth century, the variety of (published) names for this ornamental—many of them East Indian and emphasizing the plant's beauty—was reduced to a single term still used internationally, *Poinciana pulcherrima,* a name commemorating General Philippe de Lonvilliers, chevalier de Poincy, a seventeenth-century governor of the French Antilles.[10]

The story of imperial nomenclature describes well the fate of thousands of plant species that came under European perusal in the eighteenth century. In the second part of this essay I explore the exceptions to this account. One finds, despite Linnaeus's many-sided prescriptions and proscriptions, a goodly number of exotic names in European nomenclature. A prominent case of a plant not named for European botanists, royalty, or patrons of botany or exploratory navigation is *Quassia,* named by Linnaeus after the African slave in Surinam who was honored for having developed this plant as a useful medicine.[11] Though this example of a plant named for a former male slave is interesting, it may well share in the heroic individualism typically commemorated in eighteenth-century botanical names.

In the third part I look at the efforts made in the eighteenth century to develop botanical nomenclature that incorporated names from the cultures within which the plants grew. Botanist S. M. Walters has argued that the names and also the classification of flowering plants "would be substantially different" if botany had developed in, say, nineteenth-century New Zealand instead of in early modern Europe. He contends that the stability of flowering plant families, essentially unchanged since Antoine-Laurent de Jussieu, "is no evidence of their correctness; family boundaries might be seriously different, and the main purposes of classification still be adequately served."[12] As we shall see below, modern botanical nomenclature might have developed substantially differently if the system of Linnaeus's ardent opponent, Michel Adanson, had been chosen as the starting point of modern systematics. Adanson, along with others, attempted to conceptualize plants globally in his taxonomic system and often chose to retain plant names indigenous to the areas where they are found. Linnaeus complained concerning Adanson's nomenclature: "All my generic Latin names have been deleted and instead come Malabar, Mexican, Brazilian, etc., names which can scarcely be pronounced by our tongues."[13] Even within Europe there were alternative practices and viewpoints.

While I will (for lack of materials) necessarily focus on European botanical nomenclature, it is important to keep in mind that Amerindians and African slaves were also active linguists. Alexander von Humboldt reported that the "savage tribes" of the Americas (he meant Caribs and Arawaks) had names for every nation of Europe: Spaniards were called the "clothed men"; the Dutch were "inhabitants of the sea"; and the Portuguese were "sons of musicians."[14] Pierre Barrère, sent to Cayenne by the French Crown in 1722, noted that the "Negresses," who educated children (white as well as black), had introduced many words from their own country into the Creole language that dominated that island.[15] While we glean glimpses of these words from European documents, neither the Caribs, Arawaks, nor "Negresses" of Cayenne recorded botanical names in any way that has survived.

Linguistic Imperialism?

What's in a name? In Spanish territories of the seventeenth century, the University of Lima turned down a proposed new chair of medicine devoted to botanical studies on the grounds that physicians should instead study Quechua because that ancient language embedded the medical virtues of a plant in its name. Physicians, it was suggested, could learn the uses of plants more quickly by studying this language rather than by studying the plants themselves.[16] Martín Sessé, director of the Spanish Royal Botanic Expedition to New Spain (1787–1802), made similar claims for Nahuatl.[17] It was precisely this type of cultural information—whether of medical uses, biogeographical distribution, or agricultural utility—that was to be stripped from plants in the eighteenth-century naming projects.

Michel Foucault has defined the eighteenth century as the "Classical Age," the age that fashioned new conceptual grids to discipline the unwieldy stuff of nature. Within these grids, names became technical tools of reference—simple tags or neutral designators—no longer burdened by Baroque notions of resemblance.[18] A name, in other words, was to have no essential connection to the plant, but was something agreed upon by convention. The illustrious B. D. Jackson, secretary of the Linnean Society and keeper of the Kew Index in the 1880s, wrote that a plant name is merely a "symbol," and if it and the plant to which it belongs are firmly united so that the name recalls the plant without doubt, it matters little what the name might be.[19] Nomenclaturists today emphasize the playfulness of names: a fossil snake might be called Monty Python; granting agencies might be encoded into names (*Simiolus enjiessi,* when pronounced phonetically, says "NGS" in honor of the National Geographic Society); and numerous natural objects have, indeed, been named for spouses or lovers.

Though names today may be abstract and arbitrary in this way, naming practices are not. They are historically and culturally specific, growing out of particular contexts, conflicts, and circumstances, and it is the job of the historian to ask why a particular naming system developed and not another. My argument is that what developed in the eighteenth century was a culturally specific and highly unusual practice of naming plants after prominent Europeans, especially botanists. Linnaeus argued long and hard for this relatively new practice, and it was given additional justification when Linnaeus's works were made the starting point of modern botany at the beginning of the twentieth century—a period that also marked a high point in European imperial power. My claim is that naming practices devised in the eighteenth century buttressed Western hegemony and, importantly, also embedded in botanical nomenclature a particular brand of history—namely, a history celebrating the deeds of great European men.

The grand achievement of the early modern period was the invention of binomial nomenclature, emerging in nascent form in the work of the Swiss botanist and anatomist Gaspard (Caspar) Bauhin in the seventeenth century and systematically developed by Linnaeus in the eighteenth century. Binomial nomenclature refers to that system of naming whereby a species of plant is designated by a two-word name, consisting of a generic name and followed by a one-word specific epithet, as in *Poinciana pulcherrima.*[20] A plant is considered completely named when it is furnished with a generic and a specific name.[21]

There is no doubt that in Linnaeus's day reform was needed. Seventeenth-century botanists raised a cacophony of botanical names.[22] Until the end of the fifteenth century, *materia medica* were commonly Arabic versions of Greek texts in Latin translations. These texts typically provided five types of names: Latin, vernacular, apothecary, Arabic, and polynomials or phrases.[23] Take for example an entry in John Gerard's 1633 *Herbal* that begins by listing names for "Sow-Bread." Gerard provides the name in Greek, in Latin, what it is called in apothecary shops, and in the vernaculars—Italian, Spanish, High Dutch, Low Dutch, French, and English. This array of names filled nearly half the short entry.[24]

Naturalists in the sixteenth and seventeenth centuries expanded these unwieldy practices as they came into contact with new plants abroad. Francisco Hernández, collecting in New Spain in the 1570s, recorded Aztec names for the plants he encountered.[25] The military man Charles de Rochefort, working in the West Indies in the 1650s, employed the European method to collect synonyms for particular plants from many Indian languages. Writing of the *"manyoc"* of the Caribs, for example, he gave equivalent names in Toupinambous (*manyot*) and "other" Amerindian tongues (*mandioque*).[26] Charles Plumier, in his 1693 description of the Americas, gathered and recorded Taino and Carib names.[27] Hendrik Adriaan van Reede tot Drakenstein, while exploring the coast of Malabar, provided "Brahmanese" and Malayalam names for the plants he found.[28] Pierre Barrère, in Cayenne, offered names for plants in Latin, French, and "Indian."[29] Jean-Baptiste Pouppée-Desportes, working in Saint Domingue, supplied names in Latin, French, and Carib.[30]

While many naturalists happily incorporated names from other continents and cultures into the European corpus in order to accommodate a burgeoning flora, others were less sanguine about the practice. Linnaeus emphasized in his 1737 *Critica Botanica* the urgency of developing a "science of names," by which he meant laws regulating how names should be made and maintained. He judged the reigning practices a "Babel" of tongues and, with characteristic flourish, exclaimed, "I foresee barbarism knocking at our gates."[31]

In efforts to stay the "barbarisms" entering Europe, Linnaeus promoted botanical names of Latin and Greek origins. And importantly, Linnaeus promoted "as a religious duty" generic names designed to preserve the memory of botanists who had "served well the cause of science." To this point he devoted an uncharacteristic nineteen pages in his *Critica Botanica* (most entries are one to three pages long). Linnaeus energetically promoted the practice of "engraving the names of men on plants, and so securing for these names immortal renown."[32]

Linnaeus himself expected resistance to this practice, writing that, "if any of my aphorisms should provoke opposition, it will assuredly be this one." He justified these naming practices in four ways. First, bestowing a man's name on a plant "aroused the ambition of living [botanists] and applying a spur where it is suitable." Botanists, he mused, would be motivated to find new plants when rewarded by having their own names immortalized in the names of plants. Second, naming discoveries after the discoverer was a practice sanctioned in other sciences. Physicians, anatomists, pharmacists, chemists, and surgeons, Linnaeus wrote, customarily attach their names to their accomplishments. Third, and perhaps most significantly, this naming practice followed that of voyagers; as they expanded outward around the globe, voyagers named and often claimed huge portions of the earth. "How many islands," Linnaeus wrote, "have not obtained their names from their first European visitors? Indeed a quarter of the globe has received from that insignificant specimen of humanity, Amerigo, a name which no one would refuse to give it." Who, he continued, would deny a botanical discoverer his discovery? Linnaeus was well aware that his system of botanical nomenclature folded a particular history of botany seamlessly into itself: "it is necessary for every Botanist to treasure the history of the science which he is passing on, and at the same time to be familiar with all botanical writers and their names." Hence, Linnaeus considered it merely economical for the names of plants and of great botanists to be one and the same.

Men immortalized in the Linnaean system included Tournefort (*Tournefortia*), van Reede (*Rheedia*), the Commelins (*Commelina*), Sloane (*Sloanea*), and André Thouin, gardener at the Jardin du Roi (*Thouinia*). With not a little feigned modesty Linnaeus wrote of the *Linnea:* it was named by the celebrated Gronovius and is a plant of Lapland, "lowly, insignificant, disregarded, flowering but for a brief space—[named for] Linnaeus, who resembles it."[33] Few women's names appear in this 1737 list; surprisingly absent is that of Maria Sibylla Merian, whose work Linnaeus often cited.[34] *Meriania* was introduced in the 1790s by Olof Swartz, who worked extensively in Surinam.[35]

Linnaeus could have chosen many things to highlight in naming practices—for example, the biogeographical distribution of plants or the cultural uses of plants. In fact, he chose to celebrate botanists—a practice that reinforced the notion that science is created by great individuals, and in this case European men. In so doing, he inscribed a particular vision of the history of botany into the very language of this science. Linnaeus's naming system itself retold—to the exclusion of other histories—the story of elite European botany.

It is important that Linnaeus's naming practices developed at a time when naturalists were newly regulating who could and could not do science. It was a time when the informal exclusion of women was formalized.[36] It was also a time when European science was establishing its power vis-à-vis other knowledge traditions. As part of this, Linnaeus closely guarded the power to name. According to Linnaeus, "no one ought to name a plant unless he is a botanist" (and thus, de facto, not a woman or person from another culture). Linnaeus's system also served to calcify growing professional divides: "I see no reason why I should accept 'officinal' names, unless one should wish to place the authority of pharmacists unnecessarily high." Finally, the aging Linnaeus required that only "mature" botanists, not rash young men or "newly hatched botanists," name the various parts of nature's body.[37]

Naming Conundrums

A plant the voyager Maria Sibylla Merian called "*flos pavonis*" epitomizes the shift away from the multicultural naming practices of the late seventeenth century to the linguistic imperialism developing within botanical nomenclature in the eighteenth century, and we will look in some detail at this case study. The peacock flower, a bush often used as a hedgerow that blooms gloriously in brilliant reds and yellows, grows profusely in the tropics and is commonly employed medicinally as an abortifacient and lung remedy. When Europeans initially came in contact with this flower, they collected its various colorful, indigenous names. Van Reede collected "Brahmanese," Arabic, Portuguese, Dutch, and Malayalam names for the plant. The Latinized version of the Malayalam name, *tsjétti-mandáru,* is one Merian included in her text. (In Harvard's copy of the work, a Linnaean name is penciled in.) Paul Hermann, who in his youth served as a medical officer in Ceylon for the Dutch East India Company, reported the colorful "Zeylonese" (Sinhalese) name for this plant: *monarakudimbiia.*[38] Later in the eighteenth century, Michel Adanson called Merian's *flos pavonis,* which he found in Africa, by the indigenous name *Kamechia.*

But a new fate awaited Merian's *flos pavonis.* In 1694, this flamboyant flower was included within Joseph Pitton de Tournefort's abstract typology—the classification widely regarded today as one of the forerunners of modern systematics. Tournefort, director of the Jardin du Roi in Paris, placed the plant in his Class 21, Section 5, encompassing "trees and shrubs with red flowers and seed pods." As was typical of the new schema, Tournefort's classification focused on the physical characteristics of the plant, in this case the corolla and the fruit. The plant's Asian connections and its medical uses—both of which had played a significant role in earlier European accounts—were not discussed.

In the process of anchoring Merian's *flos pavonis* (Van Reede's *tsjétti-mandáru* and Hermann's *monarakudimbiia*) in the European world, the systematist Tournefort devised a wholly new name, *Poinciana pulcherrima,* celebrating his countryman Philippe de Lonvilliers, chevalier de Poincy, governor of the French Antilles, who used the plant to treat fevers.[39] Tournefort's name celebrated French colonial rule in the Caribbean rather than the plant's own virtues, its East and West Indian heritage, the peoples who used it, or those who "discovered" it or supplied Europeans with information about it—all of which were featured in other names given at one time or another for the plant. Following Tournefort's lead, Linnaeus approved of this name and added only that the plant grows in the Indies and under the sign of Saturn (because of its woody character).[40]

The story of Merian's *flos pavonis* typifies naming practices in this age of colonial expansion. In the 1794 *Florae Peruvianae et Chilensis,* documenting Spanish expeditions to those countries, the description of each genus included a note explaining the individual honored in the name. Following Linnaeus, the botanists named New World plants after Spanish botanists and influential patrons of botany: of the 149 new genera presented in Ortega's *Prodromus,* 116 celebrated the deeds of great men of Europe.[41]

The curious history of the *Poinciana pulcherrima* describes the fate of the vast majority of plants in eighteenth-century botany. But to what extent did names of African slaves in the West Indies, Arawaks, Caribs, and others who were not European botanists enter European taxonomic schema? There are in Linnaeus's system, several celebrated examples of plants named for persons who did not fit Linnaeus's general profile. Take, for example, the popular stomach tonic that came into Europe bearing the exotic name *Quassia amara.* This small tropical tree indigenous to South America was named for a freed Surinamese slave, called "Graman" (Greatman) Quassi.[42] According to Linnaeus's own criteria, Quassi should not have qualified as a worthy recipient of the "immortal fame" gained from European naming practices. Quassi was a medical man, not a botanist. In this case, however, the slave was assimilated to the peculiarly European system of celebrating individuals for heroic contributions to botany; Quassi's name was immortalized in the name of the plant as a reward for having been the first to discover its medicinal usefulness (or at least the first to make this known to the learned Europeans in Surinam). As the historical record reveals, Quassi did not develop the cure for which he was celebrated. The cure was commonly known in Surinam, and the names of the persons (Amerindian or displaced African, we do not know) from whom the cure originated were not commemorated in naming this plant and have been lost to history. Quassi served as a middleman, the individual who brought a widely used remedy to Europeans.

The Peruvian bark (*Cinchona officinalis*), widely known in the eighteenth century by its Quechua name, *Quinquina,* serves as another example of a plant—in this case, a plant of strategic importance to Europeans in tropical areas—named for a botanical outsider.[43] *Cinchona,* the source of quinine, was named by Linnaeus after the wife of the fourth Count of Chinchón, the Spanish viceroy Luis Fernandez. The countess hardly qualified for such an honor in that she was neither a botanist nor a patron of the emerging science. Though Linnaeus was taken to task for misspelling the countess's name when he coined the term in 1742, it was widely held that he rightly immortalized this woman's heroic deed, that of risking her life by trying the drug. In coining the name *Cinchona,* Linnaeus laid laurels at the feet of a woman of European extraction (celebrating European colonial rule) and cut the plant's ties to its South American home and the Incans who first came to know it or who chose to inform the Europeans about it.

With both Quassi and the countess, we see Linnaeus celebrating individuals who served as "intellectual conduits" between natives of the New World and Europeans. Quassi did not discover the drug for which he is celebrated any more than did the Countess of Chinchón, but both were instrumental in introducing the drug into Europe. Priority and discovery, so much celebrated in science, were obscured in each of these cases.

I have found only one male slave, Quassi, immortalized in eighteenth-century botanical nomenclature. As far as I am aware, no Native Americans were so honored. Several European women, however, numbered among "heroic" individuals, mostly of high birth, to be honored in botanical nomenclature. It should be noted that women of botanical stature were not so honored by the great Swede. Maria Sibylla Merian, whose work Linnaeus knew well, was not honored in Linnaeus's lifetime; nor was Mary Somerset, Duchess of Beaufort (1630?–1714), who maintained one of the grandest private gardens of the seventeenth century.

Alternative Naming Practices

Linnaeus's system encountered many forceful foes in the eighteenth century, precisely surrounding issues concerning naming. It was not clear to anyone at the time that the Linnaean system would serve as the foundation for modern systematics. According to contemporaries, the "system-madness" of this age was truly "epidemical."[44] Michel Adanson in France counted sixty-five unique systems of botany in 1763; his counterpart in England, Robert Thornton, enumerated fifty-two different systems in 1799.[45] While Linnaeus's system dominated England and Sweden, its acceptance on the continent, especially in France and Germany, was never complete. The Linnaean system, one among many, became the starting point of modern botany (as mentioned above) only in 1905.

Chief among Linnaeus's rivals was Buffon, his contemporary and director of the Parisian Jardin du Roi. Buffon opposed system building generally and ridiculed Linnaeus's system in particular for being too abstract and, most grievously, too artificial. Buffon decried the proliferation of classification schemes, each with an accompanying system of names. "To speak truthfully," he continued, "each method is only a dictionary in which one may find names arranged according to an order relative to a certain idea and, consequently, arranged as arbitrarily as if in alphabetical order." The many methods of naturalists are just so many "systems of artificial signs."[46] But, Buffon emphasized, no method will ever capture nature itself; "nature advances by imperceptible nuances so that it is impossible to describe it with full accuracy by strict classes, genera, and species." Nonetheless, Buffon acknowledged the usefulness of methods as heuristic devices: they produce a common language that leads to mutual understanding; they shorten the work, assist the memory, and serve as an aid for studying and a means for mutual understanding; finally, they provide an imaginary goal that sustains naturalists in their real "labor," which is the accurate description of natural objects. "Nothing is so rare as to discover exactitude in descriptions, novelty in details, and subtlety in observations."[47]

In respect to nomenclature, Buffon took the traditional approach of listing all known names for a particular species. He cited names given by the ancients (e.g., Aristotle and Pliny), sixteenth-century authors (e.g., Gesner, Aldrovandi, and Belon), and finally the moderns (e.g., John Ray, Linnaeus, Klein, and Brisson). Equally important for Buffon were "common" names, whether in Greek, Latin, Italian, Spanish, Portuguese, English, German, Polish, Danish, Swedish, Dutch, Russian, Turkish, Persian, Araband, "Savoyard," Old French, or Grison (Romanche). For New World fauna or flora he provided names in "Indian," "Mexican," and "Brazilian" and names used by the French living there, which were often indigenous American names rendered in French.[48]

Buffon opposed European practices of assimilating exotic plants or animals to the Old World schema. Hence, felines with striped fur found in South America should not be called tigers. Such practices led to assuming that they existed there, when in fact they do not. Names poorly adapted, borrowed, badly applied, or newly invented, he urged, confused nature's order. In direct opposition to Linnaeus, Buffon advocated using native names. Native names (as opposed to newly coined Latin names) served as a clue to the geographic distribution of species. The Greeks and Romans had no name for the buffalo, for example, because that animal is not found in the Old World. As Buffon pointed out, the exotic name "buffalo" indicated a foreign origin. Another advantage of exotic names was that they signaled relationships between animals found in different locations. Thus, one might conjecture that the Cayenne *cariacou* is perhaps the *cuguacu* or *cougouacou-apara* of Brazil because of the similarity of names.[49] Félix Vicq-d'Azyr, Buffon's successor at the Académie Royale des Sciences in Paris, also entered the fray, asking why Linnaeus restricted botanical names to Greek and Latin roots. Would it not be preferable, he queried, to use names for plants that "the naturals of different countries" have given them?[50]

Michel Adanson, born two decades after Linnaeus, was also critical of Linnaeus's highly artificial sexual system of classification, which, according to Adanson, gave "unnatural" attention to the numerical proportions of the parts of fructification, with undue attention lavished on the male parts.[51] Adanson was equally critical of Linnaeus's nomenclature. He pointed to the absurdity of Linnaeus's naming a colonial plant *Dillenia* after Oxford's Johann Dillenius rather than retaining *Sialita,* one of its traditional names. He even refused the Tournefortian name *Adansonia* for the baobab tree and cited "the vanity of botanists" as one of the three causes impeding the progress of the science. Adanson charged that requiring all natural historical names to end only in *ia, um,* or *us* merely revealed Linnaeus's unsophisticated use of Latin; such endings often served merely to lend the term a "scientific air."[52] In step with Tournefort, Buffon, Lamarck, and the French philosophes more generally, Adanson rejected the use of Latin, choosing instead to publish his scientific works in French.

But Adanson's challenge went beyond ridicule. He proposed very different naming practices. First, he argued for retaining traditional names particularly for economically important plants because nonbotanists—"physicians, apothecaries, and the people who collect herbs in the fields for medical use"—know these names and easily understand one another when using them.[53] Hence, *Jalapa,* the four-o'clock flower, should be reinstated for Linnaeus's *Mirabilis.*

For naming new plants, Adanson argued strenuously for adopting common vernacular names from any language—"French, English, German, African, American, or Indian"—unless they were too long. While botanizing in Senegal, he learned the Ouolof language and recorded numerous names in that language in his great *Familles des plantes.* Adanson championed a pragmatic approach to naming. When, for example, required to chose between synonyms, as in van Reede's *Hortus Indicus Malabaricus,* where often both a "Brahman" and a Malayalam name were given, Adanson suggested that the shorter and one more easily pronounced by the greatest number of people be chosen.[54] Adanson opposed Linnaeus's extensive revision of botanical names, charging that Linnaeus had upset and changed the majority of the best-known names in botany and medicine.[55]

Specifically attacking Linnaeus, who never traveled outside Europe's borders, Adanson wrote, "if dogmatic authors had traveled, they would have seen that in those other countries [e.g., Africa, America, and India] our European names are treated as barbarous."[56] For Adanson, nomenclature could not be fixed until a truly natural system was identified. Until that time, nomenclature should be simple and convenient. Adanson himself chose vernacular names, particularly if they had priority in European publications, such as van Reede's *Hortus Indicus Malabaricus*.[57] Adanson's inclusive, decentralized nomenclature cultivated a vision of human unity that prompted him, like Buffon, to adopt names from many of the world's languages.[58]

Linnaeus, for his part, did not appreciate Adanson's revisions, writing that his "natural" system is the most unnatural of all. Linnaean aficionados called Adanson's categories *Familia confusarum*.[59] In a letter to Abraham Baeck, Linnaeus took a typically inflated view of his own accomplishments: "Adanson himself has no experience, everything he has written he has compiled from my works, which I can prove."[60]

Why was the Linnaean system preferred over Adanson's? In this instance, personal idiosyncrasies and institutional politics came to shape the history of science. From the beginning, Linnaeus self-consciously managed his legacy. He wrote, for example, a history of botany that celebrated his own work as having set "the whole science on a new foundation."[61] Furthermore, Linnaeus fashioned his garden and position at the University of Uppsala into a vast botanical empire. Traveling little himself, he corresponded with naturalists throughout Europe and sent his many students (twenty-three of whom eventually became professors), with specific instructions and queries in hand, to America, Africa, India, Ceylon, Java, Japan, and Australia. Although Linnaeus's son was not the man his father was, he nonetheless carried on the elder Linnaeus's work and consolidated his father's fame.[62] Thus, Linnaeus's methods and nomenclature were passed on and popularized perhaps beyond their intrinsic value.

Adanson, by contrast, never managed to win an academic position and therefore had little lasting influence. He lost out in the intricate play of politics in the French academic system and was passed over for a position at the Jardin du Roi, the center of eighteenth-century French botany. Consequently, he had few students and correspondents. In the spirit of

Enlightenment reform, Adanson embedded his new botanical nomenclature in a more general language reform that swept away outmoded doubled letters and diphthongs.[63] Adanson's simplified prose proved enough of a deterrent to the general public that, unlike Buffon's wildly popular *Histoire naturelle,* Adanson's books were not widely read. Although his daughter and only child became a working botanist, she (like most women in this era) was proscribed from occupying an academic position.[64]

Others in Europe were equally critical of Linnaeus's system. Sir William Jones, founder of *Asiatick Researches,* objected stringently to Linnaeus's sexual system of classification, which so "inflame[s] the imagination" as to be completely useless to "well-born and well-educated" women. Jones also found "childish" Linnaeus's practice of naming plants for the persons who first described them, a procedure, Jones suggested, that ought wholly to be rejected. He found the names *Champaca* and *Hinna* not only more elegant but far more proper for an Indian or an Arabian plant than the Linnaean *Michelia* and *Lawsonia.* "Nor can I see without pain," Jones continued, "that the great Swedish botanist considered it as *the supreme and only reward of labour* in this part of natural history, to preserve a name by hanging it on a blossom, and that he declared this mode of promoting and adorning botany, worthy of being *continued with holy reverence,* though so high an honour, he says, *ought to be conferred with chaste reserve, and not prostituted for the purpose of conciliating the good will, or eternizing the memory, of any but his chosen followers; no, not even of saints.*" Jones took issue with one of Linnaeus's hundred and fifty such names in particular, suggesting that the *Musa* (banana) did not derive from a proper name but from the Dutch pronunciation of the Arabic word for that fruit. Jones, like Adanson, suggested keeping the common, often-indigenous Indian name for a plant. In the case of India, however, he preferred Sanskrit (the Latin of India) over names from what he referred to as "vulgar dialect[s]."[65] Beyond these aesthetic objections, Whitelaw Ainslie, an English surgeon in Madras, raised the practical point that the English names given by European botanists to some of the trees and shrubs of tropical countries were so "obscure and unfamiliar" that he had chosen to substitute the common Indian terms so that the plants could be more easily obtained from the native practitioners.[66]

In African and New World colonies, too, voices arose against Linnaeus. Spanish Creoles in New Spain, such as the priest and botanist José Antonio de Alzate y Ramírez, complained that Linnaeus's universal system obscured crucial information about a plant's location, environment, and flowering season and the soil characteristics required for cultivation. Many

Creoles objected further that Linnaeus's sexual system was abstract and did not capture important characteristics of plants, such as their usages.[67]

In the French territories, Jean-Baptiste-Christophe Fusée-Aublet, who worked in both the Île de France (Mauritius) and Cayenne, was another advocate of names indigenous to the place where the plants are found. Anti-Linnaean feeling was so strong in France that in 1772 Louis XV ordered the Jardin Royal des Plantes to be redone. Antoine-Laurent de Jussieu, director of the garden, reordered its plants using his and his uncle Bernard's "natural system" and in the process adopted many of Fusée-Aublet's plant names.[68]

These controversies have been all but forgotten today. Modern practices reinforce Linnaean principles of giving priority to European languages, names, and hegemony in botanical taxonomy and nomenclature.

Eighteenth-century botanical nomenclature was one of the instruments of empire that detached plants from their native cultural moorings and placed them within a schema comprehensible first and foremost to Europeans and responsive to their economic and political needs. With the rise of modern botany, a uniquely European system of nomenclature developed that swallowed into itself the diverse geographic and cultural identities of the world's flora.

In the eighteenth century, however, as we have seen, there were rivals to Linnaeus's nomenclature even within Europe. Had Adanson's system been chosen for the starting point of botanical nomenclature, nomenclature today might be more inclusive of the world's languages. But there are many forms of imperialism. Even Adanson used European classificatory logic, bringing exotic names along with exotic plants into a net of seemingly natural, but in fact highly European, relationships. Existing sources rarely allow historians to know what extra-European taxonomic and naming systems existed. The rich traditions of the Malabar Coast are said, for example, to be best represented today in van Reede's *Hortus Indicus Malabaricus.* According to Richard Grove, this European text remains the "only faithful textual record of the accumulated Ezhava botanical knowledge of the seventeenth century."[69] There is, in other words, nothing sacred or necessarily best about Linnaeus's nomenclature. It is the system of naming that for many historical reasons has come into common usage, and its emergence illustrates the importance of understanding knowledge making as a social process.

PART 2

Practices of Reading and Writing

CHAPTER FIVE

Novel Knowledge: Innovation in Dutch Literature and Society of the Fifteenth and Sixteenth Centuries

HERMAN PLEIJ

Old and New

Erasmus opens his 1522 dialogue about Johannes Reuchlin by having Pompilius ask Brassicanus whether he had brought news from Tübingen. His friend replied, tellingly: "Remarkable how all people long for news. To think that I heard at a sermon in Louvain . . . that everything new should be avoided." Pompilius immediately scoffed at this attitude. A man who avoided that which is new would never remove his old shoes or change his dirty underwear and would be condemned to eating unsavory eggs and drinking stale wine. Perhaps this was indeed a bit exaggerated, Brassicanus responded, since such a man would prefer fresh soup made that day to old soup prepared the day before. So then what was the news? persisted Pompilius. Unfortunately bad, replied Brassicanus, without managing to deter his friend. Nevertheless, bad news would automatically turn into good news over time: "Frankly, how could it be otherwise, if everything old is good, and everything new is bad, everything that is good *now* was once bad, and everything that is *now* bad will become good in the future."[1]

In this humorous dialogue Erasmus gives voice to conflicting connotations of "old" and "new" that had been developing since at least the fifteenth century. It is difficult from a modern perspective to imagine a culture where new was bad and old was good: consider the remarkably pervasive modern adage "stagnation is decline." But this drive toward continuous change was by no means accepted in the late Middle Ages and early modern period. Anything old was familiar and according to plan, while new ideas were inevitably suspect. After all, they upset divine intentions and might come from the devil. That is why heretical movements were labeled Die Nieuwe (the New Ones), a term that in the Middle Ages was virtually synonymous with undesirable conduct.

Inconstancy represented an earthly life of sheer pretence, an unobstructed target of the devil, a risky path to the "true life," in which destruction was replaced by immutable eternity. This Augustinian outlook equated all inconstancy with decline. Long after the Middle Ages, Western civilization remained permeated with this disassociation from anything earthly (and thus changeable), which remained the standard for civilized conduct among the elites that formed in the cities in the late Middle Ages.

Attitudes toward climate embodied this perspective. In the imagined utopia that was the land of Cockayne, as presented and performed in the literature and dramas of the late medieval cities of the Netherlands, the ideal climate was perfectly uniform. Stability was the ideal, while inconstancy boded downfall. A climate characterized by sharp contrasts was viewed in the Middle Ages as the scourge of humankind. In horror-struck terms, the famous fourteenth-century travel book of John of Mandeville, printed and reprinted well into the sixteenth century, describes such weather in the land of Tartary: "That land is seldom without great storms. And in summer there are great thunderstorms, which kill a lot of animals and people. The air temperature changes, too, very quickly—now great heat, then great cold—and so it is a bad place to live." Such an intemperate land could not be good, and its defects rubbed off on its inhabitants: "They are a very foul folk, cruel and full of ill will."[2]

Paradise, of course, was at the other end of the spectrum, where there were no seasons, no wind chasing the clouds, and likewise no burning sun and no wintry cold. Constancy was the Creator's original intent. Columbus, who never wavered during the course of three voyages from the opinion that he was frequenting the immediate vicinity of paradise, always described ideal weather conditions in terms of spring in Andalusia or elsewhere in Spain. For him, too, the most important things were balminess and constancy. Novelty and inconstancy were therefore suspect.[3]

After the Fall, humanity had become ill and infirm, as well as mortal, and unable to recognize and comprehend the elaboration and culmination of salvation. Life on earth should therefore be dedicated to a quest for God and his plan, aided by priests and the sacraments of the church and by the supreme being, his family, and the saints. This quest would yield insight into the plenitude and completeness of the divine plan and the role of humanity in it. Anything that seemed new along this journey of discovery could therefore never be more than a discovery or rediscovery of the long-standing. Change was presented only as a restoration of what had once been better or even perfect. In fact, only one new thing was good and made all the old superfluous, namely, the New Man according to Paul the Apostle. In his letters, Paul explains how the old Adam was absorbed into the New Man, as personified by the Redeemer. This was meant to serve as a universal inspiration to embrace Christendom and abandon the old life forever. To people in the Middle Ages, this meant that everything in their lives was new in the Pauline sense, and nothing else could exist.[4]

In this context, the movement started by St. Francis of Assisi was described as a restoration—not as an innovation—of the New Man. Although his movement was identified as promulgating a new and previously unheard-of life plan ("new" and "unheard of" were the terms used by his biographer Thomas of Celano, d. 1260, and were unambiguously favorable), the movement was regarded as a restoration of something that had existed in the past but had unfortunately become invisible.[5]

Still, the terms "new" and "innovation" did appear in the course of the twelfth and thirteenth centuries. New manifestations of the eternal truths could be so described, as could the guides that led one to them: a *poetria nova* or even a *retorica novissima.* Especially in the cities, with their different and constantly changing lifestyles, all kinds of things came to be designated as new, even by laypeople. The *Nieuwe doctrinael,* for example, was an instruction manual featuring new forms of etiquette prescribed by city life and written in the early fourteenth century by the local city physician from the Flanders town of Ieper, Jan de Weert.[6] At the same time, there seems to have been, by the late Middle Ages, a reluctance to label changes as "new," a trend that is increasingly evident in both Latin and vernacular texts.[7] The influential "Modern Devotion" movement (Geert Grote, Thomas van Kempen) reasserted St. Augustine's diatribe against *curiositas,* the pursuit of pointless knowledge purely for the sake of knowledge. This curious hunger for knowledge impeded true devotion, in which supreme humility, simplicity, and modesty should be unrestricted. This perspective was also advisable for reading the Bible to avoid being distracted by the "subtlety of the words." Were the *colores rhetorici,* those subdued verbal pigments, not favored tools of the devil for tempting mankind? The Sisters of the Common Life continued to be commemorated and praised in these terms. Virtually every sister was of noble mind and had never—or only as a child—succumbed to temptation by "curious things."[8]

Likewise, traditional literature cast suspicion on and ridiculed anything that pursued or even remotely hinted at originality. Theoretical treatises on drafting texts and speeches stressed, as the essence of "creative" work, the importance of including and rearranging known material for a new public. This was also emphasized in music, seen as narrowly related to poetry. For example, Joannes de Garlandia pointed out in his *De musica mensurabili positio*, from the first half of the thirteenth century, that frequent repetition not only promoted the recognition of the melody but also strongly moved the listener.[9]

Reproduction was regarded as the noblest form of art, while originality was considered a vulgar amusement attributable solely to avarice. Moreover, making something up was considered far easier than elaborating known patterns. This appears to have held true into the sixteenth century, when the erudite humanist Macropedius included an explanation packaged as a *modestas* formula aimed at cultivating sympathy among the public in his Latin school comedy *Aluta* (1535). Written in four days during a hectic schedule, Macropedius's work falters in some places: "I would have liked to improve a few sections. Realizing that producing new material was easier than revising what was already there, I simply added a few lines here and there."[10]

The pressure to present each original idea as the restoration or rediscovery of an old one even led some authors to attribute their innovative views to fictitious or actual authorities of the distant past. Jan van Boendale, a fourteenth-century municipal secretary of Antwerp and author of an original manual with pragmatic directives for adjusting to city life, availed himself continuously of this option, denying in his writings any hint of the originality so manifestly present: "Do not think that I made these things up. I have searched everywhere for material in books once written by wise authors."[11] This respect appears to have become even more powerful in vernacular literature. Around 1500 the play *Van nyeuvont, loosheit ende practike* (Of Newfangled Things, Fraud and Wrongdoing) appeared in print for a reading audience, featuring an explanatory subtitle and a woodcut illustration that symbolically depicted deception and fraud. Personified as women who were driven by the "new findings" (devious tricks), these framing devices proclaimed Lady Lortse as a new saint and yet another embodiment of the deception that proved in the course of the play to consist primarily of the broadly condemned trade in relics by mendicant friars.[12]

In sixteenth-century drama, this vile "novelty" became the standard frame of reference and in many cases a designation for all possible forms of deception as well. Ever the embodiment of total irrationality on the stage, the fool so greatly cherished in drama was referred to as Nieuloop, a frivolous hunter of idle rumors (remember St. Augustine's *curiositas*) frolicking with his hobbyhorse, Clappage (malicious gossip).[13]

This particularly conservative view of everything that even hinted at innovation seems to converge in the work of the Antwerp poetess Anna Bijns (1493–1575). In her lifetime she published three volumes of refrains (*refereynen*), which were all reprinted several times, praised extensively, and widely emulated. As an independent woman (she was a schoolteacher), she had her work published and reprinted with her name on the title page, thanks to the interventions of local Franciscan monks acting out of personal interest. At the time this was rarely the case for living authors and was in fact unique for a woman. Although Bijns lacked higher education, her deftness with language and themes such as the frenzied persecution of the new Lutheran heresy was unprecedented and unparalleled. Few diatribes in verse have ever been more vivid than her vernacular ballads. The core of her proficiency in rhetoric lies in this lifelike quality. Bijns achieved this effect through creative but meticulous application of the lowly style known as the *sermo humilis*, which has convinced literary historians to this day that she listened to those around her and copied directly from the utterances of the common folk.

Despite the novelty of her language, however, Bijns used the adjective "new" and all compound words containing the term exclusively in negative contexts. This was entirely in keeping with the long-standing suspicion of change. "New" denoted every conceivable form of heresy. Bijns associated the term especially with the reform movements of her era, irrespective of whether the aim of these movements was to break with the Catholic Church. She ascribed all such contemporary heresies to Luther and his adherents. By referring to heresies that had been known for centuries as "new," their "news" became the rediscovery of something old, even though they feigned otherwise. In addition, they circulated "new lies." Luther was therefore a "new evangelist" who gave rise to "new teachers" who disseminated "new twaddle." Opposite them were the "old teachers," who had become objects of disdain. They were inspired by the old recluses, the desert patriarchs who laid an exemplary foundation for the church during the early centuries of Christianity.

The variations on this theme were endless. Luther's misled adherents had charted "new courses"—an expression that has exclusively favorable connotations nowadays—and were spewing "new poison." They sang "new songs" or indulged in a "new dance." And their "lust for news" had lured them into this trap. Elsewhere, Bijns wrote that "everybody longs for news," since people listen with fascination to lapsed monks. They have followed Luther's cowardly example: "The people heed him in their yearning for something new." According to Bijns, Luther even led humble widows and virgins astray, as they too "felt earthly longings and became completely bound up in curiosities and lust for news."

Likewise, she portrayed the deceitful lovers who figured so prominently in her ballads as erratic drifters lured away from their true love by something new. Lack of resolve epitomized the deplorable state of humanity since the Fall; man was a fragile creature who constantly stumbled along his journey to the hereafter, tripped up by the devil. This unstable condition was manifested by the uncontrollable yearning for "news" and "novelty" that the devil used to advertise his world of make-believe. Those infected pursued earthly honor, power, and wealth. In their quest, they resorted to the most devious tricks and guile, characterized as "new findings, "false practices," and "new fabrications," as in the title of the play mentioned above.[14]

Rhetoricians' Chambers

Like these plays, Bijns's texts were known as rhetoricians' literature, which became the chief literary pursuit in the Low Countries during the fifteenth and sixteenth centuries. These *rederijkers* (rhetoricians: urban, semiprofessional poets, playwrights, and actors) gave cohesive form to the new literary life of the city. They employed subtler forms and techniques than used previously and shaped their work with specific functions in mind. These functions were in the first place aimed at legitimizing and promoting the vested interests of the city in its competition with other cities. Such aims manifested themselves in richly attired processions, triumphal entries, and competitive events. On these occasions, plays were performed in theaters and on stationary or moving pageant wagons, tableaux vivants were enacted, and triumphal arches were erected, all decorated with allegorical figures. All these displays aimed to convey with their elaborate fixtures all that the city wished to project in terms of its self-image.[15]

Not unlike the religious fraternities or urban militia companies, the chambers of rhetoricians, at least the ones in the large cities, were founded from about 1400 onward by well-to-do citizens. In larger cities, membership in these societies was highly restricted. Not only were the financial obligations heavy, but one had to have some education in rhetoric. The most highly regarded *rederijkers* would have included in their ranks semiprofessionals, one of whom was usually appointed head (*factor*) of the society (*kamer*). Within the cities the chambers functioned as schools for the education of the rising patriciate, while outside the city, their work represented the city before the sovereign and other urban centers. As a consequence, their texts are rather obscure both in form (being filled with complex structures and neologisms) and in content (consisting of difficult allegories that presuppose a broad cultural background). The intentionally elitist character of this type of vernacular literature, which constitutes a new departure, represents an attempt by the bourgeoisie to distinguish itself both from the peasant culture of the countryside and from the other strata of urban society.[16]

The skills of the *rederijkers* also served as a weapon in the hands of established urban groups and others who wished to join their ranks. The *rederijkers* endeavored to provide answers, often in the form of consolation, to the everyday frustrations and ambitions of that social milieu. The issues most frequently addressed were the whims of fortune, the dangers of foolish love, and the constant lurking of death. At the same time chambers of rhetoricians were especially well suited to serve as educational institutions in the city. Their members' public appearances in processions and tableaux vivants illustrated the same lessons they sought to convey in their dramatic and poetic work: self-control, moderation, proper behavior, and refined conversation. The pleas for these values often took the rather heavy-handed form of an uncompromising offensive against everything that failed to meet the new standards.[17]

They were in fact the progenitors of a Dutch literary language, which differed from the daily vernacular. Although prohibited as a woman from joining any chamber of rhetoricians, Anna Bijns developed close ties with various rhetoricians locally and elsewhere, undoubtedly due in part to the tremendous respect for her poetic talents. Owing to the typically bourgeois quality of this new literary institute and the fact that most members were amateurs (except for the *factors*), *rederijker* work became known in later centuries (and especially since the nineteenth century) as contrived, traditional doggerel that invoked ideas that had already become stale by the late Middle Ages. The cited use of "new" in their texts would appear to confirm this assessment. In fact, however, it demonstrates once again the tremendous problem with quantification and selection in the humanities. Within this rigidly organized hierarchy, a web of literary and social innovations emerged. Their unmistakable originality did not need to be substantiated as the alleged restoration of something from the past.

The rhetoricians' unbridled curiosity was inspired and legitimized by local education—from simple schools for reading and writing to universities. The yearning for knowledge among students and scholars, pilgrims, explorers, as well as money-hungry merchants and entrepreneurs in and from the city, was unstoppable and, by 1500, not only socially acceptable but also commonplace. The printing press in particular functioned to satisfy the hunger for news and for "renewal," and, furthermore, it created and expanded needs of this kind.[18] The thriving industry of the rhetoricians' chambers apparently became a driving force for innovation, keeping pace with the change occurring in the cities.

By the sixteenth century, fulminating against novelty and certainly the hubbub created by Anna Bijns on this subject had thus de facto become obsolete and was reduced to the nostalgic aftereffects of a massive legacy with deep theological roots that was not easily eroded by change. Such deliberate conservatism appears to have been rather widespread. Otherwise, why would somebody like Erasmus have derided it so extensively in the dialogue quoted at the opening of this essay? In this passage, Erasmus channeled his message through two erudite humanists to refute the old-fashioned views about old and new in the context of the Christian plan for salvation.

In the more official manifestations of the *rederijkers'* movement (especially the festive competitions), different views about "new" and "unprecedented" were in fact propagated early on. In addition to acquiring a more favorable connotation within their ranks, such terms even came to dominate the demands the *rederijkers* made on themselves and on others. The earliest report on this subject conveys the elite tenor of a new standard imposed by the civilizing institute par excellence—the chamber of rhetoricians—on all those aiming to excel in literature and art. This report concerns the rhetoricians' efforts on behalf of the splendid procession that the municipal authorities of Ghent held to honor Philip the Good in 1458. Such deeply solemn occasions required a fitting representation on the part of the city and were always arranged by the local rhetoricians. They had the expertise and artistic abilities to design suitable triumphal arches, tableaux vivants, pageants, eulogies, fireworks, and other ornamental displays. In addition, the entire municipal infrastructure was mobilized and subsidized to stage performances, preferably around or in front of their guildhalls and association premises and otherwise in competitions at a central site, generally the marketplace. Even there, the rhetoricians usually supervised. In 1458 all conceivable societies, from guilds to district associations, were invited to perform an *esbattement* (a short and usually comic play) at the city hall. The best would receive an award. The play was expected to be in pure (uninfected by foreign elements), new, and well-written verse. "New" had become an audacious standard that conflicted with the long-standing tradition of demonstrating expertise with known material.[19]

Apparently, however, originality had in fact become the foundation for the innovative literary pursuits of the rhetoricians, which drifted away from the medieval literature in this respect as well. An invitation to a competition, written at the Transfiguratie chamber in Hulst in 1483, called for plays "newly written and not previously heard or seen." In 1496 the eminent Violieren chamber of rhetoricians in Antwerp invited their counterparts throughout the Low Countries to submit entries for a great drama festival featuring plays "newly composed, new in content, and never previously performed."[20]

By around 1500, this criterion of a superior performance appears to have become so self-evident to the literary elite that it ceased to be stipulated in invitations issued after this date. This reflected the acceptance among these circles that *curiositas* was a foundation for knowledge, science, voyages of discovery, and apparently literary art as well. The eagerness to offer something new from the classical corpus proved useful for recommending literature and other printed texts. By contrast, few references to new or original aspects appear in the prologues to Middle Dutch literature from the manuscript period.[21]

This insistence on original material coincided with an inestimable innovation in literary endeavors: the invention of the printing press. Mechanistic replication of a text in considerable numbers yielded a product for the free market. Enticing an anonymous public with a printed text required fundamental adjustments, starting with the presentation. A title page featuring the title, explanatory subtitles, and a woodcut served to encourage potential buyers, as did laudatory remarks in a preface. Again, presenting the new and the unheard appears to have been used to boost sales; "novelty" served a commercial goal.[22]

Something "new" even came to denote an entirely new type of text. The initial designation of the novella, which was later conveyed literally in Dutch as "novelty," consisted simply of adding "new" to "history" or "story." The adaptation of an Italian novella entitled *Teghen die strael van minnen* (Against the Sting of Love), printed around 1484, was described as "a new, entertaining history." Nonetheless, this work was far from a true innovation that replaced the old version. In fact, the novella was presented as an existing story in a new edition. Only in the course of the sixteenth century did authors have the courage (e.g., Cervantes) to boast that they had conceived their novellas themselves.[23]

The extensively annotated Dutch adaptation of Boethius's *De consolatione philosophiae* presented a remarkably favorable perspective on the new and different. Both the translation into the vernacular and the explanation, published in Ghent in 1485, are clearly the work of a humanist (who has unfortunately remained anonymous). His linguistic style and views reflect contemporary innovations among the rhetoricians. The certificate of establishment (1448) for the main chamber for Flanders in Ghent (De Fonteine) indicated that rhetoricians had a duty to fight life-threatening melancholy. The translator of Boethius's work, for example, stated in his preface that literature was primarily supposed to be new and different and to have a curative effect. Did "every depressed soul not crave change or something new?" What had been initially and, according to Anna Bijns, still remained the domain of the devil had come to justify the primary struggle against him.[24]

Praise for the new, different, and unprecedented thus became a permanent feature of sixteenth-century book production. On several occasions, however, explaining this style of advertisement proved necessary. The editor of the *Buevijn van Austoen,* published in Antwerp in 1504 as a prose adaptation of an old *chanson de geste* in verse, believed that his endeavor would be innovative only if the underlying demand was presented as timeless: "metaphorically, new things have improved mankind since time immemorial, as we note daily around us. . . . Everybody loves new things, after all, and I have traced new material to present something new as a pleasant diversion."[25] Calling such material new would be deceptive, however, as it had been disseminated widely from the thirteenth century onward and appeared initially in French. The printer was undoubtedly referring to material that already existed but had not yet been used in this manner and indicated that he had prepared it for a new, urban audience.

These formulations became generalized. In the collection of allegorical fables *Der dieren palleys* (The Palace of Animals; 1520), which was as entertaining as it was informative, it was observed "that people [were] more inclined to take note of and welcome new things than old matters they had been hearing from time immemorial."[26] It became increasingly clear that printers capitalized on an interest in novelty, news, new things, and even change. The success of the Reformation in Germany and the Low Countries, the popularity of printed news reports, and the growing list of innovations in Dutch literature during the sixteenth century reflected and fostered this attitude as well.[27]

The lust for knowledge and new information was insatiable according to the printers. The reissue of *Der vrouwen natuere ende complexie,* a popular work (reprinted continuously since 1530) on the nature and physical constitution of women, had to be justified—considering the presumed audacity of the text and woodcuts—by noting the urgent questions from many people "longing to hear new and strange things and to know all about strange and new matters."[28] In 1531 a news report was presented in the same nearly apologetic terms. Apparently, the new and different still required proper justification: "wherever we go, we find everybody eager for news, whether they be sovereigns, aristocrats, merchants, farmers, laypeople, or members of the clergy."[29]

More and more members of the rhetoricians' chambers attempted to match the Latin erudition of the humanists in all respects, especially Erasmus, whose work they diligently translated, edited, and read. The type founder, printer, publisher, linguist, and writer Joos Lambrecht, who was a schoolteacher by trade, shared this view. His receptiveness to new ideas led him to explore reforming views of God's creation, and he was prosecuted repeatedly. Notwithstanding (or perhaps because of) these experiences, he had a thriving press in Ghent during the period 1536–53. Lambrecht proved to be an all-round innovator and one of the first humanists to express his ideas in the vernacular, as was happening with increasing frequency in schools and among printers. His provocative motto on his printer brands read *Cessent solita, dum meliora:* "the traditional practices should cease, now that better ones are available."

Lambrecht felt most at ease with the rhetoricians. On 20 April 1539, he printed the refrains written by nineteen chambers from far and wide as a prelude to a drama competition between these same chambers six weeks later. These statements abounded with criticism of the Catholic Church, which now and then appeared to violate its self-discipline and appeals for restoration. The church, its institutions, and its sacraments were rotten to the core. The Tienen chamber submitted a refrain answering the competition question as to who was the most foolish in the world by mocking all monastic and convent orders. Monks were accused of being lunatics with shaved crowns, hypocrites, lechers, swindlers, drunkards, and gluttons. This went considerably beyond incidental reproaches for wrongdoings and was an attack on the institution as a whole.

Lambrecht sought legitimacy for such radical positions, and to this end, he employed Roman typeface, which was the standard medium for printing Latin texts; Gothic type was used for the vernacular. Now, however, the appearance of the vernacular was supposed to have the same standing as Latin. In his preface to readers, Lambrecht explained how deeply he regretted the widespread inability to read "Dutch, German, or Flemish" printed in Roman typeface. These people mistook it for Latin or Greek. He provided corresponding alphabets in both typefaces at the end of his preface as a reference.[30]

An initiative to disseminate the *Const van rhetoriken* (Art of Rhetoric) by Matthijs de Castelein via the printing press in 1555 was similarly motivated. This book, which presents guidelines and samples for writing, reciting, and acting out texts by rhetoricians and which is based indirectly on rhetorical doctrine from antiquity, illustrates an intention to civilize the lay population. Ideas about education and the formation of the elite were similarly important, since the work was designed both for beginners and for the most experienced aficionados. The quest for improvement can be glimpsed in the woodcut at the beginning of this book, which depicts the "noble art of poetry" and features Lady Rhetorica surrounded by Quintilian, Demosthenes, Roscius, Gracchus, and Cicero. Nevertheless, the description of the author Castelein on the title page as an "excellent modern poet" indicates a break with the past. The last adjective was a new word and indicated that, in the view of the printer, this new type of poet was on a par with the greatest in contemporary and past Latinity.[31]

Rederijkers and the Printing Press

Cultural historians have maligned and scoffed at the printing press during its first century of operation. They blame the introduction of book printing for delaying cultural refinement. They allege that typography artificially extended the waning Middle Ages.[32] These allegations are aimed in particular at the literary situation in the Low Countries because of the relatively long wait there for contemporary texts to appear in print. In the fifteenth century, virtually all the texts disseminated in the north of the Low Countries (primarily Holland) were from the past, while in the south (Flanders and Brabant) very few texts appeared in the vernacular. In the south the rhetoricians and the militias continued their literary endeavors, especially for public performances, to which they contributed oral renditions of texts. This continuation of literary activity in familiar formats made demand for the new opportunities for distributing texts provided by the printing press practically nonexistent.

Simultaneously, in the far less sophisticated north, however, the foundations were established within two decades for explosive literary innovations, which proliferated from Antwerp across the Low Countries. A substantial share of the vernacular literary legacy from the past four centuries soon emerged in print. One result was that now it was much easier to compare these texts with one another. Other ongoing experiments during these early years concerned presentation techniques, costs, sales opportunities, and of course the more technical aspects of book production. Members of the clergy figured prominently in producing and purchasing these texts as part of the didactic techniques revived by adherents of the *devotio moderna* and the resulting longing to read among laypeople.

This typography-driven surge of literary activity in the north did not last long. The market was soon saturated with all those old informative texts in print. The clergy's bookshelves were full for the time being. Given this situation, it was understandable that the printers who survived the risky business in this early stage went off in search of different texts as well as markets for them. Gerard Leeu of Gouda in Holland (on whom more below) made this transition almost prematurely and moved to Antwerp in 1484. Around 1500 others followed. Antwerp soon became the epicenter of typography and book selling.

Although the old texts remained in circulation, their suppliers started to target a new public of city dwellers, whose general ambitions, expectations, and frustrations served to guide production of literature in print, including satisfactory adaptations in the vernacular. The clergy ceased to be a significant market. This meant that texts now took on the character of merchandise and as a result underwent some drastic changes, not only in form and content but also in the ways they were used by the consumer. Texts now had to attract customers in a free market, and one of the consequences of this was the development of the title page. The importance of such commercial tactics for the form and content of the text can hardly be overestimated. Rarely, however, have historians of literature devoted serious attention to this phenomenon. The fact that the object of their study, after just a few decades, always appeared with a title page—and one that usually boldly acclaimed the contents—is accepted as if it were an inevitable development. In the age of manuscripts, the prologues to texts usually included content descriptions of what was to follow. The title page, however, sharply focuses expectations, compelling the reader to approach the text in the light of what has been announced. And the titles are formulated with an eye to marketing the book to as broad a public as possible. One thing that stands out immediately is that titles of historical literature and fiction favored the names of ancient heroes.

Another radical change from manuscripts is the use of numerous laudatory formulas on title pages, in prologues, in prefaces, and even in epilogues and colophons. Other innovative features that deserve more attention are the woodcuts, chapter titles, and especially the adjustments of content made to interest a new, broader public. A chivalric novel in verse that originated in courtly culture offered middle-class culture attractive possibilities for identification, provided that it was tailored to the demands of urban ambitions and ethics. In addition, the form had to be adapted to the new communicative possibilities brought about mainly by the printing press: reading to one's self or hearing works read aloud in a small circle of listeners.

This explains why a thirteenth-century verse romance like *Heinric en Margriete van Limborch* was altered on a few points of content and rewritten as prose for the printed version of 1516, which proved very successful with a middle-class public. An important scene portrays Heinric's elevation to knighthood. He must swear that he will uphold all the traditional knightly virtues such as loyalty to his lord and the protection of widows and orphans. But the prose version adds that he must, under all circumstances, be creditworthy: "Pay generously wherever you travel, whether by land or by sea, then people will speak honorably of you." The obligatory concept of honor is maintained in the text, but it is now linked to a value exclusively associated with urban ways of thinking: the knight should be equipped not only with a trusty sword but also with a well-filled pocketbook.

We repeatedly encounter small changes like this one, together with the strong tendency to present the annexed knightly world in the new dimension of contemporary sayings and adages. These, too, were inserted in order to clarify the text's relation to the middle-class world and, in general, to strengthen the contemporary grasp of the divine plan of salvation. This justified the assertions on the title page and in the prefaces that the text would present "something new," because the adaptations were supposed to create a totally new situation in which the old matter should and could still function.

What should also be studied is the technical side of the reception. The first printers went to great lengths to make clear that their texts could be used in every conceivable (and even barely conceivable) way: for listening, reading aloud, reading to others, reading silently, following the text as someone else reads aloud; for reading selectively, discursively, pensively; for rereading, leafing back and forth through the text, and simply looking at the pictures. The unprecedented variety of suggested modes of reception also contributed to changes in the interpretation of literature and fiction, as well as in their impact.

Gerard Leeu, who worked in the small town of Gouda, was the first printer in Holland to produce a varied series of Dutch literary texts. He was fully aware of what he was doing. He had to find a wide audience in a population not accustomed to reading fiction in Dutch. People may have heard these texts—read aloud, acted by a passing troupe of entertainers, or recited from the pulpit in the form of examples—but always in a crowd, as a member of a group, aided by the performer, who emphasized, winked, accelerated, slowed down, explained when necessary, and was constantly aware of the attention of his audience. Leeu had an eye for the problems his untrained public might encounter. And that is why he provided his first editions with very practical instructions: how to read Dutch fiction in print on one's own. His interventions are all the clearer to us because he published older texts or texts known from elsewhere, and thus his adaptations and additions are immediately apparent. He was firmly convinced that he had to assist his audience to read Dutch by themselves, separated from the crowd.

Looking over Leeu's instructions today we feel somewhat embarrassed by the way he lectures us, as if we are unwilling, if not suspicious, children. But we have to put ourselves in the position of being confronted with a radically new and difficult situation for all concerned. Perhaps it may be compared with my own experiences concerning the introduction of the television set in the living room, to which I was first exposed as a schoolboy in the fifties. The room was jammed with relatives, some of whom I had never seen before. Earlier, I had been instructed to line up all the chairs we owned (and even some more from the neighbors) in five rows and to ensure that everyone was seated in complete darkness long before zero hour. No one knew exactly how to behave, and that applied to the television producers as well. They wrestled, for example, with the problem of how to start their programs, as they were no longer in the position to put out the lights themselves. And that is why the first television programs in the Netherlands started exactly at 8:00 p.m. with a loud bang on a gong, which was supposed to make every living creature within the surrounding area quiet. After that a female announcer constantly interrupted the broadcasts with motherly advice and directions concerning the regulation of the volume, the definition of the picture, and the need to take a break after one full hour of watching. Consequently, at least twice an evening the programming was interrupted for a five-minute break to give everyone an opportunity to exercise their bodily functions, which caused serious problems with the national water supply. Excitement rose to a fever pitch when the evening closed with a lady waving us goodbye; most of my relatives, including my parents, used to wave back! It is important to keep in mind that this was the behavior of adults, all middle-class and well educated. It was not until the sixties that television watching in Holland found its own form.

In 1479 Gerard Leeu took grown-up people by the hand from the very first moment they decided to open a copy of one of the first literary texts in print: *Reynaert die vos* (Reynard the Fox). To start with, the subject matter was reassuring because well known: it is quite unthinkable that there was anyone, even in Gouda in the year 1479, who had never heard of the cunning tricks of the famous fox. Probably, Leeu recognized that familiarity was essential for success: introducing a new technique would be easier if there was at least some familiarity with the object a reader held in his hands. Moreover, Leeu knew that for more than three centuries this text had been a dependable success, so that in this respect his enterprise could not possibly fail.

Leeu immediately puts his audience at ease by assuring them that the booklet not only is particularly fit for their benefit and amusement but also is certainly not intended for the world of learning, as might be expected of a printed book. The subject matter deals with the tricks one can witness daily, if not suffer from, in all social circles, including the world of merchants and the common people. This applicability allows the entertaining animal story to benefit a wide audience, Leeu continues; one can, as it were, learn how to protect oneself. But there is still the problem of understanding. Leeu shows how well he appreciates our hesitations: "He who desires to understand this book completely has to read it several times over and to contemplate all the printed words with diligence, because the author has put these words together in an ingenious way, as you will notice when reading. Should it happen that after a single reading the exact meaning or import of the text remains unclear, read it many times over, until it is fully understood." So do not give up, but keep on trying, advises Leeu, for it is quite possible, even probable, that you will not understand it all on the first attempt. Start again from the beginning and at last you will grasp the full meaning. That is the way to read fiction in your own mother tongue. And provided you follow his advice carefully, you may feel flattered by the conclusion of his preface: "To conclude, it will be very amusing and also profitable to intelligent people." So, fortunately, the act of reading is pleasant too: one might have thought that this new form of mental exercise merely yielded simple utility and no fun.[33]

Conclusion

Urban literature of the late Middle Ages and early modern period played an active role in forming, defending, and propagating urban values, which revolved around the key concepts of practicality and utilitarianism. But this does not mean that the heroes of yesteryear were not still very useful to urban society. Virgil, represented as a sort of magician, and Reynard the Fox are given a fresh lease on life in the town.[34] Reynard the Fox represents an individual with practical ingenuity who can take on the whole world and take good care of himself under all circumstances. This hero holds all the traditional values, court mores, and every other kind of accepted behavior in disregard, yet embraces them with alacrity if that is to his advantage. Reynard knows the ceremonial of legal proceedings completely, and so he can faultlessly exploit the weaknesses in the system. He is courteous whenever that will yield results, for instance, toward King Noble and his wife, Gente. In this way he softens them up for his con games.

At the same time, aristocratic culture formed an attractive reference point for the creation and propagation of the citizens' own package of virtues and behavioral codes, thus giving the required luster to this new form of society. Until long after the Middle Ages, the world of the court was to remain a model for emerging elites, who arranged tournaments in the marketplaces of the towns, read or listened to prose stories based on chivalric literature from earlier centuries, organized themselves into "round tables," and, at festivals and parties, adopted the names of famous knights of the past. To this end, however, the antique, biblical, and medieval heroes had to be equipped with the right qualities: qualities in which the new urban elites could, and wished to, see themselves mirrored. This sometimes resulted in a pseudoknightly culture in literature, street theater, and visual arts that resembled modern operetta more than it provided a faithful reproduction of actual courtly entertainment.[35]

It was in this way that a literature evolved in middle-class circles that, by means of annexation and adaptation, propagated a whole set of virtues considered essential for maintaining and extending the power that this group had acquired. The power was based on trade and industry and owed its momentum primarily to an increased striving for individualism, with its underlying assumption that one could take on the entire old world. The printing press also eagerly offered its services in the formation of this mentality by portraying countless rogues and rascals, each of whom manages to get the best of traditional wielders of power and their rigid codes by a playful use of words and wits. Dozens of *rederijker* texts present such shrewd go-it-aloners, who are mostly of simple appearance, behavior, and origin or even have physical defects: Aesop, Marcolphus, Jan Splinter, Heynken de Luyere, Aernout, Everaert, François Villon, Tijl Ulenspieghel, the Parson of Kalenberg.

These new values appeared unmistakably at this time among a "nouveau riche" nobility of ambitious merchants. In this fluid situation a more pragmatic approach to life was emerging, with eyes on the street and everyday reality. These new attitudes spread within and outside city walls through all the strata of society, whether nobility, clerical hierarchy, international merchants, or guild master craftsmen. A new power, based on personal skill and business acumen, soon openly competed with the traditional authority of the sovereign powers.[36]

It is misleading to suppose that all was new in the city. The middle-class virtues mentioned previously were certainly not "invented" in the city. Individualism, hard work, and making a career were not exclusively traits of urban society, even though they often were paraded as new in the cities themselves. A substantial number of the qualities derived from classical antiquity, many of them could be found at court, and nearly the entire list was already present in the earliest monastic environments. In every case we are confronted with overlap, for the primary characteristic of this mentality is that it adapts to and combines with other attitudes. In brief, the classical authors presented the dictates of reason and control of the emotions as guidelines for earthly life, together with instructions for the careful running of a household (*oeconomia*). The monastic orders emphasized hard work, discipline, and the related need for efficiency in measuring time. Self-sufficiency could also be found in this milieu. Finally, the individual adventurer who goes out to challenge the world and vie with fate (Fortuna) was first portrayed in the chivalric literature of courtly culture.

What was unique about the late medieval and early modern city, however, is that a highly original set of virtues appeared, forged from the classical, biblical, and medieval traditions. Urban dwellers borrowed from vernacular, as well as elite, traditions, from old and new, and were continually in search of useful elements that could be adapted to reinforce, embody, and foster the city's own interests and ambitions. And this passion for annexation and adaptation was presented in terms of novel knowledge, novelty, and renewal, particularly in the literature produced in new institutions such as the *rederijkers'* chambers, which assembled, tested, and propagated this new collection of urban virtues.[37]

CHAPTER SIX

Watches, Diary Writing, and the Search for Self-Knowledge in the Seventeenth Century

RUDOLF DEKKER

The word "revolution" is often used to describe the development of science in the sixteenth and early seventeenth centuries. This may be an adequate term, but this revolution did not take place in isolation. Like its political equivalents, the Scientific Revolution was embedded in a broader process of cultural change. In this essay, one essential aspect of the Scientific Revolution, the great improvement in techniques to measure time, is linked with an important cultural development, the invention of the modern diary. While the search for knowledge of the outer world intensified around 1600, there was also a growing need for knowledge of the inner world. Although by no means the work of great individuals alone, the Scientific Revolution is associated with famous thinkers like Galileo and Isaac Newton. Likewise, the growing interest in self-knowledge had a broad basis but is also linked to some famous authors, of whom the most important was Michel Montaigne. His essays set a new standard for self-analysis. The autobiography of Benvenuto Cellini and the diary of Samuel Pepys are among the most famous landmarks in the development of new literary tools for researching the self. Producing knowledge and producing self-knowledge clearly went hand in hand, and the two cannot easily be separated. In egodocuments like those by Cellini and Pepys, the authors not only tried to learn more about themselves but also described and analyzed their immediate environment. Like astronomers, they sought to describe and explain reality. There even was a direct link between both developments: the invention of new timekeeping devices directly influenced the development of the modern diary, in which the author kept track of his or her life.

Stuart Sherman has recently drawn attention to the connection between the birth of the modern diary and the invention of the first accurate clock, relating the way in which Samuel Pepys kept his diary to the introduction of the pendulum clock, which was invented by the Dutch mathematician and astronomer Christiaan Huygens.[1] In the seventeenth century daily writing became a common practice, and this new diurnal form would influence culture in various ways, contributing to the development of, for example, the shipping journal and the daily newspaper. The link between the revolution in time measurement and the rise of the modern diary can be found within the Huygens family itself, as can be learned from an analysis of the extensive diary kept by Christiaan Huygens's brother Constantijn Jr. The Huygens brothers embodied the technical and cultural renewal of the seventeenth century. Science and culture, as is clear in their case, sprang from the same source—the wish to know the world. While technical devices like the pendulum clock contributed to knowledge of the material and public world in general, diaries—kept in growing numbers—produced an ever-increasing amount of knowledge about the mental and private worlds of their authors.

The Huygens Family

Three generations of the Huygens family have played an important role in politics, culture, and science in the Dutch Republic. The father of the brothers, Constantijn Huygens Sr. (1596–1687), held the prominent post of secretary to the Prince of Orange. He was a key figure in Dutch culture of the Golden Age and well known as a poet, classicist, musician, composer, and connoisseur of art; he also took great interest in natural science. In 1627 he married his cousin Susanna van Baerle. Their eldest son, Constantijn Jr. (1628–97), also was appointed secretary to the Prince of Orange; after the Glorious Revolution of 1688, when Prince William of Orange became king of England, he usually spent winters in London, summers in Holland, and during the spring was often on military campaigns with the king. Huygens's second son, Christiaan, devoted himself full time to natural philosophy. The youngest son, Lodewijk, was the least scientifically gifted of the three and held a government post in a Dutch town. There was also a daughter, Susanna.

For Constantijn Huygens Sr. and Jr., writing was a profession, and as secretaries to the Princes of Orange they wrote enormous numbers of letters, memoranda, and other official papers. In writing, they also maintained contact with friends, artists, and scholars throughout Europe. Several members of the family wrote and published poems and plays. There was also a tradition of private writing within this family.[2] Constantijn Sr.

was a prolific writer of all kinds of egodocuments. He kept a diary, of which only a fragment has survived.[3] He wrote travel journals and made notes in his almanacs, which are now in the Dutch Royal Library.[4] His tendency toward self-observation is also revealed in a well-drawn self-portrait. He kept extensive notes on his children, which contained acute observations from their birth onward. When he was a mere thirty years old, he wrote an autobiography to serve as an example for his children.[5] When he was in his eighties, he produced another autobiography, again written in Latin verse.

Constantijn Huygens Sr. stopped writing notes on his children when—as he wrote—he regarded them to be old enough to keep their own diaries. It is indeed very likely that his three sons obeyed their father. However, only the diary of the eldest son, Constantijn Jr., has survived. He started writing a travel journal during his grand tour in 1649 and 1650.[6] Thereafter, his diary covers long periods during the years from 1673 to 1683.[7] Some earlier parts of the diary may have been lost, but we still have a continuous text, in which he made daily entries, starting in 1688 and ending in 1696, the year before his death. Christiaan wrote much, mainly scientific tracts and letters (his complete works number twenty-two volumes), but unfortunately only two short travel journals have survived.[8] The youngest brother, Lodewijk, left two surviving travel journals.[9]

The Huygens Family and Science

Technology, practical mathematics, and natural philosophy flourished in the Dutch Republic.[10] The invention of the telescope and the microscope, which improved the measurement of space and time, and the subsequent advancements in the art of cartography were all symptomatic of the cultural shift that took place in the republic. Constantijn Huygens Sr. was a *homo universalis* who was skilled in many arts and sciences and followed new developments keenly. He combined the two spheres in his poetry, for example, in his laudatory poem to Antoni van Leeuwenhoek, in which he praised Leeuwenhoek's "glass keys"—his microscope—which "unlocked the secrets of nature."[11] Elsewhere he praised the telescope: "what a small ball shall the whole world become!" In another poem he compared the human body to a clock and the rhythm of the heartbeat to the ticking of clockwork ("Balance Wheel"). Huygens was an admirer of contemporary inventors and practical mathematicians of his time, such as Cornelis Drebbel, who had developed a craft that supposedly sailed underwater in the Thames, and Simon Stevin, who built a cart with a sail that is reported to have reached high speeds on the beach of Scheveningen.

An indication of the family's appreciation for technology is to be found in a 1668 portrait by Caspar Netscher of Geertruyd Huygens, Constantijn Huygens's sister, posing with a watch in her hand. It was customary to allude to time passing in portraits as a memento mori, usually by the depiction of an hourglass.[12] Mechanical watches, invented in the mid–sixteenth century, were still a rare possession at the time. The watch in the portrait could have a religious meaning. However, since this symbolism is already expressed in the portrayal of Father Time in the background, it is likely that the watch was included as a technical showpiece.

The portrait that Constantijn Huygens Sr. commissioned in 1627 from Thomas de Keyser also has a telling detail. Huygens is portrayed in his function of secretary, handing over a letter to a messenger. His desk is crowded with paper, inkpot, a quill pen, and, on closer observation, also an open timepiece. At that time watches were not part of the usual equipment of a secretary. Title pages of instruction manuals for secretaries, already widespread by then, only showed the writer with pen, inkpot, and paper. Twenty-five years later, on the frontispiece of *Secretaris d'à la mode,* for the first time a watch is added to these paraphernalia (fig. 6.1).[13]

Their father was quick to realize that his two eldest sons possessed great capacities. Constantijn Jr., born in 1628, picked up Latin quickly. After his studies in Leiden, he was appointed secretary to the Prince of Orange, just like his father. He was known as an authority on art, wrote Latin verses, and was an excellent draughtsman.[14] He also had a great interest in the investigation of nature and assisted his brother regularly with experiments. Being very adept at grinding lenses for telescopes and microscopes, he customized them for his brother and at a later age wrote a treatise on the art of grinding lenses.

Christiaan Huygens, one year younger, was active in many areas of natural investigation.[15] He constructed telescopes and discovered Saturn's rings and one of its moons. With his microscopes, he experimented with air and vacuum and was one of the first mathematicians to work on the calculus of probability. In particular, he made a great contribution to the improvement of the technology of time measurement. Until the middle of the seventeenth century, clocks remained inaccurate instruments because, among other problems, stored energy could not be transferred efficiently to the moving hand by means of weights and pulleys. In 1659 Huygens designed a clock in which the transport of energy was regulated by the constant movement of a pendulum. (The patent for building pendulum clocks according to this principle went, however, to a clockmaker in The Hague, Salomon Coster, in 1659.) For the first time a clock could have

Fig. 6.1. Frontispiece of Jean Puget de la Serre's *Secretaris d'à la mode* (Amsterdam: Jacob Benjamyn, 1652).

a minute hand that worked with precision. Huygens published his findings in *Horlogium oscillatorium*, and his father promptly wrote a poem praising his son's invention.[16] On a portrait from around this time, Christiaan was painted with his pendulum clock, as a parent would have himself painted with a child, and in a letter to his brother Lodewijk, he wrote that he viewed his inventions as children of the mind (fig. 6.2).

Fig. 6.2. Christiaan Huygens, engraving by G. Edelinck, 1686; Huygensmuseum, Hofwijck. Huygens's inventions, his books and clock, stand on the table to his left.

More precise time measurement was essential to position finding on ships at sea. A clock that could give precise local time during the journey made calculation of longitude more accurate. The problem with longitude determination had impeded the exploration of the world for centuries, and in 1610 the Dutch States General announced a reward of 15,000 guilders to the person who could solve the problem. Huygens had built his pendulum clock for use at sea, hoping to receive the reward. Initially, his clocks were tested at sea with success. In 1665 Huygens published a manual for his sea clocks, *Kort onderwys aengaende het ghebruyck der horologien tot het vinden der lenghte van Oost en West.*[17] The clocks were not good enough during bad weather, however. To solve this problem, Huygens invented the spiral spring as an alternative to the pendulum, a discovery for which he received a patent in France in 1675. The spiral spring could also be used in watches, improving the reliability greatly. Initially, watches were considered jewelry rather than instruments, but now they were functional as well.[18] Christiaan Huygens came to be considered the first great horologist. And that was great praise, especially in a time when God came to be viewed as the great watchmaker of the universe.

The Diary of Constantijn Huygens Jr.

The diaries of Constantijn Huygens Jr. and of Samuel Pepys have much in common, if only because both wrote often about sex.[19] Constantijn Jr., however, mainly wrote about the sexual adventures of others, particularly Dutch regents and courtiers of the Prince of Orange. His diary is full of gossip and a gold mine for the history of sexuality.[20] He also wrote much about his wife and son, as well as about his domestic servants.[21] The reader is informed of his professional practice as secretary in a manner that would not have been possible using only official documents.[22] Moreover, Huygens wrote of many other subjects, from feasts and games to magic and witchcraft.[23] He was generous with the trivia that give a diary local color. His diary, now kept at the Dutch Royal Library, is in his own handwriting; it comprises seven volumes, with entries that cover the period between 1673 and 1683[24] and daily entries from 1688 to 1696.[25] In the years before 1682 Huygens usually wrote in French; after 1688, mainly in Dutch. The diary was, like that of Pepys, private and not meant for others to read. Once in a while, he would use a letter code, which, since he was a secretary, took very little effort, and once he used a cipher code (9 July 1689). This is another similarity to Pepys, who used the shorthand in which he was skilled. Nevertheless, Constantijn Jr. took great care to ensure that no one would read his diary. Once he noticed that there was a section missing and feared it had fallen into the wrong hands. A few days later he found the missing pages, to his great relief, as he noted in his diary. Constantijn Jr. kept his diary in a locked writing cabinet that also served as a traveling desk, which he took with him when he accompanied the king during military campaigns. It was solidly built and had a secret drawer for money (12 September 1694). Once, he lost his key and could not get to his papers and a locksmith had to be called to open the cabinet (12 February 1691).

Like Samuel Pepys, Constantijn Huygens Jr. made daily notes with great regularity. Over the entire nine-year period between 1688 and 1696 he missed only 25 days. In all those cases he would later note that he had forgotten to write (see, e.g., the entry for 28 November 1694). We have this type of information because he wrote his diary first as a draft, which he would write out in a clean version at set times. Then he would notice these omissions. It is only the final version that has survived. Once in a while he would mention in his diary that he took his draft notes to write out a clean copy. On 4 June 1694 he wrote, "In the evening I wrote a part of these notes, to catch up," and on 26 March 1696: "I copied a part of this journal this morning." Constantijn Jr. had a well-trained memory. As secretary to Prince William he frequently had to memorize dictated letters, without being allowed to makes notes, to write out later. Only once does he mention that he had to return to the king to inquire about a forgotten detail. At another time, he wrote in his diary that he had forgotten the names of two members of a certain party (24 February 1690). On another occasion he later added to a story he had noted down: "This turned out not to be true" (27 July 1696). Once he also corrected the chronology and added in the margin that a certain meeting had taken place a day later (19 March 1694). Additions at a later date, such as on 9 June 1693, are rare. The diary served in the first place as an *aide-mémoire,* and once he describes in detail a room in which he was lodged "simply for the sake of memory."

Characteristic of Huygens's diurnal temporality are the many observations made on the spot. "Looking out of the window I saw boys throwing snowballs at carriages passing by," he writes on 16 February 1692. Or, on campaign, he notes that he saw a lizard walk through his tent or that he found a louse in his coat (26 August 1695). Another characteristic of Huygens's diurnal temporality is that each day contains an entry even on the days when nothing of note happened, such as "I didn't have visitors the entire day" (24 February 1693) or "I was not out and nothing happened" (29 November 1695). On 22 May 1691 he wrote: "I did not get out in the afternoon, got to reading." Constantijn Huygens Jr. wrote, just as Pepys did, to the rhythm of time measured with the newest technology. His daily notes always followed a strictly chronological pattern. Usually he used the labels "morning," "noon," and "evening." The flow of time in the diary is uninterrupted and even. It recalls the linear, objective notion of time that was introduced by Newton and replaced the traditional, cyclic notion of time.[26] Occasionally, Constantijn Jr. gave a more precise indication of time, for example, when he awoke at one o'clock in the morning and could not fall back asleep (8 September 1695), or when he spoke to someone for "three quarters of an hour" (19 July 1694). In short, the connection between the Huygensian chronometry and the diurnal form of writing can be made between the two Huygens brothers. Christiaan, the horologist, and Constantijn Jr., the diary writer, were dealing each in his own way with a new notion of time they themselves were producing.

The connection between diary and newspaper, suggested by Stuart Sherman, is also traceable to Constantijn Huygens Jr. He regularly made note of what he read in the newspaper. On 12 November 1695, after his arrival in Holland, he states that his return from London is mentioned in Dutch newspapers: "My return is reported in the newspapers." It is the first diary entry in which the writer reacts to being mentioned in the newspapers. None of his almanacs have survived, but presumably he made as much use of this type of printed work as had his father. It is certain that he provided Queen Mary with almanacs, for on 1 May 1691 he received a visit from a lady-in-waiting, Miss Vijgh, who, on behalf of the queen, thanked him for the almanacs he had had delivered to her.

The well-organized diary of Huygens stands in contrast with many earlier diaries, like that of the Frisian farmer Dirck Jansz, one of the earliest examples in the Netherlands. Jansz was born in 1578 or 1579 and kept a diary between 1600 and 1636, but very irregularly. He made dozens of notes each year, but not in chronological order, and he obviously possessed a different attitude toward time.[27]

Constantijn Huygens Jr. and Science

In his diary Constantijn Jr. regularly shows interest in the investigation of nature, and it turns out that he had quite a reputation in this area. On 18 July 1692 he wrote that the king had said to a visitor about him: "il est le plus grand astronome du monde!" It is one of the few times he uses an exclamation mark. On several occasions he mentions meetings with like-minded persons and conversations about science. For example, on 18 October 1690 he was visited by the daughter of Cornelis Drebbel, the famous inventor who was admired by his father. She told him many details about his legendary submarine, for instance, that a "pipe with quicksilver" provided the air supply.

Only a few days after his arrival in London during the Glorious Revolution of 1688, Constantijn Jr. received a visit from someone who said he had known his brother in Paris and offered to introduce him to natural philosophers such as Boyle, Newton, and Wallis (8 January 1688). Then the conversation turned to a long discussion of the merits of Chinese astronomy.

Two weeks later, Christiaan himself arrived in London and introduced his brother to members of the Royal Society of London for Improving Natural Knowledge. The society had been founded thirty years before, with Christopher Wren and Robert Boyle among its first members and Robert Hooke as its curator. The society met weekly to discuss the investigation of nature. Christiaan Huygens had been chosen as a member in 1663 and had since been in contact with English members. The brothers' father, Constantijn Sr., also corresponded with the society, apprising them of his son's discoveries.

Constantijn Jr. maintained regular contact with the society. On 16 January 1692, for example, he described a dinner with eight to ten members in an inn, Pontac, as the society did not yet have its own housing. Among those present were President Robert Southwell, Vice President Thomas Henshaw, Hans Sloane, Patience Ward, and William Stanley. During this dinner Huygens raised a question: why was the *Philosophical Transactions* published by the society "so meager and small" of late? They told him that once the war with France was ended, it would be thicker again. The lack of interest shown by the editors Hale and Hooke was also blamed. On 18 December 1696 Constantijn Jr. attended a meeting at which Halley lectured for no less than two hours. Once in a while Constantijn Jr. would drop in on the bookshop of the printer of the *Transactions*, Smith (as on 3 and 4 March 1694 and 26 December 1695). He always bought copies when they came out, as he did with other books on natural topics. When on 20 February 1693 Stanley showed him the recently published *History of the Air* by Robert Boyle, he "promptly sent a servant to buy the book in the city," as he noted that same evening in his diary. His scientific interest was great, but often he was too busy to attend meetings. On 3 December 1690, for example, he mentioned that Stanley had forcibly tried to take him to a Royal Society of London dinner, for which he had no time and no desire.

In those years Constantijn Huygens Jr. kept track of Isaac Newton. On 10 July 1689 he wrote that his brother Christiaan went to the king with Newton to "recommend him for a vacant regency for a college in Cambridge." They did not succeed. In 1694 he heard from Christiaan that Newton had gone mad for a while and had suffered from "phrenesia" for eighteen months. Whether or not Constantijn Jr. ever met Newton himself remains unclear. Presumably he did know Robert Boyle personally. He tells several stories about him in his diary, such as on 10 September 1690, when he had heard from someone that Boyle had ingested enough arsenic to kill a hundred people, but that he, with a drop of a preparation, had vomited it all out and felt fine again. On 1 July 1689 Constantijn Jr. visited one of the founding fathers of the society, Christopher Wren. He spoke most frequently with the mathematician Nicolaas Fatio de Duilliers, as courtier and his equal, who, among other things, improved the watch with the use of diamonds. On 3 September 1689 Constantijn Jr. went to Greenwich to visit the famous observatory and to meet the director, John Flamsteed. This entry is probably the first reference to the word "observatorium" in the Dutch language.[28] It is also fairly certain that the diary contains the oldest mention of the word "laboratorium" (3 December 1690) in Dutch.[29]

Constantijn Huygens Jr. was always enthusiastic when he met others with an interest in natural matters, as he was on 3 March 1690, when a Scottish lord asked him during the king's *coucher* if his brother or his father was the inventor of the pendulum. He also received visits from inventors and technicians who hoped that he would put in a good word for them with the king. For example, on 1 February 1690 he received a visit from an engineer who wanted to introduce fire engines in London. On 24 January 1692 Constantijn Jr. heard Halley, known mainly as an astronomer, tell of his "ability to stay under water for an hour and longer, having an invention to let the used air out of his diving-clock as new air from a container is piped back in." On 4 February 1695 he received a visit from an engineer named Tompson who said he was working on a new "invention" by which a diver could get air through a pipe and then go to a depth of twenty fathoms, "but that from there the pressure of the aire was so great that one couldn't stand it." Such vivid stories speak to the imagination.

Natural investigation and alchemy were strongly related, as shown by a note made on 20 June 1689, when Constantijn Jr. heard from his brother Christiaan that Robert Boyle had received a visit from a man "who, in front of him, made an ounce of gold from lead with a powder that was red and clear." Constantijn Jr. noted in the same entry that he had heard that the man was later arrested in France. A little after that, on 5 July 1689, Constantijn Jr. heard a similar story, about making gold from quicksilver. The inventor—or alchemist—had passed away in the meantime and had taken his secret with him to his grave. Then there was an old friend from the army with whom he spent much time, the engineer Willem Meester, who produced explosives to blow up forts and prepared ships that could blow up enemy ships and coastal works (3 September 1694).[30] Meester also made fireworks for parties (11 April 1694), but this apparently did not appeal to Constantijn Jr., who seemed to show more interest in their utility (as he noted often in his diary) than their potential for entertainment.

Clocks and Watches

The measurement of time occupied Christiaan Huygens for his entire life, and this interest was shared with his brother Constantijn. After Christiaan had invented a well-regulated spring for watches in 1675 while staying in Paris on a pension from King Louis XIV, the French king soon received a particularly fine specimen. Soon after, such a watch was made for the Dutch stadtholder William of Orange, despite the ongoing war between France and the Dutch Republic. In April 1675 Christiaan wrote a letter to his brother with instructions on how the prince should carry the watch, namely, in a sachet on his belt and not in the pocket of his trousers. Two months later the watch was sent from France.[31] Constantijn Jr. made a note of its arrival in his diary on 17 July 1675: "At eight o'clock in the evening Mr. Boreel arrived with the mail from Paris. His Highness immediately asked if he had not brought his watch. He said yes and gave him the small box containing it. Willem Meester and I got it moving and His Highness showed how happy he was with it and received great pleasure from watching it move." In those years William III and Louis XIV fought against one another on the battlefields, and it is remarkable to learn that one brother was instructing the stadtholder while the other was instructing the king in the use of this new instrument.

Constantijn Jr. wrote much about his watches—he possessed several—in his diary over the years from 1688 to 1696. Having barely arrived in London with the stadtholder, Constantijn Jr. paid a visit to clockmaker Lownes on 13 January 1688 to have his watch repaired. He was clearly impressed with the English clockmakers and noted: "He made 'repeating watches' small and large, the large ones very neat and handsome." A few days later, on 22 January, he and his friend Meester paid a visit to the best clockmaker in England, Thomas Tampion: "Went to see Tampion, the watchmaker, with Meester in the morning. He showed me a repeating pocket watch." In the following years Tampion remained the clockmaker with whom Constantijn Jr. maintained regular contact for the purchase and repair of clocks and watches, but also to have an occasional conversation. Constantijn Jr. was important to Tampion because he could recommend him to the court. On 10 March 1688 Constantijn Jr. wrote that he had picked up a pocket watch from Tampion that "sounded on the hour and quarters" to show it to the king.

Constantijn Jr. was impressed by this clockmaker, whose workplace looked like a factory, as is apparent in the entry dated 30 September 1690: "I went to see Tampion during the morning, and had him make a silver watch and repair my golden one. I noticed he had at least 20 apprentices, each in a room, all working for him and he provided for all of them." The store was a meeting place for those interested in technology. On 18 July 1689 Constantijn Jr. and his brother Christiaan stopped by to purchase a "ring-sundial." A sundial was still an important instrument among the arsenal of time measurers and was essential for setting clocks. On 12 December 1690 at Tampion's shop he met Flamsteed again, the director of the observatory in Greenwich, who was also a regular visitor there.

Smooth-running clocks were of great importance, especially for determining longitude at sea. As did the Dutch Republic, the British government offered a reward to the one who could solve the problem. In all of Europe clockmakers were trying to claim the reward promised by the London Admiralty with the perfect clockwork to measure the hour. Constantijn Jr. received firsthand information about the attempt made by the French clockmaker Thuret: "In the morning I received a visit from Dr. Stanley, who told me, having been at the Royal Society meeting as a member, . . . that Thuret, watchmaker from Paris, had also been there, claiming to have the invention of the longitudes" (29 November 1689). The Frenchman's clock could not, however, withstand the test. It would take many decades before the first reliable sea clock was made.

Another favorite theme of conversation among clockmakers was patents for different findings and improvements. Christiaan Huygens had already had problems with obtaining the patent for the pendulum, and later he would have trouble over the English patent for the spiral spring. Inventors and clockmakers were in competition with one another frequently and were not above stealing ideas from each other. In 1675 a conflict broke out about the English patent on the spiral spring between Christiaan and Robert Hooke, who claimed to have made the discovery earlier. The Royal Society of London intervened but were not able to bring the two men to an agreement. On 9 December 1690 Constantijn Jr. made his rounds to the London clockmakers once again. First he visited Lownes, where he had seen a "repeating clock" that would run for eight days and sound every hour; it cost sixteen pounds sterling. Then he went to see Daniel Quare, where he saw a similar clock said to cost twenty-eight pounds. He also saw a "repeating pocket watch" that cost sixty pounds. This watchmaker told him all about the conflict he had had with Tampion about the patent for the "repeating pocket watch."

These clockmakers also made other instruments, such as barometers. On 14 December 1694 the king summoned his secretary to show him his new barometer, made by Quare, which was—and this was new—portable.

Tampion could not let such a success by the competition go unmatched. On 14 April 1695 he visited Constantijn Jr., who noted he was carrying a portable barometer. Tampion was on his way to the king to ask for a patent similar to the one Quare had received. On 6 June 1695 Constantijn Jr. visited Tampion's store and saw the instrument: "He showed me his newly made barometers."

When Constantijn Huygens Jr. was in the Netherlands, he went to see Dutch clockmakers, such as Van der Cloese in The Hague (23 and 25 November 1693). He had taken a clock from Oosterwijck with him to England, as he noted on 27 October 1695: "Went by Tampion and retrieved my pocket watch. Sent him my standing clock made by Oosterwijck, which needed some repairs." In October 1694 Constantijn Jr. was in Holland and visited his brother Christiaan at the family country house in Hofwijk "and saw there his newly invented pendulum clock" (8 October 1694, 21 May 1694). Christiaan worked with several clockmakers and constantly invented improvements.

Constantijn Huygens Jr. had an ever-changing arsenal of standing and hanging clocks, pendulums, watches, and sundials. He was constantly purchasing new clocks and watches and exchanging old ones. All of these continually had to be readjusted and repaired. There were periods in which Tampion and Constantijn Jr. saw one another weekly, such as in the winter of 1691–92 when Constantijn Jr. had bought a new "pulling clock." Sometimes Tampion came by to adjust the clock, and on other occasions Huygens brought his clock to the store. A series of such entries ends on 27 February 1692: "In the morning I went to Tampion, under the impression that my pulling clock was not working properly, but found her to be wound down completely." ("Pulling clock" is the English term used by Constantijn Jr., who in his diary now and then throws in some English phrases.) Constantijn Jr. possessed a watch with a key, a detail that personalized this type of clock, but such a small key could easily get lost. On 1 June 1691 he noted: "I bought a new key for my pocket watch, having lost the old one."

At the royal court Constantijn Huygens Jr. was known as the expert on watches. During his campaign in 1690 King William sent a servant to borrow Huygens's watch when his own broke: "The king had sent Baersenburg in the morning to borrow my watch, his being out of order" (20 July 1690). During the military campaign in the southern Netherlands the following year, King William appealed to his secretary again: "I was in the dining room in the afternoon when the king seated himself, spoke to me about his watch, and borrowed mine again" (8 August 1691). In return the king showed his secretary his newest acquisitions, as on 14 December 1694: "He said he had a repeating watch from Quare which was better than that of Tampion."

Finally, it was Constantijn Huygens Jr. who introduced the phrase "time management" into the Dutch language. On 1 September 1690 he said to Queen Mary that she "managed her time better than I do." This was a polite compliment, and not completely true. In his diary it is apparent that Constantijn Huygens Jr. carefully managed his time, making a report of his activities at the end of each day. The two Huygens brothers, Christiaan and Constantijn Jr., were in fact both specialists in time management in their own ways: one as an inventor of clocks and the other as a modern diary writer.

Conclusion

Christiaan and Constantijn Jr. Huygens were both original and creative in their own right, but their innovations are also part of the Dutch culture of the seventeenth century. The modern diary and the precision clock both resulted from more general changes: the rise of modern time awareness and the diurnal form. There are several similar developments, for example, the change in character the almanac underwent around 1650. Almanacs had been printed for more than a century but now became more practical books with useful information, such as market hours, travel times, and departure times of towboats and mail carriages. Blank pages for notes were added. Furthermore, the circulation was much larger than before. Almanacs from then on played a more and more important role for more and more people in planning their lives.[32]

The rise of the shipping journal also occurred in the middle of the seventeenth century. On the ships belonging to the Dutch East India Company, the skipper and the helmsman were required to keep a daily register. Starting in 1650, lined journals with preprinted columns were handed over to them at departure and were to be returned on their arrival. Such journals were implemented on warships not much later. In the archives of the Dutch East India Company and the Dutch Navy the number of journals increases from the middle of the seventeenth century. After 1670 helmsmen of the fleet kept journals. Journals were probably kept less frequently in the merchant fleet, but there, too, more and more seamen kept some sort of journal.[33] In a sense, Constantijn Huygens Jr.'s diary had the same purpose as a ship's log, in his case determining his own position within the turbulent world of the royal court.

These developments around 1650 fit within the cultural climate of the republic as a whole in the first half of the seventeenth century. Investigating, recording, and describing are also reflected in Dutch painting, as Svetlana Alpers, among others, has shown.[34] A favored subject of Dutch painters in this time was the still life in which both writing instruments and watches are depicted, with a diary often added to the scene as well. Such a still life was painted in 1668 by Maria van Oosterwijck (plate 2), whose art Constantijn Huygens Sr. praised in a laudatory poem. All the objects in her painting are replete with religious symbolism, but they can also be interpreted as a reflection of the close connection between time measurement and diary writing that arose during this time. This painting, the numerous diary entries surveyed in this essay, as well as the very act of keeping a diary, speak to the ways in which all the objects of timekeeping depicted by Maria van Oosterwijck structured not only individual lives but also large-scale enterprises of overseas commerce and empire. But more than this, timekeeping, in all its written and mechanically measured modes, shaped understandings of the individual self at the same time that it helped form a new picture of the cosmos.

CHAPTER SEVEN

The Moral of the Story: Children's Reading and the Catechism of Nature around 1800

ARIANNE BAGGERMAN

"No! Go on: it is something else to know that there is some unknown, planets and stars, mock suns and mock moons, water and land, sea and rivers, humans and animals, birds and fish, insects and plants; something else to contemplate all these carefully in the prescribed manner. To observe, dear pupil! has great advantages over hearing and reading. You must, whenever possible, see everything."[1] With these words the eighteenth-century author J. F. Martinet exhorted his young readers to lay aside their books and go outdoors to see nature, to feel, smell, taste, and—armed with microscopes, telescopes, and other equipment—test it. Wisely, he only did so at the conclusion of the roughly 1,600 pages of his *Katechismus der natuur* (Catechism of the Natural World), when all of creation, from a grain of sand to the human body, had been duly described. In the preceding text the author was less drastic, but he repeatedly encouraged his readers not to limit themselves to reading his book. Their book knowledge should be supplemented by active research out-of-doors. And the actual reading of the book could also be regarded as a walk. It was written in the form of a dialogue in which, during a long hike through various regions of the Netherlands, the master drew the attention of his pupil to the natural world in all its aspects—from the structure of snowflakes, birds' feathers, and grains of sand to the formation of sand dunes, the disposition of the stars, and the structure of the human body.

This encyclopedic knowledge, in combination with the way it was presented, served to foster an awareness of the miracles of creation in young readers, to enrich them with the power of amazement at the natural world, to enable them to enjoy it, to find peace there, but especially to learn a wise lesson. Whether it was the ebb and flow of the tides, the way in which spiders weave their webs, or the path of the earth around the sun,

Fig. 7.1. "O what a marvelous view!" In J. F. Martinet, *Katechismus der natuur,* 5th ed., 4 vols. (Amsterdam: J. Allart, 1782–89), vol. 3.

essentially it boiled down to the same physico-theological moral: nature in all its manifestations was the best proof of God's omnipotence and infinite wisdom.

One of the first lessons to which pupils were treated was about how the term "nature" was to be understood and how the word could best be used. The term should be understood as "the created works of God." Statements like "nature works, nature does this and that, etc." were to be discouraged: "Rather say, the Creator of all things works, he does this and that, etc."[2]

Two pages later, the desired vocabulary was extended in a passage in which the pupil summarized an exposition on the ingenious efficiency of a grain of corn, which, although small, was powerful enough to eventually feed millions of people: "You say: God's works are great," and the reply is "Yes! There is nothing more courageous, greater, or more masterful, even in the rough and uncivilized chunks of nature. . . . Observe but the leaf of a tree." After this, it will come as no surprise that Martinet's account of the structure of leaves, complete with foldout pages, could only lead his pupils to the following conclusion: "I must confess that the creation of the slightest things in nature (such as a leaf of a tree) by far exceeds the best things made by the hands of men, both in artfulness and in neatness." This reaction comes not from the fictional pupil proposed in the book, however (who for obvious procedural reasons was not overly smart), but from a real reader: the twelve-year-old Otto van Eck, who commented thirteen years after the appearance of the last volume of *Katechismus der natuur,* on 17 November 1792.[3]

The crucial role of books in the transfer of knowledge was long taken for granted by historians. The contribution of books to knowledge making was so obvious that their function did not merit further research. More recently, however, new developments in both the history of science and the history of the book have led to the questioning of the relationship between books and knowledge, which now turns out to be much more complicated and much less self-evident than previously thought.

In the history of science, most studies have focused on famous scientists and their books. The greatest of these books are even mentioned in surveys of world history, like that by R. R. Palmer and J. Colton. In their much-used handbook two works have been included in the very selective list of important events in world history: Copernicus's *Revolutions of Heavenly Orbs* of 1543 and Newton's *Principia* of 1687.[4] The history of the book initially also concentrated on the famous and exceptional. These two preferences can be elegantly combined in studies on first editions of the works of authors like Shakespeare and Rousseau, for instance. Over the last few decades interests have changed in both fields of research. Today the focus is no longer solely on the summits of high culture, but also on the broad context in which such works originated. The role of print and reading is no longer studied only within the elite, but also among the popular classes. Last but not least, the exclusive attention to the production of books by authors and printers has shifted to the consumption of books, focusing on the readers. The next step for both book historians and historians of science would seem to be the reconsideration of the interaction between production and consumption.

By approaching books about science with new methods developed in the field of book history, we can learn much about the transfer and spread of knowledge. Martinet's *Katechismus der natuur,* already introduced, will here serve as a case in point. The scientific status of this book is easy to determine. The book does not aim to extend scientific knowledge, but it is a comprehensive and well written compendium of knowledge around 1780. It is much more difficult to find out how the book was received by its readers. Who read the book and why did they choose it? How did they read it and react to it? To find answers we have to track down that rare species: the "real reader."

Although the cultures and practices of reading have been high on the agenda of book historians since the 1980s, the real reader has suffered from neglect. Attempts to identify the readers of books have so far been limited to studies of the putative audience of forewords and prefaces, a method of research which only uncovers the "intended reader."[5] Much effort has also gone into research based on probate inventories, usually leading to the discovery of another type of reader: the "deceased reader."[6] Finally, research has been done into booksellers' account books, which has yielded knowledge about the "buying reader."[7] None of these readers has informed us as to their motives: Why did they own a particular book? Did they actually read their books, or did they merely purchase them to fill their bookshelves? They also remain profoundly silent as to their reading habits: were the books read aloud in company, or did the readers quietly ensconce themselves in a corner? More importantly, there is no answer to the question of how readers interpreted texts—according to the letter, or did they give them new meanings?

Still more important is the question of the horizon of expectation of past readers. The historical reader is no longer seen as a passive consumer, manipulated by writers and publishers and dependent on the logic of the text. According to the French historian Michel de Certeau, the text, by its internal logic, determines the limits of the reader's interpretation; but within these boundaries a reader can imbue a text with meaning of his own invention. Reading was never a form of passive consumption but rather a form of "silent production": "the drift across the page, the metamorphosis of the text effected by the wandering eyes of the reader, the improvisation and expectations of meanings inferred from a few words, leaps over written space in an ephemeral dance."[8]

Questions as to the "silent production" of knowledge, as well as questions about the motives of readers and reading habits, can be answered only through personal testimony: letters, diaries, and autobiographies, sources that not only are rare but have also been insufficiently investigated. Historians have only just begun to discuss the possibilities they offer and the specific methods needed for determining their meaning. So far historians have only discovered a few readers, scattered in time and space. One is the sixteenth-century Italian miller Menocchio, whose bizarre interpretation of the creation story was discovered by Carlo Ginzburg.[9] Robert Darnton has written about a French silk merchant, Jean Ranson, who read and reread the work of his hero, Rousseau, as if it were the Bible.[10] The reading habits of the Englishwoman Anna Larpent have been exhaustively documented by John Brewer, as have those of the Dutch schoolteacher David Beck, analyzed by Jeroen Blaak.[11]

Interesting as the results may be, these studies have been regarded with distrust by fellow historians. Analyses based on egodocuments—roughly understood to include diaries, autobiographies, personal letters, personal travel journals, and other forms of private writing—do not comply with the strict requirement of representativeness still upheld within this field of scholarship. Yet should representativeness indeed always be the touchstone of historians? Carlo Ginzburg has answered this question with a loud and clear no.[12] Another approach was proposed by myself some years ago, that of reading over the shoulder of the historical reader—in other words, a close reading of both the private manuscript and the printed text itself that was read and commented upon by the diarist.[13]

The lack of studies based on the private reactions of real readers is caused not only by scholarly fear of a theoretical vacuum but also by the paucity of sources. Egodocuments are rare, and diarists who extensively commented on their reading matter are even rarer. Presumably that is the reason why studies on real—nonprofessional—ways of reading always start with a certain reader and follow his or her literary preferences and not the other way around.[14] To find the reactions of real readers to a certain book, hundreds of diaries have to be perused. An additional condition of success is that the book in question has to have been popular, as only then can more than an incidental reader's reaction be expected. So far, such extensive research has not been undertaken. In the case of Martinet, however, I could in part build upon an earlier research project, a recently compiled inventory of Dutch egodocuments written between 1500 and 1914. All Dutch libraries and archives have been screened in the course of an intensive search for such diaries, autobiographies, travel journals, and other egodocuments, both in manuscript and in print, which has resulted in a catalog with brief descriptions.[15] During this research it was possible to trace four readers who wrote about their encounter with Martinet's popular

Katechismus. This small pool of readers responding to the same book carries no statistical weight, but the standard of evidence at least surpasses that advocated by Carlo Ginzburg, who maintains that even "just one witness" is sufficient to make a case.[16]

Thanks to this research it was possible not only to ask the kind of questions historians are always so eager to ask but also to find answers to them. In his introduction Martinet has revealed the intended reader: children and young men and women. But what was the impact of his book on its intended young readers? Did they agree with—or at least understand—Martinet's conclusions? Did they develop the new attitude toward nature that Martinet advocated? Before entering into such depths of reader response, we need to establish the background of Martinet's concept of nature.

The History of the Book of Nature

Martinet opened his work with the words "philosophy is a catechism to strengthen faith."[17] What he called philosophy was in fact something more limited, physico-theology, or natural religion, as it is called in English, an intellectual movement going back well into the seventeenth century, and at that time a way of reconciling the contradictions between science and Christian faith. Experimental science aided by its new tools, the telescope (invented by Newton and perfected by Christiaan Huygens) and the microscope (invented by Antony van Leeuwenhoek), revealed a surprising world, which surpassed the limited universe of Descartes's mechanistic philosophy. The bond between God and nature, which Descartes had severed—animals without souls, creation without God, nature without secrets—came into its own again with the physico-theologians. Descartes's arrogant attitude toward nature was replaced by an honorable admiration for a marvelous universe, too large for man to comprehend, or, to quote the physician and biologist Swammerdam, "too high for my dull mind to grasp."[18] In this view, natural scientific research was essentially a religious undertaking: research into God's invisible hand to strengthen religious beliefs. Nature was seen as a perfect mechanism, designed by God—the "master artisan"—a fine-tuned mechanism, as the physico-theologians would say. All living creatures from ant to human had their specific function in a greater whole, forming a "great chain of being," invented by God in his infinite wisdom and power. From an endless number of possible laws, God had chosen those that were best for man and all other parts of this chain.[19] Experiencing nature as a sinister and threatening contingency made way

Fig. 7.2. "The Creator worshipped in His creatures." Frontispiece of J. Van Westerhovens, *Den Schepper verheerlykt in de Schepselen* (Haarlem: C. H. Bohn, 1771), vol. 1.

for the "enlightened" and relieved feeling for nature that Alexander Pope famously described:

> Nature and Nature's laws lay hidden in the night:
> God said, "let Newton be!". and all was light.[20]

Humans lived, as Leibniz said, in the best of all possible worlds.[21]

The main message in Martinet's *Katechismus der natuur,* in other words, was anything but new. His book, within and outside the Netherlands, was preceded by a surge of physico-theological writings, which had been typified by one of his contemporaries as "the flooding of the Nile before Egypt." Nor can it claim the honor of being the first Dutch physico-theological children's book. N. A. Pluche's famous sixteen-volume *Spectacle de la nature* had appeared in a Dutch translation forty years earlier.[22] However, it was Martinet's *Katechismus,* and not Pluche's *Schouwtoneel der natuur,* that triggered an avalanche of catechisms of nature for children, in all sorts of shapes and sizes, appearing well into the nineteenth century. Martinet's success in the Dutch market, relative to Pluche, can be partly explained by the fact that the work was more compact by four volumes and thus more affordable than the sixteen-volume *Schouwtoneel.* A more important difference, which has not drawn much attention, was the typically Dutch character of the work. The scenery into which Martinet lured his readers and the scenic vistas he opened to their minds' eye were all recognizably Dutch. The scenery of the Netherlands was not dull, Martinet assured his readers, but offered unlimited variation. From the "many stagnant waters . . . in North Holland," where mosquitoes thrived, he led his readers to the expansive meadows between Amsterdam and Enkhuizen, the "planted orchards and rich wheat fields" in the Betuwe, the "grand spectacle of the open sea, which near Scheveningen and elsewhere seems to disappear into the clouds," the dunes built up by God to protect the western seaboard—"born of crashing floods and flurrying winds, the raging shocks of the fiercely roaring sea serenely to receive"—and the "green forests and the gray rolling heathland with the yellow sands, . . . of the hills of the Veluwe.[23] Especially in a genre that wished to stress the acquisition of knowledge by starting close to home, such recognition was important.[24]

In his preface, Martinet stated that he had written the book to fill the gap with which he had been confronted in his lessons to young people during his ministry in Zutphen. At these well-attended gatherings he taught groups of boys and girls not only their Heidelberg Catechism but also the "Catechism of nature."[25] His pupils' eagerness to acquire this knowledge can be seen from one of his letters to a university friend to whom he addressed a request for demonstration materials: "I am in need of large quantities, for every week I teach 38 young ladies of the first rank their Catechism. . . . For this large company, which has eyes and wants to see, I perform physical experiments to compensate for the dryness of instruction. And they have to see many things from Natural History. They have requested me to do so, they are like sparrows, avid for anything I care to offer."[26]

In contrast to the practical demonstrations that Martinet's pupils enjoyed in Zutphen, his *Katechismus der natuur* in fact contained mainly bookish knowledge. It made an attempt, however, to bridge the gap between the readers and the book of nature as much as possible. The work was situated in the Netherlands on purpose, richly illustrated and written in the form of a lively dialogue between the enthusiastic master A., modeled on the author himself, and one of his insatiable "sparrows" of his catechism lessons, the pupil De V.[27] The catechistic character of the book was reinforced by the rules of living with which the book closed, which were urgently requested by the pupil to enable him to go on enjoying nature to the fullest and to the greater glory of God after the departure of his master. The personal bond between master and pupil—in other words, between author and reader—is emphasized even more by a portrait of the author printed, not on the first, but on one of the last pages of the work, as an aid to the pupil during his further life. This portrait is supposed to have been drawn by the pupil De V., who had begged his master for this privilege.

In reality it was of course not the pupil De V. who signed this rather sophisticated portrait but someone with the same initial: the famous Dutch artist Reinardus Vinkeles. From the portrait the author looks out at the reader with a look of understanding, not with his book in his hand but holding his hat. With his hand he points explicitly to the window, through which we can see the picturesque river landscape near his hometown, Zutphen, with rainbow and dramatic clouds. The teacher has fulfilled his role as mediator. He trusts that his pupil has developed the right state of mind to look at nature through his eyes: "Go on! You must see everything!" However, though nature may be more important, the author dominates the scene. This is confirmed by the caption underneath the portrait, where pupil De V. exclaims that the sight of his teacher will be a source of inspiration during the rest of his life: "Seeing the representation of your person will impress on my heart when I shall be away from you."[28] In his foreword the author had already expressed the hope that his book would generate a new attitude toward nature. He sees a future "Fatherland" populated by "a people that are better educated in the wonderful knowledge of creation."[29]

Could those expectations be exaggerated? Martinet's *Katechismus* was a huge commercial success. On the basis of the high sales (six thousand copies in the year of publication alone), reprints (at least six), abbreviated versions, and imitations of and supplements to Martinet's *Katechismus*, it would not seem too bold to assume that Martinet's ideas found

Fig. 7.3. "And I will, while looking at your Portrait, say this, this is the Face of my Tutor and such things I learned from Him." In J. F. Martinet, *Katechismus der natuur*, 5th ed., 4 vols. (Amsterdam: J. Allart, 1782–89), vol. 4.

wide circulation.[30] The abbreviated version for younger children—*Kleine katechismus der natuur voor kinderen*—which appeared in 1779, also ran to six reprints. The spread of this work was not limited to the Dutch Republic alone, and Martinet's *Katechismus* was translated into French, German, Malay, and English and released by publishing houses in Leipzig, London, Philadelphia, Dublin, and Batavia.[31] Further evidence of its popularity is to be found in a number of eighteenth- and nineteenth-century sources. Intensive research into household inventories has revealed that the book was a standard possession of nineteenth-century families. The list of top-ten authors (those occurring more than ten times in the inventories) compiled by Marie van Dijk shows Martinet's *Katechismus* in second place.[32] According to the customer ledgers of Van Tijl booksellers in Zwolle, Martinet's work was among the bestsellers. Even before any reviews had been written,[33] sixteen inhabitants of that town acquired one or more volumes of *Katechismus der natuur* in 1778–79.[34] They were a rather heterogeneous group. Among them was the mayor, important merchants, senior officials, and academics, as well as junior officials, shopkeepers, a milliner, and a charcoal burner.[35] Han Brouwer, who researched the customer ledgers, concluded that the book, in contrast to other titles sold by Van Tijl, inspired "unity among all classes."[36]

Research into reading societies yields the same perception. The book circulated both in an Alkmaar society consisting mainly of merchants and shopkeepers and in the more elite Leiden society called Utile Dulci.[37] A nameless society in Hindelopen also had it among its humble possessions. Jacob van Lennep and Willem Hogendorp came across the *Katechismus* during their walking tour of the Netherlands in 1823 when they had lunch in the "bawdy inn" Het Gouden Anker at Hindelopen: "Here there was a reading society; their list of books was posted on the wall, and this precious museum consisted of twenty-six works, such as *Inleiding tot de geographie; Historie der kozakken; Geschiedenis der landing in Noord-Holland; Star; Letteroefeningen; Catechismus der natuur; Predikatie bij gelegenheid der inwijding van de kerk te Molquerum;* and so on."[38] The irony underlying this passage is a telling reference to Martinet's popularity. In a remote corner of the Netherlands where the streets were deserted ("we had already walked the length of two streets and not seen a creature excepting a cockerel and a dog") and the library was situated in a dilapidated inn, the reading matter must have been limited to absolute essentials—among which we find the *Katechismus der Natuur.*

Nevertheless, for an analysis of the effect of this literature on youthful readers around 1800, we will have to dig deeper for sources more eloquent than household inventories, catalogs, or customer ledgers.

Moving Images: Readers in Action

The first witness is Jacob Nieuwenhuis, born in 1777, the son of a shopkeeper who sold hunting equipment and fireworks. In 1848, when Jacob wrote his autobiography, looking back on a long life, first as a minister, later as a professor of philosophy, he clearly remembered when he first became acquainted with Martinet's work.[39] On his fourteenth birthday, in 1791, the book had fallen into his hands like manna from heaven. Jacob extensively described how, at the time, he would get up at four in the morning to go into his parents' garden with his friend Berkhout. There they would do their homework in the garden house, play in the garden, and do their "hobby studies" until it was time to go to school. Their hobby consisted "of making all sorts of notes and summaries of what we had read and found remarkable." Before Martinet's *Katechismus* came to hand, Jacob and his friend had to make do with copying hardly legible remedies from the papers of Berkhout's father. This source was completely exhausted by the two boys: "even prescriptions for dropsy, epilepsy, rabies, difficult delivery, suckling tricks, and many others, we copied carefully. You never knew whether then or later, both in our own households and for the beneficent and charitable healing of others, it might come in handy."

The present of the *Katechismus der Natuur* therefore must have seemed heaven sent: "I cannot remember that any other book presented to me gave me so much pleasure as this one. It was Martinet, morning, noon, and night. It was always there beside me as I worked for school; and just the sight of it made me work harder so that I would gain more time for reading Martinet." Jacob turned out to be Martinet's—indeed any author's—ideal reader, for not only was he inspired by him, but sixty years later he acknowledged his debt to the author (still very much alive in the mind of his reader) who had opened his eyes to the wonders of the natural world and had been a beacon to him in his later career as a theologian, a scientist, and a philosopher:

> It was as if Martinet had opened up all creation for me, and his pleasant, pious spirit, well versed in human nature, inspired the whole work and filled me with respect for nature, with deep awe for the wisdom, omnipotence, and goodness of the Creator, perceptible in all his works. Blessed Spirit! if you could look down on Earth and know the utility that all your writings have engendered; if you could read my heart, could perceive the impression you made on my spirit and the zeal for natural science with which you filled it; then, even now, after more than half a century, you would be party to the gratitude of an old man of seventy years, who is happy to bear witness to the infinite debt he owes to your works!

One could ask if Martinet, as a guru, did not overshoot his goal in this case. Nieuwenhuis's mental horizon seems to be much less dominated by the "goodness of the Creator" than by Martinet himself. When Jacob looks up to heaven, he sees, not God looking down, but the author of his favorite children's book.

Jacob Nieuwenhuis can be seen as Martinet's paper pupil, De V., grown up and with undiminished enthusiasm and all the fluency of speech necessary to serve as a propagandist. Among the three other "readers in action," there is also a real "De V." She was not a figure made of paper but one of the "38 young ladies of the first rank" whom Martinet taught both the traditional Reformed catechism and the catechism of nature, one of the hungry sparrows who could never get enough of studying the phenomena of nature. This was Anna Aleyda Staring, daughter of the mayor of Zutphen. We learn more about her later passions in the autobiography written by her son, Maurits VerHuell. In this document he extensively discusses his mother's passion for nature, said to have been inspired by "the reverend Martinet, by whom my mother was confirmed in the faith."[40] Only when his mother was staying at the family's country house near Doesburg was she really in her element: "There my mother was cheerful and happy and ever admiring the beauty and wonders of her natural surroundings, which she would point out to us with warmth and emphasis."[41] Martinet was a passionate adherent of "the theology of utility"—the efficiency of nature in all its aspects—and the idea apparently took root in his pupil. We find it reflected, for instance, in her attitude to the house cat, Snoek. In her opinion this cat was "the epitome of cat virtues: courageous in a fight, cunning in the catching of rats and mice, meek and docile and faithful like a dog."[42] We also find evidence of her, at the time, unusual attitude toward thunder and lightning. She did not seek shelter indoors but would take her children outside to enjoy the spectacle: "There she made us remark how majestically the pregnant thunderclouds approached, how lightning flashed down and thunder rolled, what sharp shadows lay across the scene, how softly the crowns of the high oak trees would rustle, how calmly nature awaited the downpour of salutary rain, how childish it was to be afraid, since out of a hundred bolts of lightning often not one would reach the ground."[43]

On the many long walks taken by Maurits with his mother and sister, not only Snoek the cat but also Martinet was their constant companion. Thus, Anna would urge her daughter to make a collection of birds' feathers, "in imitation of the famous writer, the reverend Martinet." She herself, following Martinet's example and assisted by the little Maurits, started a collection of mosses: "Thus, the noble woman succeeded in instilling a passion for natural beauty in the hearts of her children. . . . Neither a remarkable shadow nor glittering fall of light on heath or woodland would

Fig. 7.4. "Industrious John. Little John favors book reading to catching Butterflies." In J. F. Martinet and A. van den Berg, eds., *Geschenk voor de jeugd* (1781), vol. 1.

go unnoticed by her, not a birdsong went unheard."[44] In turn, Maurits was to develop into a passionate collector of butterflies. In later life he achieved fame as an entomologist who illustrated his work with drawings showing meticulous biological detail.[45]

An eye for detail—to see God's omnipotence in the square millimeter—was also developed in Jan Pijnappel, when he was eleven years of age. During the daytime he read Martinet "with great pleasure," as he noted in his diary in January 1806.[46] In the evening he saw everything himself with the microscope: "We saw (everything with great amazement and pleasure) several beautiful things; for instance, a slice of a lime tree, the prickly plants growing in the dunes, several seeds, the dust on a fly's head, also a complete fly, the funnel in the throat of a bee, the beginnings of the skin of a plant, and more magnificent and amazing things."[47]

Otto van Eck was only one year older—twelve—when he first mentioned reading Martinet's *Katechismus* in his diary. His notes are much more elaborate than those of Jan Pijnappel and give us the opportunity to observe a reader of Martinet in action during the course of several weeks. Because he was a very diligent writer (his parents obligated him to keep a diary and tell them everything about his thoughts and his reading experiences), we know that this young reader did not read the book in the intended order. He picked his own way or followed the advice of his parents, to whom he regularly read aloud from the work. The order in which he read seems to have been determined, not by the author, but by the events happening around him. Thus, he read about the structure of grains of sand when garden work on the country estate was at a high point.[48] The garden was being transformed from a classicist to a romantic English landscape garden with meandering streams, which necessitated a lot of digging by the gardeners.[49] Otto wrote proudly that he had helped the workmen carry the wheelbarrows of sand. With the aid of Martinet's *Katechismus,* the already-interesting heaps of sand the laborers generated took on a deeper significance: "Tonight I read Martinet on sand, which (when regarded with the naked eye) may seem very ordinary, but when a magnifying glass is used it appears that each grain is an artful triangle or round ball."[50] The structure of snowflakes was on the program in the middle of the winter, and Otto read Martinet's discussion on people's natural tendency to be strongly attached to life, in spite of the promise of perfect happiness in the hereafter, not surprisingly after a period of illness.

In the beginning Otto's summaries of what he had read in Martinet stayed close to the original text. Later on, however, more interpretation was added. Take for instance Otto's already-quoted confession on 17 November 1792: "This evening I read in Martinet about the greatness of God's works. I must confess that the creation of the slightest things in nature (such as a leaf of a tree) by far exceeds the best things made by the hands of men, both in artfulness and in neatness."[51] Otto's conclusion, however, differs greatly from Martinet's. Otto does not conclude that this is proof for the existence of God. Instead, Otto writes that the insignificance of human beings is an excuse for his own failings: "It is also (I think) very logical that God, who created all men, has more order in his work than those created by him in theirs."[52]

Another example of this personal form of "appropriating" Martinet's text was evoked by Martinet's attempt to free children of their fear of death by emphasizing that attachment to life is part of the natural order and at the same time stressing that they should trust that all good and faithful people would go to heaven. This reasoning was not completely successful in Otto's case. Otto was glad that his attachment to life was natural. Nevertheless, he added: "It is, I think, understandable that a human being, even if he believes in happiness in the hereafter, is attached to life, because death separates him suddenly from all his best friends and destroys all his plans, and the promise of eternity is after all obscure, and therefore I dare to confess that I would prefer to live on for a while, although I have not the slightest doubt about the promise of Jesus Christ."[53] On closer inspection, his arguments appear to be a cocktail of Martinet's reasoning and that of the philanthropic pedagogue J. B. Basedow, whose *Manuel élémentaire d'éducation* was read by Otto during the same period.[54] The conclusions drawn by Otto are, however, of his own creation.

This same process took place when, a day later, Otto read in Martinet about sunlight, how perfect it is, and how it is both a gift of God and another proof of his existence: "It gives the earth fertility and food. . . . This is an honor to God and a testament to the workings created by his hand."[55] Otto's rather free interpretation and conclusion were as follows: God allows the sun to shine on both "good people and bad people, although the latter do not deserve this; however, now and then God allows that in this world the bad people prosper, but he is just, and they will after their death receive their deserved punishment, and they cannot escape God's judgment."[56]

The moral that the injustice of the equal division of sunlight between the good and the evil would be put right in the hereafter was nowhere to be found in Martinet's work, nor in the work of Basedow. Otto could not have found this message in the Bible either. The fact that God makes the sun rise for both the good and the bad, in Matthew 5:45, has a much less judgmental moral: "But I say unto you, 'Love your enemies, bless them that curse you, do good to them that hate you, and pray for them which despitefully use you, and persecute you; That ye may be the children of your Father which is in heaven: for he maketh his sun to rise on the evil and on the good, and sendeth rain on the just and on the unjust.'" Otto's derailed reception was probably caused by other books that he read during the same period: enlightened children's fiction, like *Adèle and Theodore,* by Madame de Genlis, *Contes moraux,* by Madame de la Fite, and *Little Grandisson* and *Little Clarissa.*[57] In these strongly moralistic tales, good was always rewarded and evil inevitably punished.[58] It would appear that Otto had the tendency, when the familiar moral was lacking in a certain work, to read the message into it himself.

On other points, however, Otto understood Martinet's message well, maybe even too well. Martinet's call upon his readers to prefer the lessons of nature to those written in books found, for example, fertile soil in Otto's mind. An analysis of Otto's diary has shown that his enthusiasm for the natural world was much greater than that for his daily reading program, including Martinet's *Katechismus der natuur.* Otto read the book daily, but

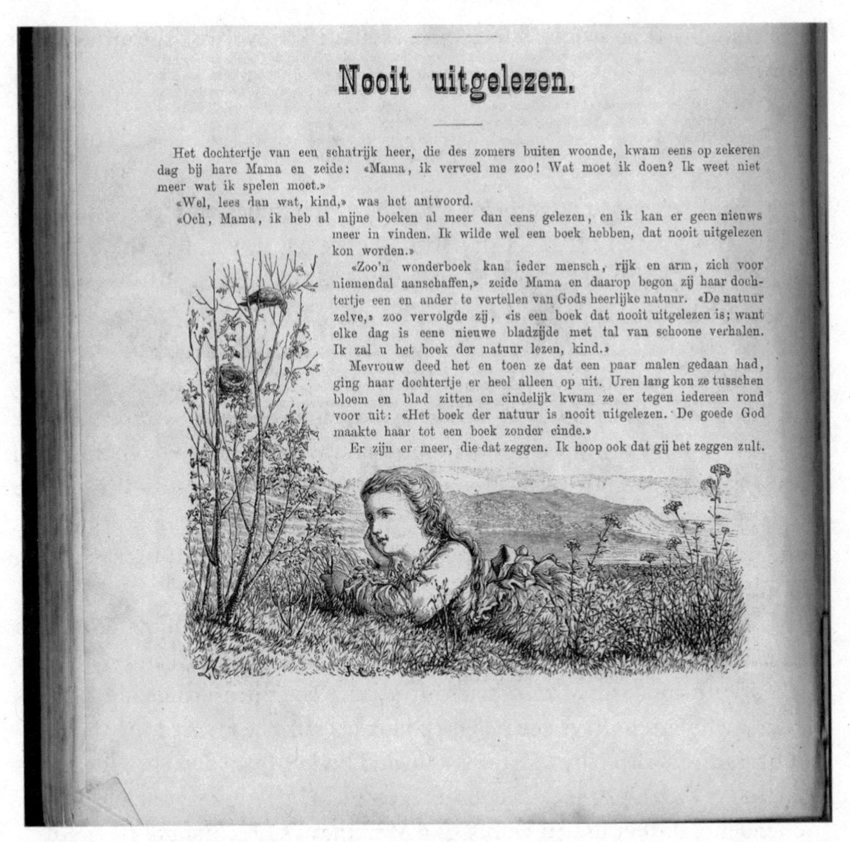

Nooit uitgelezen.

Het dochtertje van een schatrijk heer, die des zomers buiten woonde, kwam eens op zekeren dag bij hare Mama en zeide: «Mama, ik verveel me zoo! Wat moet ik doen? Ik weet niet meer wat ik spelen moet.»

«Wel, lees dan wat, kind,» was het antwoord.

«Och, Mama, ik heb al mijne boeken al meer dan eens gelezen, en ik kan er geen nieuws meer in vinden. Ik wilde wel een boek hebben, dat nooit uitgelezen kon worden.»

«Zoo'n wonderboek kan ieder mensch, rijk en arm, zich voor niemendal aanschaffen,» zeide Mama en daarop begon zij haar dochtertje een en ander te vertellen van Gods heerlijke natuur. «De natuur zelve,» zoo vervolgde zij, «is een boek dat nooit uitgelezen is; want elke dag is eene nieuwe bladzijde met tal van schoone verhalen. Ik zal u het boek der natuur lezen, kind.»

Mevrouw deed het en toen ze dat een paar malen gedaan had, ging haar dochtertje er heel alleen op uit. Uren lang kon ze tusschen bloem en blad zitten en eindelijk kwam ze er tegen iedereen rond voor uit: «Het boek der natuur is nooit uitgelezen. De goede God maakte haar tot een boek zonder einde.»

Er zijn er meer, die dat zeggen. Ik hoop ook dat gij het zeggen zult.

Fig. 7.5. "You can never finish the book of nature." The book of nature in a Dutch nineteenth-century children's journal: "'Nature itself,' said her mother, 'is a book that you can never finish. I will read to you the book of nature, child.'" In *Almanak voor 't Jonge Volkje* ('s Gravenhage: Joh. Ijkema, 1884), no. 2, p. 42.

only under pressure from his father: "So that father may be satisfied with me in this matter, tonight I took up *Manuel d'instruction* and the *Natuurlijke historie* by Martinet."[59] Otto was inspired to go outdoors and observe nature with his own eyes, as stated in Martinet's exhortation at the end of his book, long before he reached this chapter of the book. He was no bookworm at all but a real child of nature who preferred to play with his goat, to take care of his little birds, to help the gardener at the country estate, or to take long walks through the countryside with his father, who informed him during these walks about the miracles of nature, as Martinet did with his pupil De V., and as Anna Staring did with her son, Maurits. It is exactly this element that would survive the scientific revolt of the nineteenth century.

Otto, Jacob, Anna, and Jan belonged to the first generation that was thoroughly educated in the book of nature. However, it was also the last. Science evolved further, toward an evolutionary worldview. Otto was not to witness these developments, for he died prematurely in 1797, aged seventeen. Jacob Nieuwenhuis died, blind and demented, half a century later. He wrote up his memories two years before the publication of Charles Darwin's *The Origin of Species,* which was to undermine physico-theology. The concepts of the survival of the fittest and of nature as dynamic and ever changing were hard to reconcile with the idea that order in nature pointed to a creator whose higher intentions were inherent in every organism. At the same time, the idiom "the book of nature," as if nature were a completed entity, fell into disuse.

The same fate befell Martinet's *Katechismus der natuur.* In Willem van den Hull's memoirs, written in 1841, this book was still automatically linked with modernity. In order to illustrate the open-mindedness of his father—who might have seemed orthodox because he favored the theological literature written by Van der Kemp and Brakel—van den Hull mentions Martinet's *Katechismus* as a counterexample. His father was an enthusiastic reader of "other useful works like Martinet's Catechism of the natural world."[60] During the same period but at a much younger age, the future engineer and politician Adriaan Gildemeester read the same book—at least he tried to read it. His memories of "days long gone" date from 1891, more than half a century after his first confrontation with the work of Martinet. Even so he managed to remember the exact date (14 June 1834), the name of the teacher who obliged him to read this book (his teacher of religion Bähler), and his own inability to understand the meaning of the book. This work was far above the grasp of the five-year-old child and was recalled only because of the stupidity of his former teacher: "That our Bähler in the meantime did not have a clue about appropriate literature for a gentleman of almost six years of age is expressed by the book I got from him on the 14th of June 1834. . . . It was called Catechism of the natural world by J. F. Martinet." In 1891 the book was no longer seen as manna from heaven, as it was in Jacob Nieuwenhuis's mind in 1791, or even as an indication of the progressive attitude of the elderly generation, as it was in Willem van den Hull's memory in 1841. "Nowadays" it was supposed to be "old-fashioned."[61]

There is one important aspect of Martinet's work that has survived into the present, for instance, in the pioneering works of the Dutch biologists Eli Heimans and J. P. Thijsse and of the famous American naturalist and pantheist John Burroughs—icon of the present-day ecology movement.[62] Their works were published in the late nineteenth century and written in reaction to the dry, systematic approach of the decades that followed Darwin's publications. They were not out to prove the existence of God. Burroughs even seems to choose the opposite position when he declares:

> "We must get rid of the great moral governor, or head director. We must recognize only Nature, the All; call it God if we will, but divest it of all anthropological conception. Nature we know; we are of it; we are in it. However, this paternal Providence above Nature—events are constantly knocking it down. . . . We want no evidence of this God. . . . The universe is no more a temple than it is a brothel or a library. The Cosmos knows no God—it is super Deus.[63]

Yet, there is an important similarity between those late-nineteenth-century authors and Martinet. Like this eighteenth-century author, they pleaded for practical instruction about the world around us; young people especially were to be made familiar with the natural world in an evocative way. They, too, took their readers by the hand and inspiringly walked them through parks, woods, fields—or, like Burroughs, through the wilderness. Just like *Katechismus der natuur*, these books were meant to stimulate the reader to seek the great outdoors on a journey of discovery, but above all to enjoy nature. The caption under Martinet's portrait could be easily exchanged with the famous words of Burroughs: "Each of you has the whole wealth of the universe at your very door."[64] And just like Martinet, this author refused to step out of the picture when he underlined his own importance as a mediator between nature and his readers: "People admire my birds, but it is not the birds they see, it is me."[65] This is one aspect we saw in Jacob, the young reader who had traded God for Martinet. In Otto's and Anna's reception of Martinet as well, it was not the religious moral of the story that made the greatest impression. Nevertheless, in the end, their passion for country life and their romantic love of all things bright and beautiful, "the walking variety" of natural education, endured.

Because Martinet tried to influence his readers not only in their thinking but also in their feelings and behavior, his book offered an opportunity for a wider form of reception research than usual. As we have seen, walks in the countryside, attitudes to natural phenomena, and even an aversion

CHAPTER EIGHT

Method as Knowledge: Scribal Theology, Protestantism, and the Reinvention of Shorthand in Sixteenth-Century England

LORI ANNE FERRELL

Historians have been hard-pressed to explain why English men and women converted to Calvinist Protestantism in the late sixteenth and early seventeenth centuries. A confession built upon the idea of God's omnipotent and exacting justice, expressed pastorally in decrees of predestination, appears harsh and mystifying. And so a recent generation of social historians insists, for example, that Protestantism was imposed upon rather than embraced by the English people, with early modern monarchical and ecclesiastical initiatives producing (in Christopher Haigh's infamous formulation) "a protestant nation but not a nation of Protestants."[1] And revisionist cultural historians have often found it easiest to tout the persistence and strength of England's traditional Catholicism by proclaiming English Calvinism's weaknesses, declaring it far too confounding and complex to satisfy a laity whose previous beliefs had been buttressed by a reassuring round of familiar rituals and formulas.[2]

And yet convert the English people did, often with an enthusiasm that demands a plausible explanation for the attraction of early modern English Calvinism. The theme of this volume, knowledge making, offers the opportunity to examine this question. With the help of some rather unusual evidence from the late sixteenth century, I will argue in this essay that Calvinism, this most stringent form of Protestantism, offered its English practitioners a chance to process new and complicated religious information using two methods familiar from older religious and intellectual practice: manipulation and memorization. These visible and interactive techniques allowed committed and enthusiastic Protestants to form a community of cognitive elites within the larger community of a national church. Moreover, in allowing them to transform some very powerfully instilled habits of practice, it also allowed them to transform some equally powerful habits of thought.

In the 1580s, England's late medieval culture, saturated in religious imagery and perfectly suited to its condition of widespread illiteracy, surrendered, more or less gracefully, not only to the forces unleashed by the new ideas of the Protestant Reformation but also (and crucially) to the *look* of those new ideas clad in typeface and graphics. A world drenched in sensuous color gave way to one reified in black-and-white. This transformed culture enshrined the idea of *The Word* and, consequently, privileged the verbal over the pictorial message: a worldview undeniably Protestant and recognizably early modern.[3]

This was not, however, a worldview we would accurately describe as "iconoclastic": the historian Patrick Collinson's famous term "iconophobic" takes us closer to its character but still does not quite do it justice. Catalyzed by Calvin's doctrines and built upon the tropes of Renaissance learning, England's emergently hybrid religious culture is perhaps best described as *iconoskeptic.* In the late sixteenth century, the English people did not deface or break images; they were nonetheless profoundly suspicious of the idolatrous tendencies such images provoked—and profoundly aware of the sensual attractions of a painted saint or a sculpted apostle. Their concerns led them, not to eschew all religious imagery, but instead to explore the world of the nonverbal with facility and rare creativity. As a result of this redirection of imaginative energy, English Protestantism developed its own distinctively visual pedagogical culture with a unique and striking aesthetic, inspired by the elegantly patterned architectonics of English religious print.[4]

Calvinist anxieties regarding religious idolatry and modes of salvation were well served by the medium of print, with its increasingly innovative approach to graphic design and intrinsic appeal to an expanding lay class of educated adepts. In this respect, moreover, the theology print market was no different from the secular print market. In the last decades of the sixteenth century, we find English presses busily churning out "how-to" books on mathematics, surveying, rhetoric, navigation, and foreign languages. These are texts loaded with visual instrumentation and kinetic possibilities: graphs, glyphs, charts, maps, diagrams, and *volvelles.* The style seems designed to capture the attention of an emergent "middling" class ("pregnant wits," in John Dee's pithy phrase),[5] if by class we reckon, not solely in terms of early modern economic ambition, but also in terms of early modern cognitive ambition.

Sharing a competitive market with books on the secular, useful arts,

English Calvinist texts used the same style to teach new theological ideas and spark a kind of soteriological ambition. These strategies have gone relatively unstudied by historians of early modern religious culture. To be sure, much attention has been paid to the *philosophical* or *rhetorical* impact of Ramism on late-sixteenth-century Protestantism, but little or no interpretive attention has been paid to the *visual* impact of Ramist tables, or indeed to the visual impact of non-Ramist, Protestant graphic design in this, the first great age of print graphics. And no attention whatsoever has been paid to the way such designs could spark desire in, as well as be persuasive to, lay as well as clerical readers. This is to miss an essential—perhaps *the* essential—element of early modern Protestant pedagogy. As I have argued elsewhere, Calvinist theologians like William Perkins and writers like Thomas Middleton used pedagogical graphics literally *to shape* dauntingly new and complex ideas into demonstrable, understandable doctrines. Their graphic arguments deserve our scrutiny.[6] In the theological diagrams of William Perkins's *A Golden Chaine* (fig. 8.1) or the complex spacing of Thomas Middleton's gospel harmony *The Two Gates of Salvation* (fig. 8.2), we glimpse the attractions inherent in a "look of proof":[7] reformers' attempts to represent theological ideas with all the simple optical elegance, and divine veracity and rationality, of geometry.[8]

This Protestant representative impulse even extended to what we might consider the very building blocks of ideas themselves: not *The Word*, exactly, but *words* plain and simple. Dr. Timothy Bright's *Characterie: an arte of shorte, swifte, and secret writing by character* was published in 1588, near the end of a memorable decade in the history of the Protestantization of England. While a historical oddity (not unlike Dr. Bright himself), *Characterie* provides its modern readers a useful—if unusual—lens through which we can identify and interpret this aspect of emergent Calvinist pedagogical style.

Here I must make a disclaimer seemingly at odds with the aims of an essay collection centered on the *making of knowledge*. Dr. Bright's creation, generally considered the first English shorthand system, certainly should *not* be evaluated for its utility—its vaunted capacity to facilitate (as he promised) "short, swift and secret" writing—for indeed this particular style of character writing almost certainly fell lamentably short of all three goals. *Characterie* was tedious to learn and difficult to practice. Requiring enormous feats of memorization, it was nearly impossible to execute gracefully—or unobtrusively. Its small rank of enthusiasts was instantly identifiable by ostentatious, sometimes-obnoxious behavior in public venues. Out of vogue by the second decade

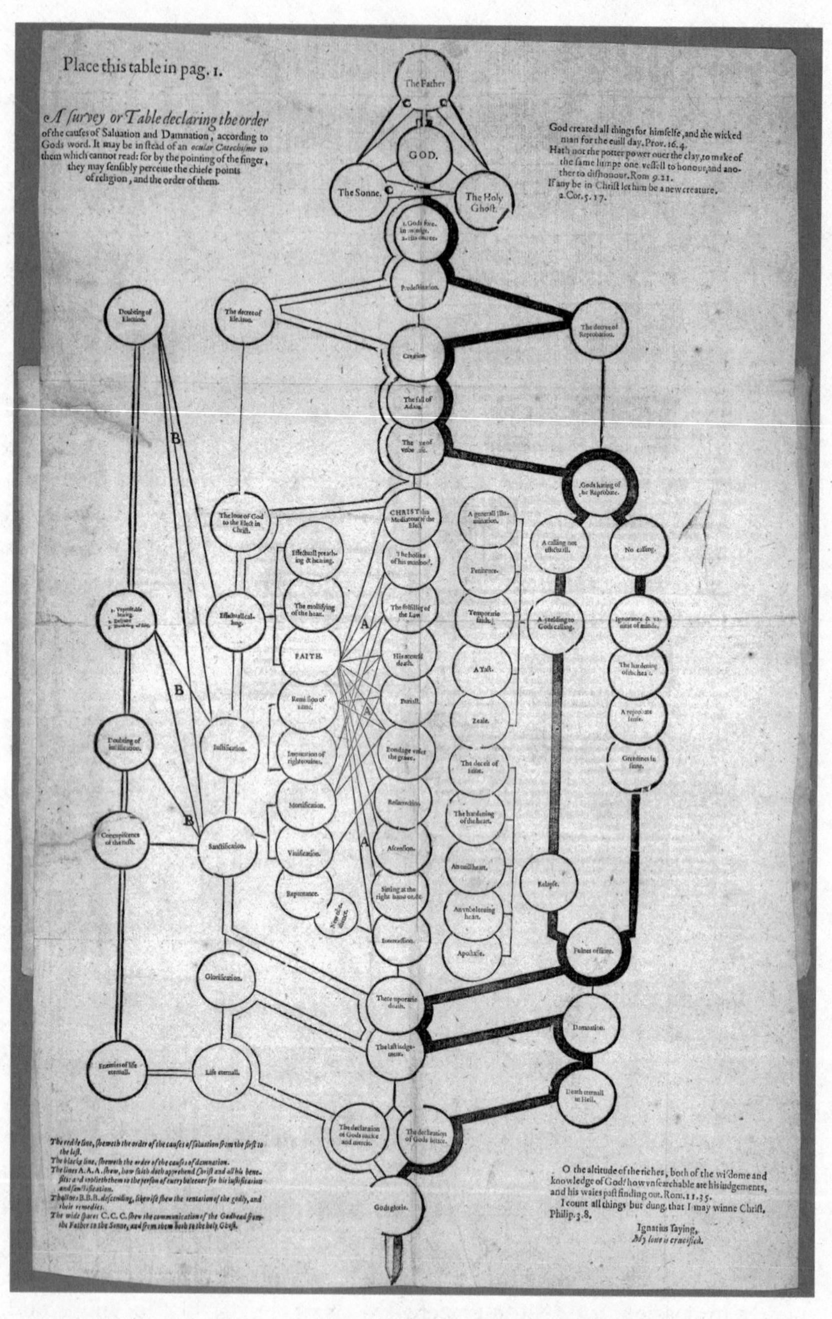

Fig. 8.1. William Perkins, *A Golden Chaine* (London, 1600), STC 19646, folded table tipped in at A1. RB 30016. Reproduced by permission of The Huntington Library, San Marino, California.

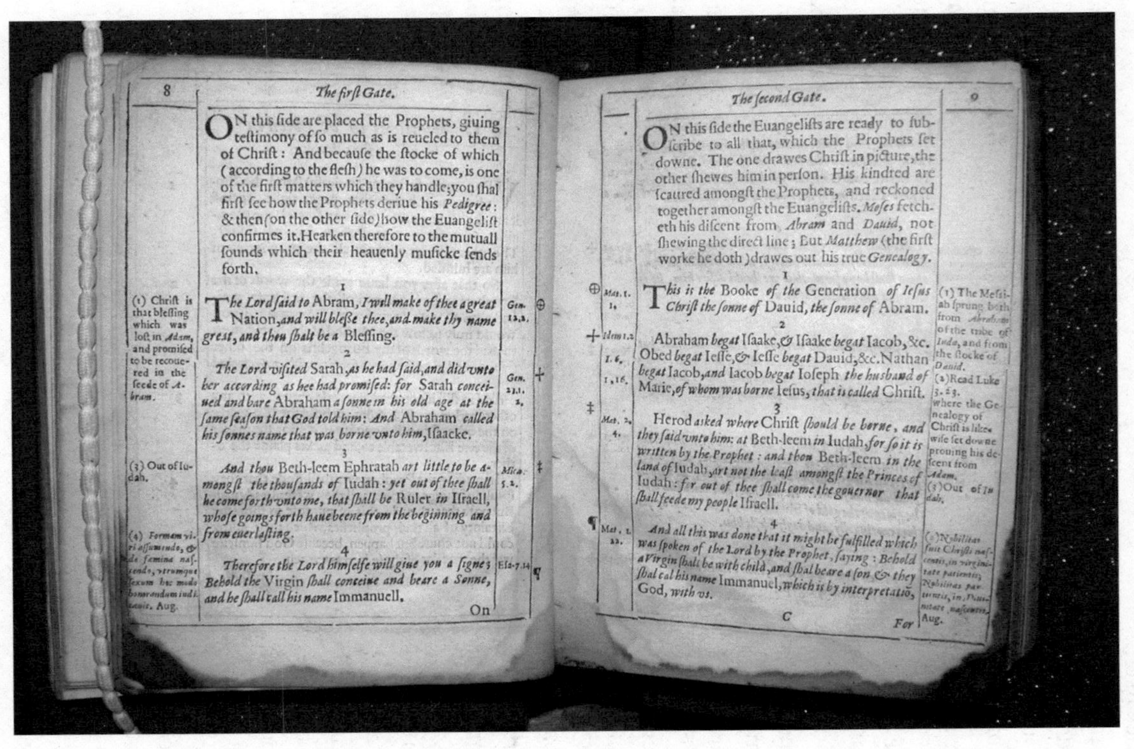

8 The first Gate.

ON this side are placed the Prophets, giuing testimony of so much as is reueled to them of Christ: And because the stocke of which (according to the flesh) he was to come, is one of the first matters which they handle; you shal first see how the Prophets deriue his *Pedigree*: & then (on the other side) how the Euangelist confirmes it. Hearken therefore to the mutuall sounds which their heauenly musicke sends forth.

1

The Lord said to Abram, *I will make of thee a great* Nation, *and will blesse thee, and make thy name great, and thou shalt be a* Blessing. — Gen. 12.2.

(1) Christ is that blessing which was lost in *Adam*, and promised to be recouered in the seede of *Abram*.

2

The Lord visited Sarah, *as he had said, and did vnto her according as hee had promised: for* Sarah *conceiued and bare* Abraham *a sonne in his old age at the same season that God told him: And* Abraham *called his sonnes name that was borne vnto him*, Isaacke. — Gen. 21.1. 2.

3

And thou Beth-leem Ephratah *art little to be amongst the thousands of* Iudah: *yet out of thee shall he come forth vnto me, that shall be* Ruler *in* Israell, *whose goings forth haue beene from the beginning and from euerlasting.* — Mica. 5.2.

(3) Out of Iudah.

4

Therefore the Lord himselfe will giue you a signe; Behold the Virgin *shall conceiue and beare a Sonne, and he shall call his name* Immanuell. — Esa. 7.14.

(4) *Formam viri assumendo, & de femina nascendo, vtrumque sexum hic modo honorandum iudicauit.* Aug.

On

The second Gate. 9

ON this side the Euangelists are ready to subscribe to all that, which the Prophets set downe. The one drawes Christ in picture, the other shewes him in person. His kindred are scattred amongst the Prophets, and reckoned together amongst the Euangelists. *Moses* fetcheth his discent from *Abram* and *Dauid*, not shewing the direct line; But *Matthew* (the first worke he doth) drawes out his true *Genealogy*.

1

This is the Booke *of the* Generation *of Iesus Christ the sonne of* Dauid, *the sonne of* Abram. — Mat. 1. 1.

(1) The Messiah sprung both from *Abraham* of the tribe of *Iuda*, and from the stocke of *Dauid*.

2

Abraham *begat* Isaake, *&* Isaake *begat* Iacob, &c. Obed *begat* Iesse, *&* Iesse *begat* Dauid, &c. Nathan *begat* Iacob, *and* Iacob *begat* Ioseph *the husband of* Marie, *of whom was borne* Iesus, *that is called* Christ. — Idem 1.2. 1.6. 1.16.

(2) Read Luke 3.23. where the Genealogy of Christ is likewise set downe prouing his descent from *Adam*.

3

Herod *asked where* Christ *should be borne, and they said vnto him: at* Beth-leem *in* Iudah, *for so it is written by the Prophet: and thou* Beth-leem *in the land of* Iudah, *art not the least amongst the Princes of* Iudah: *for out of thee shall come the gouernor that shall feede my people* Israell. — Mat. 2. 4.

(3) Out of Iuda.

4

And all this was done that it might be fulfilled which was spoken of the Lord by the Prophet, saying: Behold a Virgin shall be with child, and shal beare a son, & they shal cal his name Immanuel, *which is by interpretatio,* God, *with vs.* — Mat. 1. 23.

(4) *Nobilitas fuit Christi nascentis, in virginitate parientis; Nobilitas parientis, in Diuinitate nascentis.* Aug.

C For

Fig. 8.2. Thomas Middleton, *The Two Gates of Salvation* (London, 1609), STC 17904.3, pp. 8–9. Reproduced by courtesy of the Master and Fellows of Emmanuel College, Cambridge.

of the seventeenth century, it was replaced by a more accommodating system.

In short, what we may have here is a stenographic analogy to Puritan practice itself.

Limited and futile as Bright's system eventually proved to be, from its odd vantage on the outposts of late-sixteenth-century Protestant culture it nonetheless illuminates a longer-lasting religious and pedagogical phenomenon: Calvinist—or Puritan—*style.* The art of shorthand as "invented" by Timothy Bright represents (even if poorly) certain important elements, visual and ideological, that identified Calvinist style at the end of the sixteenth century: it was designed to be graphically distinctive, demonstrably organized, and mnemonically oriented. These characteristics were necessary to the theological and cultural phenomenon known as "experimental predestinarianism," which required believers meticulously to examine daily experience, remembering and sorting through sensory and psychological data in order to ascertain the temporal state of their eternal souls. This method of mapping, graphing, and organizing personal information in books and personal diaries embodied—and, in so doing, instantiated—the anxieties, or, perhaps better said, the *ambitions,* of diligent Calvinists.[9]

We could say, therefore, that Bright's recovery of the art of shorthand—and, more important, his creation of a new writing system and a new form of communication—both responded to and created a distinctive community of learners. In his *Characterie,* Timothy Bright offered the cognitively ambitious Protestants of the late sixteenth century a chance to acquire a new, exacting, and challenging skill to facilitate their learning of a new, exacting, and challenging theology. To ask whether the system actually succeeded is, then, to miss the point. For what Bright primarily created was not so much a viable system but an elegant textual acknowledgment of the needs and desires of Protestant practitioners.

To locate this practice more precisely, I return to its heyday: the 1580s, that decidedly un-Anglican decade when a first generation of university-trained Calvinists succeeded in converting England to Protestantism. The late sixteenth-century Church of England had proclaimed itself part of the Reformed tradition inspired by John Calvin. Like any state church, though, England's was designed to be all things to all people—or, to put it more accurately, most things to most people. An institutional propensity to mediocrity might have tempered the Calvinist force of its key doctrinal formulations, but it could not dampen the pastoral ardor of its bright-eyed clergy. (It would take another generation and a couple of unsympathetic early Stuart archbishops to accomplish that dispiriting end.) And so, historians of Protestantism (most notably, Patrick Collinson) identify 1580 as a distinct watershed, the year England became not merely administratively but actually, culturally Protestant.[10]

English Calvinism's salad days were short-lived. I will, however, contextualize Bright's unpromising system, not in a failed Calvinism, one seen through the retrospective lens of its subsequent (mis)fortunes, but in an activist Calvinism caught on the rise, when reform on its terms seemed not merely possible but entirely probable. This was the moment when Reformed theology acted most creatively, imprinting the confident signs of its success on English culture. And for a brief moment, the work of Dr. Timothy Bright may be seen to have played a part in creating the zeitgeist.

But historiographical appreciation has heretofore eluded Timothy Bright (we could argue that this allies his fate, oddly enough, to that of English Calvinism altogether). Bright has never attracted the notice of historians of religious culture,[11] but if you peer carefully into a few documents you can spy him: in Francis Walsingham's Paris home during the 1572 St. Bartholomew's Day Massacre;[12] or on 1 October 1585, standing by Laurence Chaderton on the day Emmanuel College received its statutes from Walter Mildmay.[13] These associations alone might allow us to place the man within the ranks of the hotter sort of Protestant in late-sixteenth-century England. To map Bright's ideas onto late-sixteenth-century English Protestant thought requires a strong magnifying glass, however, and a willingness to extrapolate religious data from supposedly secular sources.

Bright's Calvinism, or his attitude toward the Elizabethan church, its rituals and liturgy, is not documented in the places historians of religion tend first to check. A second-career minister, Bright left to posterity no sermons, no overtly theological treatises; we have no record of his response to episcopal visitation, although we know he signed the subscription articles of 1562 upon taking orders in 1591; he wrote only one book of obvious religious significance, and that not original, but an abridgement.[14] Fact is, Bright was no rousing vocational success—in medicine, his first career, or in ministry, his last. He appears to have been the dreamy, scheme-y type, more apt to be found hatching ideas to peddle, noodling in his impressive collection of learned books, or doodling successive drafts of his character-writing project than ministering to the sick in either body or soul.[15]

So, let us consider how a Protestant like Timothy Bright spent Protestant England's watershed decade: increasingly negligent of his medical practice and writing books.[16] Bright wrote three notable books in the 1580s.

In 1586, after penning a number of minor medical textbooks, he published what is probably his best-known work, *A Treatise of Melancholy* (best known, that is, because it is considered the forebear to Robert Burton's famous *Anatomy*). He followed that two years later with *Characterie*, which he dedicated to the queen, receiving for his pains a fifteen years' patent granting him exclusive publishing rights to promulgate his shorthand system. Shortly thereafter, Bright wrote an abridgement of Foxe's *Actes and Monuments*, quit the medical profession (or, to be quite precise, was fired from it), and took holy orders and two clerical livings in the 1590s.

The text by Bright that best exemplifies that remarkable admixture of the skillful, the sacred, and the affective that characterizes late-sixteenth-century Calvinist style is not a religious but a medical treatise. The key to Bright's theology can be found in his *A Treatise of Melancholy: Containing the causes thereof, and reasons of the strange effects it worketh in our minds and bodies: with the physick cure, and spiritual consolation for such as have thereto adjoined an afflicted conscience* (1586).[17] Addressed to a miserable friend known only as "M," this work opens with a typical sixteenth-century analysis of depression: humors out of balance; colorful depictions of bodily secretions; methods of treatment that include not only the usual bloodletting and purging but also the wearing of "curious and precious" jewels, the eating of "domesticall," rather than "wild," meats, and the abandonment of "working of [the] brain by any study or conceit" (243–75).

Halfway through, however, the treatise shifts perspective, devoting most of its last 150 pages to the diagnosis and treatment of the condition of *spiritual* depression. It is here we can detect Bright's reliance on the language of Calvin's *Institutes* and the structures of late-sixteenth-century experimental predestinarianism. Bright's concern in *A Treatise of Melancholy* is to ensure that the humorally imbalanced condition of melancholy not be confused with the spiritual condition of sorrow for sin. Such sorrow could signal the repentance leading to regeneration in those elected to salvation, and so to confuse these sensible motions of the spirit with mere physical imbalance would be deleterious. Bright stresses that the treatments he prescribes for simple melancholy would do nothing to ease the person suffering from an acute awareness of his or her sinfulness, a condition he portrays with a hair-raising specificity worthy of a Puritan preacher. "*No* medicine, *no* purgation, *no* cordial, *no* treacle or balm are able to assure the afflicted soul and trembling heart" of the self-aware sinner, writes Bright (184).

Worse still, it *was* possible for humoral imbalance and apprehension of sinfulness, affliction of body *and* soul, to commingle in the same person. Here Bright makes detailed use of Calvin's doctrines in his medical treatise. Melancholy, cautions Bright, could cause a morbid man to inquire too closely into God's secret decrees (199); it could lead to that irreparable condition of highest concern to the author of the *Institutes.* Bright devotes several detailed chapters to the discussion of how to examine one's own mental condition while at the same time avoiding undue curiosity and its deadly consequence, despair. His meticulous delineation of the relationships between conditions of physical and spiritual discomfort makes *A Treatise of Melancholy* a text to be placed not only alongside Burton's *Anatomy* as an example of the early modern medical-instructive text but also alongside Perkins's *A Golden Chaine* as an example of the Protestant experiential-instructive text.[18]

With *A Treatise of Melancholy,* then, Bright found his métier: the writing of the Calvinist how-to book. His next, *Characterie: an arte of shorte, swifte, and secret writing by character* (1588), also combines the teaching of a useful technique with more than a hint of religious purpose. Like all writers of books on the useful arts in this period, Bright was keen to establish the originality and necessity of his creation, part of the politics of the patent system. His dedicatory epistle to Elizabeth I is revealing. Throughout, Bright refers to his system as an "invention" and stresses its national origin, describing charactery as "English yeeld." "It is like a tender plant," Bright explains to Her Majesty, "young and strange" (A5v–A6r).[19] The horticultural imagery allows Bright to promote the queen's interest in nurturing such a delicate shoot. The metaphors of husbandry stress her duty to make wise use of her enterprising subjects, and they point out to us the economic, as well as authorial, importance Bright placed on his being the inventor, or "creator," of a useful art.

Alas, what was so original and creative about Bright's system is exactly what made it so short-lived as a viable practice: its reliance on words reduced to "characters" rather than truncated alphabetically to render brevity. In *Characterie,* words are depicted figuratively, in a language system based on visual metaphor. And while Bright claimed to have created a language that transcended the limiting specificities of phonetically comprehended words, even a cursory glance at the requirements of such an art—which depended on whole-word memorization rather than glyphic substitution—is enough to convey its daunting demands.

But Bright's claim does allow us to contextualize this overly difficult skill in a number of late-sixteenth-century Protestant concerns. First, it seems that Bright aspired to create a universal language with broad ecclesiastical purposes like Latin, one that would unite all Protestants in a single tongue. His style of writing, he bragged, "excell[ed] the writing by letters and alphabet, in that Nations of strange languages may hereby communicate their meaning together in writing, though of sundry tongues" (A3r–A4v). Certainly Bright's personal interest in biblical translation into charactery bears out this contention. Two years earlier, he had produced a manuscript transcription of the Genevan text of Paul's Epistle to Titus in characters, arranged in vertical columns like Chinese script.[20] In the end, the published *Characterie* did not utilize vertical columns; in its dedication, however, Bright explicitly connects his creation to Chinese character writing: "It is reported of the people of China, that they have no other kind [of writing], and so traffic together many Provinces of that kingdom, [otherwise] ignorant one of another's speech" (A4v).

The passage reproduces, almost word for word, the section on Chinese character writing in Bernardino de Escalante's *Discourse of the Navigation Which the Portugales Do Make to . . . the East*, first published in English translation in 1579.[21] Like many Protestants in this period, Bright was captivated and inspired by narratives of Spanish exploration. His near plagiarism of Escalante might allow us to contextualize character writing in the realm of international Protestantism, especially when we consider the year of *Characterie*'s publication, 1588.[22] His concern with *regional* unification, however, is even more telling: for it is likely that, designing charactery as a tool to evangelize, Bright was actually thinking about matters domestic.

To understand his purpose, we must start with his method: *Characterie* originates in a base vocabulary, or "table," of five hundred words/figures, organized in alphabetical order. "Thou art first to learn the Characterie wordes by heart, and therewith the making of the figure of the Character," Bright instructs on page 1 of his textbook. Once the five hundred characters are learned, and the writer has established them in a mental memory bank, the system moves into the realm of abstract thought. Using a mnemonic technique Bright calls "referral," the characterist is instructed to relate all words heard or read in the act of transcription back to the base vocabulary (A8r). These nonbasic words will be either of "like signification" or "contrary" to the base words. The relations between these words to be transcribed and the base character must then be indicated with small marks placed around the corners of the base character. Such belies Bright's first assertion, that he had invented a "short" art. The time-consuming work of mental referral works against such a claim.

Two examples will take this point further: hearing or reading the words "river," "sea," "stream," or "creek," the characterist in all cases makes the figure for "water"—at which point the final transcription character can either stand as simple H_2O or include in some ingenious fashion the initial of the original word so that the characterist will recall the original accurately. Or, confronting the word "evil," the characterist, remembering that there is no character for "evil" in the Charactery Table, must instead transcribe the figure for "good" and then indicate the initial letter *e* of its antonym "evil" to the right of the base figure—literally signifying "the word that means the opposite of good that starts with an *e*" (fig. 8.3). The work of transcription would necessarily require the characterist not only to master a lengthy set of memorized words and their corresponding figures, and a corollary mental dictionary of synonyms and antonyms, but

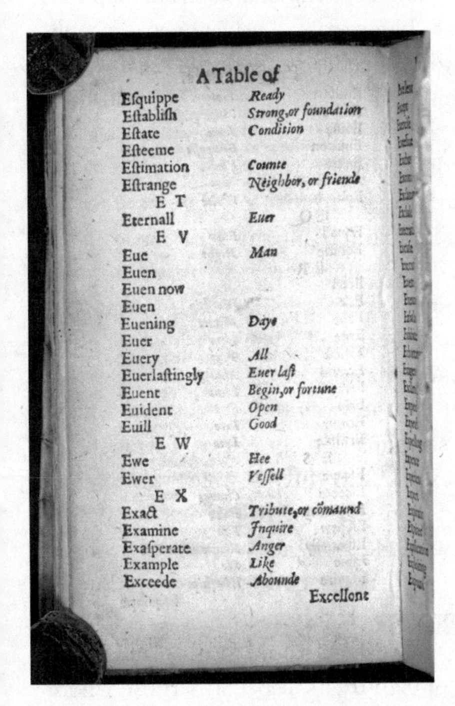

A Table of

Esquippe	*Ready*
Establish	*Strong, or foundation*
Estate	*Condition*
Esteeme	
Estimation	*Counte*
Estrange	*Neighbor, or friende*
E T	
Eternall	*Euer*
E V	
Eue	*Man*
Euen	
Euen now	
Euen	
Euening	*Daye*
Euer	
Euery	*All*
Euerlastingly	*Euer last*
Euent	*Begin, or fortune*
Euident	*Open*
Euill	*Good*
E W	
Ewe	*Hee*
Ewer	*Vessell*
E X	
Exact	*Tribute, or comaund*
Examine	*Inquire*
Exasperate	*Anger*
Example	*Like*
Exceede	*Abounde*

Excellent

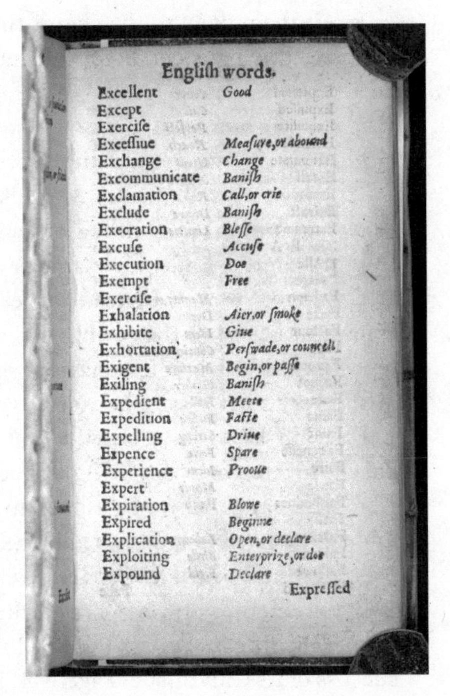

English words.

Excellent	*Good*
Except	
Exercise	
Excessiue	*Measure, or abound*
Exchange	*Change*
Excommunicate	*Banish*
Exclamation	*Call, or crie*
Exclude	*Banish*
Execration	*Blesse*
Excuse	*Accuse*
Execution	*Doe*
Exempt	*Free*
Exercise	
Exhalation	*Aier, or smoke*
Exhibite	*Giue*
Exhortation	*Perswade, or counsell*
Exigent	*Begin, or passe*
Exiling	*Banish*
Expedient	*Meete*
Expedition	*Faste*
Expelling	*Driue*
Expence	*Spare*
Experience	*Prooue*
Expert	
Expiration	*Blowe*
Expired	*Beginne*
Explication	*Open, or declare*
Exploiting	*Enterprize, or doe*
Expound	*Declare*

Expressed

Fig. 8.3. Timothy Bright, *Characterie: an arte of shorte, swifte, and secret writing by character Invented by Timothe Bright, Doctor of Physike* (London, 1588), STC 3743, from appendix 2, "Table of English Words." The many handwritten annotations are marks of diligent use. Rare Books Division, The New York Public Library, Astor, Lenox and Tilden Foundations.

also to do the complicated work of relation ("referral") and simplification word by word when transcribing text and in the moment when transcribing speech. The drawbacks to such a system are obvious, belying Bright's second assertion that his was a "swift" art.

In terms of cognitive process we detect, perhaps, the true impact of "characteral" thinking on the Puritan mind. Bright's system required a mastery of the arts of memory and a facility with word association that could serve the work of Protestant evangelization at home. A select group of undergraduates dedicated themselves to learning Bright's system and used it to transcribe sermons for not only personal use but also unauthorized distribution and even publication.[23] Contemporary reports thus show that charactery also failed in its mandate to be a "secret" art. In fact, the practice of charactery appears to have been as far from furtive as could be.

Worse still, if not skillfully practiced, the art was lamentably (if understandably) error prone. This did not make aspiring characterists popular, as it left them open to the charge that charactery fed private ego, not public need. One preacher, Stephen Egerton, forced to publish his sermon in a corrected edition, suggested in an acidic preface that his words had been taken down previously by one who "respected the commendation of his skill in charactery more than the credit of my [Egerton's] ministry."[24]

So the art of charactery, while falling short of its goals, succeeded in others not promised in the subtitle of its manual. No matter how designed to Protestantize England, charactery did not do so because it *worked*. Not as a system, anyway. Instead, our evidence points to narrower but no less important results—the establishment of a rarified, visible Protestant elite. Just imagine the public behaviors that a mastery of charactery conveyed: at the very least it was, I suspect, yet another way to prod the lukewarm, Laodicean, less-godly multitude at sermons. By scribbling furiously in weird little ciphers, screwing up faces while trying to "refer" an expression like, say, "Balm of Gilead" to the austere requirements of Bright's table of basic characters, and all the while keeping up with the preacher's exegesis, the self-evident godly surely were as much figures to emulate as they were figures of fun.[25] The work of this self-selected cohort, the leaven in the loaf of institutional English Protestantism, would be essential to the success of its Calvinist evangelism. For Calvinists, especially those whom we might be comfortable calling "Puritans," were the Protestantizers of English culture.

This sheds a slightly different light on the practice of character writing in this period. The evidence we have of sermons transcribed in charactery and later published represents the tip of a very small iceberg, but one still worth noting, for it allows us to locate shorthand in a larger realm of Puritan cultural practice. Even the godly Egerton grudgingly admitted that, if practiced with Christian humility (a quality not often associated with Oxbridge undergraduates, granted), charactery cultivated "judgment, memory, and dexterity of hand." His praise of these three "necessary" qualities, two cognitive and one of motor skill, reminds us of the close connection between things mental, things tactile, and things spiritual in late-sixteenth-century Calvinist style. Here, then, I would like to suggest another context for Bright's "art," which is the *pleasures* it would have promised: of cognitive stimulation and rhetorical dexterity and theological adeptness—all parts of the kinetic, imaginative, complex world of what we have misleadingly termed "Puritan plain style."

So I would like to make a few suggestions about how Bright's charactery allows us to admit the notion of a Calvinist aesthetic. In charactery, Bright has constructed skills—and underpinning those, a cognitive method—that are not unlike those skills and methods William Perkins would have assumed or promoted in an audience for the mnemonic charts and graphic theological designs in his *Golden Chaine. Characterie* combines the mental orientation of the medieval and Renaissance *ars memorativa*—fostering an associative, discerning power of recollection[26]—with a streamlined visual representation of The Word and a spiritual mandate for clarity and precision in language. From this vantage, it appears that *Characterie*'s cognitive method (like *Chaine*'s) is a visible revision of "place theory," connecting Renaissance rhetoric to the "art of Logic" that we often associate with late-sixteenth-century Puritan preaching culture.[27]

This brings us to Bright's final book project in the 1580s. In 1589, bypassing the Stationers' licensed privilege by a skillfully underhanded and somewhat-imaginative deployment of his royal patent, Bright published an abridgement of John Foxe's *Actes and Monuments.* In his preface, Bright explained how his condensation of the "Book of Martyrs" was designed to complement, not supplant, Foxe's great work: "The diligent man that loveth knowledge," he states, in language the Puritan preacher Egerton would echo in 1603, "will use [the abridgement] for his memory, and the [original] for his judgment." "Large Volumes and Abridgements both have their use," Bright opined, "and if Abridgements withhold from reading large volumes, because men find the contents of them in brief: even as much are large books an hinderance to themselves, in discouraging the negligent and slothful by their length." Here Bright echoes the prefatory remarks of the editors of the Geneva Bible, who claimed that the scholarly aids in that version were aimed at readers who would otherwise plead that the text's difficulty kept them from its wholesome and necessary message. In essence, then, Bright's final contribution to Protestantism's watershed decade maintained the principle set down in *Characterie:* the message emerges from the method.

This, then, is a good place to consider how contextualizing Bright allows us to open up the categorical possibilities for the cultural historiography of English Protestantism. In this collection dedicated to knowledge and its making in early modern Europe, this essay proposes that Protestantism, especially Calvinist Protestantism, was most often taught as a skill or "art," and not treated as a response to revelation, in post-Reformation England. As a result, the teaching of sacred skills was closely allied to the teaching of secular ones. Finally, I would also like to stress that Protestant practices, which centered on the printed word, actually extended the realm of The Word into that of pedagogical aesthetics. Style *was* substance, which, arguably, is what defines any culture, even religious culture. Even *Protestant* culture.

Historians of Puritanism are not shy about using the term "culture." But we tend to think of "Puritan culture," rather crudely, as what folks *did:* gadding to sermons or sounding off about images or studying the Bible. But when cultural history is confined to cultural *practice,* we tend to look for—and find—only behaviors to describe. Certainly Timothy Bright's *Characterie* allows us to talk about all sorts of Protestant behavior in the late sixteenth century and gives us plenty of scope for cultural interpretation of the usual kind. Ruminating on how an Elizabethan physician with more time for books than patients might end up writing a medical text for Calvinists, a shorthand book for evangelists, and a *Reader's Digest Condensed Book* for lazy Puritans, we can enroll Bright in a Puritan association with worthies like Walsingham and Chaderton. We can uncover a cracking good tale of publication rivalry involving the book trade, the Puritan pulpit, and the undergraduate common room.[28]

And we should look at the texts, but not only to find behaviors to document. Surely cultural history also must concern itself with the problem of thought formation, and in order to do that we may need to repudiate the historical, religious, and literary commonplaces that have long consigned "Protestant aesthetics" to the realm of the oxymoronic. We can analyze texts written by and for Protestants, like Bright's *Characterie,* with an eye to the way they *look,* the creative and intellectual impulses that gave rise to them, and the way that—out of these—Protestant principles might craft a mental, as well as a visible, praxis.[29]

And to do this, we must begin by remembering that, while "Calvinist" may not always mean "Puritan," Puritan praxis originated in Calvinist doctrine. Analyzing the tenets associated with the practice of experimental predestinarianism or sermon evangelism will actually allow us to discuss Protestant theology as form, as we consider the beautifully architectonic structure of Calvinist texts—their intellectual, rhetorical, and grammatical requirements—and how these might provide a blueprint for a characteristic behavioral, as well as material, culture. To incorporate a formal study of theology into the new cultural history might just allow us to redefine the cultural historical notion of "style" as *thought creatively expressed and realized*—not only in documented behavior but also in visual representation. And then we might start to construct a method by which we can identify a "Protestant aesthetics"—the beauty of holiness as expressed in Calvinist style.

CHAPTER NINE

Boyle's Essay: Genre and the Making of Early Modern Knowledge

SCOTT BLACK

Robert Boyle's "Proemial Essay" (1661) announces a reformed methodology for natural science and attaches this to a shift in literary form, placing the genre of the essay both at the center of his natural philosophy and at the cusp of the emergence of modern science. This paper considers the implications of that choice, exploring Boyle's adoption and adaptation of the essay. Boyle's "Essay" is both a significant moment in the history of early modern science and a chapter in the story of the essay's evolution; I argue that the latter is integral to the former.

Recent historians of early modern science have rewritten the neat stories of heroically modern innovators to include the messy details of laboratory work, the assistance of servants, the material tools, and the emergence of modern practices out of—not just against—purportedly residual practices. A literary history organized as a study of genres offers one more strand to this complex story, another way to trace the myriad local adaptations that are the actual texture of change. A history attuned to genres offers a means of specifying the contemporary categories that organized ways of knowing. By recognizing the work of these categories in the production of knowledge—the particular practices they enabled, the skills they solicited, the ways they named and structured specific conceptual possibilities—we can see one site of interplay between innovation and convention, adaptation and continuation, and begin to appreciate the knotty loops of use, reuse, and even misuse that define at once how knowledge is practiced and how knowing works.

Scholars have recognized Boyle's "Proemial Essay" as an important methodological statement, describing it as the "rules for the literary technology of the experimental programme" and a "general defense of his approach."[1] But while the importance of the "Essay" has been generally noted, the significance of Boyle's choice of the genre has not been adequately characterized or fully contextualized. By situating the "Essay" and its advocacy of essays in their local literary context, I argue that Boyle's adoption of the genre enables a practice of reflection that is indispensable to his natural philosophy. In *The Sceptical Chymist* (1661), Boyle expresses a wish that his work provoke both further thinking and further experiments: "whether the Notions I propos'd, and the Experiments I have communicated, be considerable, or not, I willingly leave others to Judge; and This only I shall say for my Self, That I have endeavour'd to deliver matters of Fact, so faithfully, that I may as well assist the lesse skilful Readers to examine the Chymical Hypothesis, as to provoke the Spagyrical Philosophers to illustrate it."[2] Essays were integral to the work of assisting "readers to examine," underwriting Boyle's willingness to "leave others to judge," a central condition of his epistemological reform.

The essay in seventeenth-century England was an instrument of moral philosophy. An established practice of writing in the service of reading, it was described by one early English essayist as a "draught of reading."[3] In locating Boyle's "Essay" within its generic context, I argue that Boyle adopted an available tool rather than invented a new one and so join those recent scholars who focus on the continuities between Boyle's work and earlier, "premodern," practices.[4] Such work puts early modern natural philosophy more squarely within a continuous stream of adapted ideas and evolving practices—philosophical, experimental, social, and, I want to add, literary—an amodern flow rather than a modern eruption or rupture of the kind imagined by both seventeenth-century moderns and contemporary postmoderns alike.[5] Instead of a break, I see what Boyle calls "a way of thinking" as a bridge that spans his moral and natural philosophies. Boyle construes this mode of reflection as a kind of reading, which he transfers from books to the world and from devout to "physical" uses. In doing so, he participates in a larger evolutionary shift and shows one route (though certainly not the only one) through which humanist moral philosophy is transformed into experimental natural philosophy. The particular hinge of this emergence, or perhaps just merge, is Boyle's adoption of an old tool, the essay, for a new way of knowing.

It was not an obvious choice. One might expect essays from his brother, Francis (according to a tutor "not soe much giuven to his booke as my most honored and affectionate Mr Robert"), as essays were lazier and looser than scholarly modes of writing.[6] In his *Moral Essays* (1690), Francis Boyle rehearses the genre's typical rationale, calling them "the small Issue and Recreation of my own private Thoughts and Meditations."[7] The generic repertoire of the essay had been well established for a century: a short, private, recreational form for writing leisurely thoughts and reflections. Earlier in the century (1628), Owen Felltham professed himself a gentleman, not a scholar, and thereby excused himself to write essays instead of studious and strict—and pedantic—works.[8] Essays are immature, premature, and amateur. They are "Faint breathings of a minde burthened with other literary employments, neither brought forth with Care, nor ripened with Age," says John Hall of his *Horae Vacivae, Or, Essays* (1646); they employ a "cursory and imperfect manner (for hee that expects exactnesse and method in an ESSAY, wrongs both the Author and his owne expectation)."[9] At the end of the century (1699), Lady Grace Gethin's work was published posthumously as *Misery's Virtues Whetstone . . . Written by Her for the most part, by way of Essay,* and described as "written for the most part in hast . . . her first Conceptions . . . and set down just as they came into her Mind, as never designed for any others view but her own." Such writing, "by way of essay," named a way of recording "private undigested Thoughts and first Notions hastily set down, without Method or Order, and designed only as Material or a Foundation for a future structure to be built thereon."[10] This cursory, inexact, and imperfect kind of writing—private, leisurely, undigested—seems an inauspicious ground for an epistemological reformation, yet this is precisely what the more scholarly and more famous Boyle advocates.

Boyle explains "why I have cast [my works] into Essays, rather than into any other form" in the "Proemial Essay": "I must freely acknowledge to you, that it has long seem'd to me none of the least impediments of the real advancement of true Natural Philosophy, that men have been so forward to write Systems of it, and have thought themselves oblig'd either to be altogether silent, or not to write less than an entire body of Physiology." Systems require writers to address every part of a subject, even those about which they have nothing to say, making them "idly . . . repeat" others, or "say any thing on them rather than nothing, lest they should appear not to have said something to every part of the Theme, which they had taken upon themselves to write of." You have to fill in the blanks even if you have nothing to add. And further, if systems make writers do things they do not want to do, they discourage readers from doing anything at all, inhibiting further inquiry because they speciously suggest that natural philosophy has been "sufficiently explicated" and there is nothing left to do but passively to "learn what [philosophers] have taught, and thankfully to acquiesce in it." In contrast, a more imperfect form of writing solicits a more active kind of reading, as writers are free to write only "Thoughts

Physiological Observations and Reflexions into Experimental Essayes," historians have paid more attention to the former, the observations, than the latter, the reflections. Shapin and Schaffer construe Boyle's essays as a "literary technology" of representation that solicits readers as "virtual witnesses" to experiments.[14] Their account presents this literary technology as representational in two senses: it represents the facts isolated in the laboratory as a kind of word picture, and it portrays a trustworthy narrator who stands for the reliability of those facts, a representative to guarantee the representations.[15] Such writing offers the "expository means by which matters of fact were established and assent mobilized," an abstraction of the activities of the lab, the work of observation and experiment.[16] Boyle's literary technology is said to have been crafted to secure the assent requisite to constituting matters of fact, an account that reduces reading to simple validation.[17]

More recently, historians have argued that this literary technology was neither as new nor as strategic as Shapin and Schaffer suggest.[18] Principe argues that the style of Boyle's experimental reports can be explained by his early interest in French romances and their epistemological claims situated in a tradition of alchemical "transmutation histories."[19] But if he suggests local contexts that provide Boyle the means to do the work of Shapin and Schaffer's literary technology, Principe gives no account of the genre specifically named by Boyle. Similarly, Sargent has resituated Boyle's experimentalism within several traditions—philosophical, legal, Christian—and offers a nuanced account of Boyle's "writing" as part of the "dynamic learning process" of his experimental philosophy.[20] But she too overlooks Boyle's particular generic context. Principe's contexts do not include the one named by Boyle, and Sargent's "writing" is not contextualized; the literary context of Boyle's "Essay" has fallen through the cracks.

When critics do think about the genre, they think in French. Paradis, for instance, argues that Boyle models his experimental essays on the "French familiar essay." Like Shapin and Schaffer, he explains Boyle's essay as a "mechanism designed to create and sequester facts." In this account, Boyle shifts the focus of Montaigne's "authentically individualist" *Essais* from the internal to the external, transforming Montaigne's "uniquely personal speculations" into "a passive instrument of observation."[21] While I agree that Boyle adapts the essay for his own purposes, I do not think Montaigne is as expressive, nor Boyle as representational, as Paradis says. He seems to abstract Montaigne's reflections from the dense thickets of quoted passages to which they respond, an account that drastically empties the *Essais* of most of their actual content and misconstrues Montaigne's project of thinking through his voluminous, and voluminously cited, reading. In contrast to Paradis, I read Montaigne's *Essais* as a record of his responses to texts, a commonplace book gone wild. Boyle's essays are like them in this sense, registers not just of facts but of reflections as well.[22] It is this practice of response that the genre of the essay facilitates.

Boyle's "Essay" has been mischaracterized and miscontextualized. Paradis situates it exclusively in a French milieu, Montaigne's *Essais,* at once suggesting a generic context (the French essay) that is dubious and eliding the one (the English essay) that is in fact most relevant. Boyle recommends "that form of Writing which (in imitation of the French) we call Essayes"—not a French genre but one that is called by a French word.[23] Seventeenth-century French writers did not write "essays"—the title was too closely identified with Montaigne to be claimed by others—and actually considered the genre an English one.[24] Further, one of the few French writers who did call his work *Essais,* Pierre Nicole, sounds more like Boyle than Montaigne, or at least articulates a generic expectation that Boyle also shares. Nicole offers an account of "what's marked out by Essays": a practice of writing various matters as they present themselves "without undergoing the trouble of disposing and ordering them according to Method" and without the "necessity of filling methodical Works with an infinite number of things, which have no other benefit, than that of being requisite to *Order.*"[25] For Nicole, as for Boyle, the essay is a way of addressing complex fields on one's own terms, and without having to answer the demands of method, saying things you do not want to say while leaving out things you do.

There was not much of a seventeenth-century French genre of the essay, but there was a large and thriving English one. The genre, though, did not, as Paradis claims, "create and sequester facts." Indeed, while the Royal Society's early *Philosophical Transactions* evidence many literary technologies that sought to secure facts ("observations," "experiments," "relations," "accounts," "inquiries," and "descriptions"), the "essay" is not one of them. And when one does find an "essay," it is used to do something else. "An Essay of Dr. John Willis, exhibiting his Hypothesis about the Flux and Reflux of the Sea" is called an essay because Willis offers a hypothesis without experiments to back it up. Willis says he does not have time for "prosecuting the inquiry and perfecting the Hypothesis," and so he gives "some general account of my present imperfect and undigested thoughts," "an Account of my thoughts, as to this matter, though yet immature and unpolished."[26] Essays are not a genre in which one reports experiments but rather in which one offers "conjectures" (as Oldenburg says in his preface to Willis's essay) and solicits responses and assistance from others.[27] Willis's essay is a response to Galileo's hypothesis about the tides, and Willis calls it an essay because it is to be adjusted as more particular information becomes available.[28] And he asks others to respond in kind to his essay:

> And what I say of Galileo, I must in like manner desire to be understood of what I am now ready to say to you. For I do not profess to be so well skilled in the History of the Tides, as that I will undertake presently to accommodate my *general Hypothesis* to the *particular cases;* or that I will indeed undertake for the certainty of it, but onely as an *Essay* propose it to further consideration; to stand or fall, as it shall be found to answer matter of Fact.[29]

Rather than a literary technology of representation that solicits facts, essays name here a tool of hypothesis and conjecture that facilitates "further consideration."

Situating the "Proemial Essay" in its generic context reorients Boyle's "literary technology" from representation to reflection, and it also reorients his "social technology" from identity to skills. Shapin and Schaffer's account of Boyle's literary technology neglects one side of his "Essay," and that practice of reflection is the (absent) literary aspect of the "social technology" of gentlemanly codes that Shapin later explores. Explaining that "the English experimental community had relocated gentlemanly codes into the practice of natural philosophy," he writes:

> It was common in early modern society to contrast the society of gentlemen with that of scholars according to the different values they respectively placed upon truth and good manners. Polite writers condemned traditional scholars because they would sacrifice the good order of conversation to the imperious demands of truth and accuracy, while the scholar might justify himself through variants of the ancient trope used to identify oneself as "a friend of Aristotle but more a friend of truth." Yet changed conceptions of the nature of scholarly practice in the seventeenth century—especially in England but also elsewhere—increasingly reordered and respecified the characters of the scholar and the gentleman. It was now urged that the end of philosophy—the search for truth—might best be acquitted by deploying features of conversational practices that had traditionally belonged

> to gentlemanly and not scholarly society. Lowered expectations of philosophical accuracy, a more reserved way of speaking, a less passionate attempt to claim exact truth for one's claims were justified on explicitly epistemic as well as explicitly moral grounds.[30]

Shapin argues that a social practice reorders epistemology as the practices of "polite writers" serve to ground the new science in gentlemanly conversation, which produces a truth organized by "a more reserved way of speaking." Experimental science was, of course, as much a textual as a social project, so when Shapin slides from polite writing to a reserved way of speaking, he is not making an exact, scholarly claim but employing a loose, gentle locution. What he means is a more reserved way of *writing*, but the slip is instructive, signaling a missing dimension to Shapin's story: a way to explain how social practices answered epistemological problems. Rather than the precise and pedantic models of truth of scholars, the experimentalists' approach was organized by the civil conversation of gentlemen.

> The unwarranted "confidence" and quarrelsomeness of the school-philosopher were juxtaposed to the humility and modesty of the experimentalist. "Diffidence" in asserting truths and the professed willingness to alter one's views were mobilized into emblems of disinterestedness. The presentation of claims as certain and exact, by contrast, was identified as the mark of a scoundrel. One was invited to recognize the genuine experimental philosopher by his civility, decorum, and display of Christian virtues.[31]

Shapin's concerns are centered on "emblems," "marks," and "displays" of identity and authority, on philosophers instead of philosophies.[32] But in tracing the ways social solutions are given to epistemological problems, he replaces the latter with social problems.[33] Social solutions, though, are solutions because they provide epistemological answers to epistemological problems.

Social motivations are only part of a complex story. I want to follow Sargent's suggestion that "the epistemic dimension of a social practice could provide a valuable resource for the introduction of an innovative idea in science" by arguing that gentlemanly practices underwrote experimental natural philosophy by installing a practice of reading at its core, not just a means of self-fashioning as its context.[34] Essays enable a practice of reflection, and this practice is at the heart of the "gentlemanly codes" that define Shapin's social technology. Throughout the seventeenth century, essays were a "Gentleman-like Learning": "A cursory Knowledge, [which] though it be not exact enough for the Schools, is more pleasant, and perhaps more useful then to over burthern the Brain with Books, which may be called a Gentleman-like Learning, or one who is in *Omnibus aliquid.*"[35] In essays, writers and readers are "left at liberty" to think for themselves and are bound, not by the exigencies or exactitudes of systems, but only by their own interests, pleasures, and uses. Writers are at liberty to write only what they know, "a little bit of everything," modeling and enabling an inexact but also unconfined skill of reflection which is not only an "emblem" of disinterestedness, an identity that does social work, but a practice of thinking.

Rather than a social technology of gentlemanly identity that guarantees facts secured by a literary technology of representation, the essay is a gentlemanly practice of reading—a tool of literacy. Boyle makes the practice of inexact and unconfined reading that Culpeper calls "a Gentleman-like Learning"—a gentlemanly skill, not just the sociological fact of being a gentleman—the motor of his reformed epistemology. In essays you must do it for yourself, if not by yourself, and this skill of participatory reading and collaborative exploration was as integral to Boyle's project as the reporting and witnessing of experiments. Within their local generic context, experimental essays look less like a technology of writing and representation than a tool of reading and reflection. Rather than a way to secure facts or fashion identities, they are a way to solicit the practice of reflection on which Boyle's project equally depends.

Accounts of Boyle's writing have been governed by the classical debate between philosophy and rhetoric, between language as exposition and language as persuasion. At a certain level of abstraction, Shapin and Schaffer's account looks rhetorical (focused on questions of authority and ethos), and Sargent's philosophical (focused on epistemic questions)—though Sargent does recognize the public, social aspects of Boyle's project, and Shapin its epistemic ones.[36] But at the same time, they share an emphasis on the "philosophical" side of the "Proemial Essay" and construct their accounts around Boyle's "matters of fact"—though they understand these in different ways.[37] In their discussions, each neglects the rhetorical aspect of the "Proemial Essay": "the 'florid' style to be avoided was a hindrance to the clear provision of virtual witness: it was, Boyle said, like painting 'the eye-glasses of a telescope.'"[38] Boyle does say that rhetoric hinders virtual witnessing, but he does not say it should be avoided. Rather, he recommends it as one aspect of his experimental essays:

> as for the style of our experimental Essays, I suppose you will readily find, that I have endeavour'd to write rather in a Philosophical than a Rhetorical strain, as desiring, that my expressions should be rather clear and significant, than curiously adorn'd . . . certainly in these Discourses, where our design is only to inform Readers, not to delight or perswade them, Perspicuity ought to be esteem'd at least one of the best Qualifications of a style; and to affect needless Rhetorical Ornaments in setting down an Experiment, or explicating something abstruse in Nature, were less improper, than it were (for him that designs not to look directly upon the Sun itself) to paint the Eye-glasses of a Telescope, whose clearness is their Commendation, and in which even the most delightful Colours cannot so much please the eye, as they would hinder the sight.

I read the phrase "in these Discourses, where our design is only to inform Readers" as indicating particular places in these discourses (not the whole of them). In such places of "setting down an Experiment" or an observation, clarity is more effective, and language serves as a lens. You do not, of course, look at but through a lens, and any pleasure given by colors painted on it distracts you from its purpose, to see through. But having made this point about the proper use of philosophical language, Boyle shifts gears. Perhaps with an awareness of how he has made his point, employing a vivid and effective figure to recommend a slim-figured use of language, Boyle catches himself and steers clear of a pedantry he does not want to slip into:

> But I must not suffer myself to slip unawares into the Common place of the unfitness of too spruce a style for serious and weighty matters; and yet I approve not that dull and insipid way of writing, which is practis'd by many Chymists, even when they digress from Physiological Subjects: For though a Philosopher need not be sollicitous, that his style should delight its Reader with his Floridnesse, yet I think he may very well be allow'd to take a care, that it disgust not his Reader by its Flatness, especially when he does not so much deliver Experiments or explicate them, as make Reflections or Discourses on them; for on such Occasions he may be allowed the liberty of recreating his Reader and himself, and manifesting, that he declin'd the Ornaments of Language, not out of Necessity, but Discretion, which forbids them to be us'd where they may darken as well as adorn the Subject they are applied to. Thus (to resume our former Comparison) though it were foolish to

> colour or enamel upon the glasses of Telescopes, yet to gild or otherwise embellish the Tubes of them, may render them more acceptable to the Users, without at all lessening the Clearness of the Object to be looked at through them.[39]

At times language should strive for clarity, the servant of things, a lens, but it is no contradiction to recognize that, even in recommending this, one does not have to be so rigorous, or dull, as to avoid figures altogether. Boyle prefers philosophical to rhetorical language in reporting experiments, not because he thinks language should just be philosophical, but because when using it in this way it should be. Language does many things. There are times when figures darken lenses (when one should be looking through language). And there are times when language adorns, when it is fit "to delight or perswade" readers. This is one of those latter times. Boyle's "Essay" on the essay is, of course, a "reflection." By his own account it is appropriate here to take the liberty of employing the ornaments appropriate to reflections. And indeed he finishes his point with a surprising and witty extension of his figure, taking up his telescope and looking at it again—though not through it. This engaging use of rhetorical language, though, does more than adorn; it makes the point. Sometimes ornaments are mechanisms too, and the figure may do more than Boyle here says. Perhaps he means what he does as well as what he says.

Boyle turns to the figure, and to figures more generally, not just because it does not matter if tubes are painted but because ornaments can clarify. At the end of the passage, Boyle opposes "darkening" to "adorning," suggesting that rhetoric be used only to delight when it can do no harm. But the image that colors also explains. That is, figures give "strength" as well as "light" to passages.[40] Instead of imagining language as a lens that does its real work by getting out of the way as much as possible, Boyle's figure here suggests that language is a complex tool. Lenses are no less crafted than tubes, but they are constructed to do a different kind of work, to make possible a particular ability. One does not have to choose between a lens grinder's diagram and a tube engraver's wit and whimsy to explain what a telescope *really* is. Rather, one must see how they work together to allow it to do the complex things that it does.

Just as telescopes are complex tools with several parts, lenses and tubes, so experimental essays are complex tools with several parts, experiments and reflections. Boyle does not banish rhetoric but recommends it as one aspect of his experimental program. You use rhetoric for reflections, which do a different kind of work than lenses; a writer "may be allowed the liberty of recreating his Reader and himself" when he makes "Reflections or Discourses" on experiments. This "liberty of recreation" repeats Boyle's characterization of the essay, the genre in which readers are "left at liberty" to think for themselves, and writers to write only what they understand and "can write well." The essay sponsors this work of literacy, of reading and reflection, not the work of representation, securing either facts or identities.

Reflections are as integral to Boyle's project as experiments; "when once a Man is in the right way of making Inquiries into such Subjects, Experiments and Notions will reciprocally direct to one another."[41] In *The Christian Virtuoso* (1690) Boyle more fully explicates this relationship between experiments and reflections.[42] He situates his experimental philosophy between "School-Philosophers" and "mere Empiricks." The former, "in the framing of their System, make but little use of Experience . . . superstructuring almost their whole Physicks upon *Abstracted Reason.*" In contrast,

> those, that Understand and Cultivate Experimental Philosophy, make a much greater and better use of Experience in their Philosophical Researches. For they consult Experience both frequently and heedfully; and, not content with the *Phaenomena* that Nature spontaneously affords them, they are solicitous, when they find it needful, to enlarge their Experience by Trials purposely devis'd; and ever and anon Reflecting upon it, they are careful to Conform their Opinions to it; or, if there need be just cause, Reform their Opinions by it.[43]

If reflecting on *experiments* distinguishes experimental virtuosi from systematizers, *reflecting* on experiments distinguishes a virtuoso from "a mere Empirick, or some vulgar Chymist . . . who too often makes Experiments, without making Reflection on them, as having it more in his aim to Produce Effects, than to Discover Truths."[44] The Christian virtuoso brings experiments and reflections together.[45] A "true Naturalist" combines an uncommon "curiosity and attention" and "competent knowledge," which together make one an "attentive and intelligent Peruser" of the world.[46]

When Boyle recommends the essay, then, it is to solicit a skill of attentive reflection that is construed as reading. Such reading serves as a hinge between Boyle's moral and natural philosophy, both of which depend on a "way of writing" that evolved out of a humanist practice of reading. This process involved two shifts, one of the objects of reading (from books to the world) and one of the ends of reading (from moral to natural philosophy). Boyle neither began nor completed these long-term conceptual shifts, but his *Occasional Reflections* (1665) offers one point of transfer in which a practice of Protestant reading begins to develop into a desacralized natural science (an end point Boyle's own work neither reaches nor aspires to).[47] "I have endeavoured to Display the Usefulness of that *way of thinking* I would invite to," Boyle writes, "an attentive frame of mind," "an attentive observation of the Objects wherewith [one] is conversant."[48] Such attention transfers a skill of reading from books to the world. An occasional reflector makes "the world both his Library and his Oratory"; "whereas Men are wont, for the most part, when they would Study hard, to repair to their Libraries, or to Stationers Shops; the Occasional Reflector has his Library always with him, and his Books lying always open before him, and the World it self, and the Actions of Men that live in it, and an almost infinite Variety of other Occurrences being capable of proving Objects of his Contemplation."[49] Boyle argues that this fosters an even more active kind of reading. Reading explicit instructions in books requires only an antlike docility, while reading the world requires an extraordinary beelike and transformative—perhaps even chymical—attention.[50]

In addition to this shift of the objects of reading from books to the world, Boyle also opens a space for various kinds of reading, a transfer (not fully undertaken by Boyle himself) from devout to secular reading. He recommends occasional reflections as a pious practice, training "Heavenly Mindedness, which is a Disposition and a readiness to make Spiritual Uses of Earthly things."[51] But he also says that devotion is not the only purpose of such reflections: "there is no necessity of confineing occasional meditations, to matters Devout, or Theological."[52] In a remark that is refreshingly clear-eyed, especially in this context, Boyle notes that it may be the insistence on theological or moral uses that has kept such reflections so little cultivated: "I would not confine Occasional Meditations to Divinity it self . . . but am ready to allow mens thoughts to expiate much further, and to make of the Objects they contemplate not onely a Theological and a Moral, but also a Political, an Oeconomical, or even a Physical use."[53] Here Boyle anticipates his natural philosophical work, which applies the skill of reflection originally articulated as a practice of moral philosophy to reflect on new objects in new ways.[54]

The particular link between the two is a "way of thinking" that *Occasional Reflections* construes as a practice of reading—one modeled and enabled by a "way of Writing" that is characterized, implicitly and explicitly, in terms of the essay: "immature Productions," "far short of being an

Exact and finish'd Piece," "written for my own private Amusement," and requiring "not any other than a loose and Desultory way of writing."[55] The *Occasional Reflections* are "Green Fruit" of the sort Boyle names in his exemplary "Occasional Consideration of a Fruit-tree," "green and immature Essays."[56] In a blueprint, or an echo, of his advocacy of essays as the tool for his natural philosophy (the *Occasional Reflections* were written before the "Proemial Essay" though published after it), Boyle recommends them in the same terms he uses in the "Proemial Essay": "such Reflections, being of the nature of short and Occasional *Essays,* may afford men the opportunitys, of saying the Hansomest things they know, on several Subjects, without saying any thing Else of them, or filling above a *Sheet,* or perhaps a *Side* of Paper at a Time."[57] Such short and sweet (honeylike, transformative) writings form at once the link between Boyle's moral and natural philosophies and the core of each.

In calling this means of reflection "essays" Boyle uses a contemporary term in a standard way and situates his text in a well-established generic context. "Essays" signified a collection of reading notes. For example, John Uffley's *Wits Fancies: or, Choice Observations and Essayes* (1659) is small and immature (a "little Treatise," "the Maiden flowers of my young age"), and in it he "endeavours to write truly those things that (by his own Experiences) he knows."[58] Uffley's *Essayes* are completely undigested and read like—no doubt are—a published commonplace book, fulfilling all too well the characteristics of the genre that Jonson scornfully derided in the 1620s:

> Some that turn over all books, and are equally searching in all papers, that write out of what they presently find or meet, without choice: by which means it happens that what they have discredited and impugned in one work, they have before or after extolled the same in another. Such are all the essayists, even their master Montaigne. These, in all they write, confess still what books they have read last, and therein their own folly so much, that they bring it to the stake raw and undigested; not that the place did need it neither, but that they thought themselves furnished and would vent it.[59]

Jonson's complaint is not with the practice—he kept his own commonplace book of "Discoveries"—but with the publishing of these necessary but properly private exercises. By momentarily ignoring Jonson's indignation, we can discern a generic profile of the essay as a way of "writing out" what you find in your reading. Throughout the century, "essay" named this practice of writing in the service of reading, a way to register raw and undigested thoughts. William Master's *Logoi eukoiroi, Essayes and Observations Theologicall and Morall* (1653) are thoughts out of school, "unstudied thoughts" that are "committed to paper" to help him "recall" his occasional considerations more profitably.[60] Similarly, though with a higher tone, Lady Chudleigh describes her 1710 *Essays Upon Several Subjects* as her way of processing the "new and useful Hints" provided by her thoughts: "the Notices they give me, I strive to improve by Writing; that firmly fixes what I know, deeply imprints the Truths I've learn'd."[61] Essays were part of a practice of moral philosophy, a way to improve thoughts by "firmly fixing" and "deeply imprinting" what one thinks in order to reduce "knowledge into practice, and live those truths we have been learning."[62]

Essays do not enable the transmission of a body of knowledge as much as the diffusion of a skill of attention that is relayed by readers through books to other readers. Such essays cannot tell you exactly what to do (that would defeat their purpose, which is to allow your own engagement with the material). Instead, they offer necessarily particular examples of that work of reading both books and the world, a skill of close attention which is as much the point of Boyle's project as the observations and experiments through which it must pass. The "Essay on Nitre," with its "Experiments and considerations tending to countenance or illustrate the Reflections therein set down," offers a rationale for regarding the essay as the proper form of such reflections. "I am sufficiently sensible of my having not yet been able to look into the bottom of it; and that very sense of my own ignorance, help'd to keep me from lengthening your trouble in this Essay, lest by solemnly endeavouring to countenance my Conjectures, I might be thought Dogmatical in a hasty Scrible, where 'tis much more my design to awaken and engage your Curiosity, than acquaint you with my opinions."[63] In writing essays, "hasty scribbles," one announces one's own ignorance, one's inability to see to the bottom, in order to solicit help, "to awaken curiosity." Boyle offers a positive rationale that answers Jonson's complaint. You publish such "unpolish'd and unfinished" work not to claim knowledge but to affirm its absence and to seek assistance in pursuing it.[64]

Forgoing authority, essays at once model and enable a "naked way of writing" that solicits the habit of careful attention at the core of Boyle's new way of knowing, a practice of writing articulated in terms of the essay in *Occasional Reflections* and put at the center of Boyle's experimental program in the "Proemial Essay." In both works, essays underwrite

Boyle's desire for a community of knowers bound by a practice of attention: "they that would compleat the Good Fortune of these Papers, may do it more effectually, by Addicting themselves, (as considerable Persons have been of late induc'd to do) to Write Occasional Reflections (how excellent so ever they may prove) then by being Kind to These; since having written them, not to get Reputation, but Company, I cannot but be Unwilling to travel alone: and had rather be *out-gone* than *not* at all *follow'd,* and Surpass'd, than not Imitated."[65] A similar statement ends the "Proemial Essay": "And you will easily pardon me the injury which for your sake I do my own Reputation by this naked way of writing, if you, as well as I, think, those the profitablest Writers, or at least the kindest to their Perusers, who take not so much Care to appear Knowing Men themselves, as to make their Readers such."[66] It may be that Boyle means exactly the opposite of what he says and performs a pose of modesty that disguises an actual interest in reputation. But to believe what he says costs exactly the same as to question his disinterest. And even if we choose distrust over trust, we have done exactly what Boyle asks, undertaken the work of reading his claim for ourselves, not just endorsing his authority. This work of reading does not, it seems to me, refer to another, hidden level of authority, or a further trick of authorship, but rather trades both for a different kind of knowing, one organized by a chain of readers reading other readers, or, same thing, writing essays.

The continuity within Boyle's work is part of a larger continuity. Principe has argued that experimental reports were "a locus of continuity, not of disjunction," between a less modern Boyle and a less ancient alchemy.[67] The same is true of Boyle's adaptation of the essay, his use of a "residual" humanist practice of moral philosophy for his "emergent" experimental science. Anthony Grafton has argued that the old story of humanism eclipsed by science masks a considerable and interesting complexity. Rather than a history of ruptures and radical epistemic breaks, Grafton tells a story of coexisting practices that build bridges across the dubious gaps introduced by modern stories, both seventeenth-century ones and our own.[68] Among such continuities, Grafton names the notebook method—a practice of reading with pen in hand, writing in the service of reading: "the student, armed with a notebook and a set of *loci,* places, or categories, in which to store material for rapid retrieval, set out as confidently in 1630 as his counterparts had one or two centuries before to break up the classics into bite-sized segments and organize them for aggressively confident re-use."[69] The essay evolved out of this humanist method of reading, "a draught of reading" organized by the peculiar exigencies of one's own needs, uses, and pleasures. The practice did not fade away. It was as integral to experimental natural philosophy as to humanist moral philosophy, as can be seen in Henry Oldenberg's preface to the third installment of the Royal Society's annual *Philosophical Transactions:*

> Neither have we discouraged or refused Essays of some famous Philosophers, learned Philologers and Antiquaries; whose Disquisitions, Readings, and Reasonings, have extended farther than their Experiences; since by such bold Excursions and Sallies many valuable Truths may be started out of their recesses. *Architects* do require some variety and store of Materials for the further satisfaction of their Judgment in the Choice: And the *Sculptor* must pare off somewhat his richest Marbles, Onixes, Diamonds, *&c.* before he can perfect the Portraicture. Such liberty an exact Philosopher must claim in his Extracts from Men of much Learning.[70]

The essay was one hinge between the practices of humanism explicated by Grafton and those of experimental philosophy illustrated by Oldenberg. It was an old tool that names the "liberty an exact Philosopher must claim in his Extracts" and was as much a part of the public *Transactions* as it was of a student's private notebook.

The essay underwrote a practice of reflection that was as important to Boyle's natural philosophy as to his moral philosophy, and as integral to his natural philosophy as experiments. Placing Boyle's use of the essay more exactly within its contemporary generic context reminds us that "literary technologies" are not only ways of writing but also ways of reading. Boyle makes the skill solicited by the genre—a particular literacy defined as responsive and not foundational, interactive and not self-fashioning, open-ended and not conclusive—the motor that drives his experimental philosophy. He uses the essay to organize his new way of knowing as a form of literacy instead of logic, and to replace reason with reading.

Boyle's "Essay" offers not only a representational literary technology that guarantees facts and identities but also a tool of literacy, specifically a practice of gentlemanly *reading* that is as indispensable to Boyle's project as the "social technology" of gentlemanly identity. I want to end by suggesting that literary history should not be a subset of sociology, grounded exclusively in categories of social authority and identity, or a subset of intellectual history, organized by abstract models of writing and discourse. Rather, the precise historical specificity of thoughts and practices may be best located in a study of the particular genres that self-consciously organize historical practices. If we are looking for continuities, they will be mediated by adaptations of available tools, somewhat like Shapin's model of a relocated gentlemanly practice. At the same time, as Sargent argues, an account of social practices needs to address the way they answer, not just replace, the problems they were adapted to solve. Gentlemanly practices provided epistemological solutions because they solicited a skill, not just an identity, around which to orient a new way of knowing. A literary history organized by genres—the coevolving tools and skills of literacy—may offer a way to mediate histories of ideas and their social contexts by offering a history of the tools of ideas that enable the skills named by particular local identities.

Boyle's adoption of the essay works more like a Rube Goldberg machine than a great leap forward across epistemic chasms. It is best explicable in terms of small, local, and convenient readjustments and adaptations—in terms, I suppose, of the essay itself. The "history" of the essay, that is, recursively takes the form of the very evolutionary process the genre makes possible. Essays are the registers and effects of evolving uses, variously reused and even misused by many people exploiting the genre for their own myriad exigencies. From the perspective of the history of science, the essay is the name of a tool available to be picked up, one that fit, and was fitted to, its procedures of experiment, participation, and perpetual correction. From the perspective of the evolution of the genre, experimental natural science newly applies the skills enabled by the essay, ones that emerge from a humanist practice of reading and that, in turn, are redeployed to support a new way of knowing—a new chapter in an old story.

The essay's evolution from a practice of humanist moral philosophy to an instrument of experimental natural philosophy offers a mediate link across the epochal divides proposed by stories of modernity (in which Something Happens That Changes Everything). "Emergent" scientists were "residual" humanists, and the practices of the latter informed the former, not paradoxically but integrally. "Residual" skills enable "emergent" practices. This kind of development looks like a contradiction only from the perspective of a progressive history organized by gaps and fissures that, one suspects, are introduced by historians so that they can be bridged by their histories. But one could start with the fact of crossing and recognize that if thinking stitches patterns across even the most rigorous conceptual separations, it may best be conceived as a practice of literacy, less a problem of writing things down than taking them up.

PART 3

The Reform of Knowledge

CHAPTER TEN

Making Sense of Medical Collections in Early Modern Holland: The Uses of Wonder

CLAUDIA SWAN

In this essay I respond to the broad question of how knowledge was produced in early modern Europe by offering an analysis of a cluster of natural historical collections amassed around the turn of the seventeenth century in the Netherlands. I aim to tie agents of Dutch medicine and pharmacology to a broader, pan-European history of natural history and medicine circa 1600 while also venturing claims about how the knowledge in which these individuals trafficked—medical knowledge, natural history—was linked to material objects.[1] With one exception, the principal subjects of this account lived and worked within a five-kilometer distance of each other, in the university town of Leiden. As pharmacists, botanists, medical doctors, and professors of medicine, they shared a professional interest in medicine as well. What further links these men is their common interest in collecting; each of them owned or oversaw a sizable collection that encompassed natural items, man-made artifacts, and ethnographic goods. Bernardus Paludanus (Berent ten Broecke, 1550–1633), the town doctor of Enkhuizen and the only nonresident of Leiden, amassed voluminous collections of *naturalia* and ethnographic materials, which lured numerous European visitors to the northern port town where he lived. Christiaen Porret (1554–1627), French-born longtime resident of Leiden and prominent pharmacist, owned a massive collection on a par with a *Kunst-* or *Wunderkammer* of a princely sort. Dirck Cluyt (1546–98), a pharmacist from Delft who moved to Leiden to manage the university garden in 1592, Carolus Clusius (1526–1609), first director of the Leiden University garden, and his colleague Pieter Pauw (1564–1617), professor of medicine at Leiden, were all professionally associated with the growing collections of natural and man-made goods (from crocodiles, blowfish, and exotic plants to maps, prints, and mummies) housed at the university.

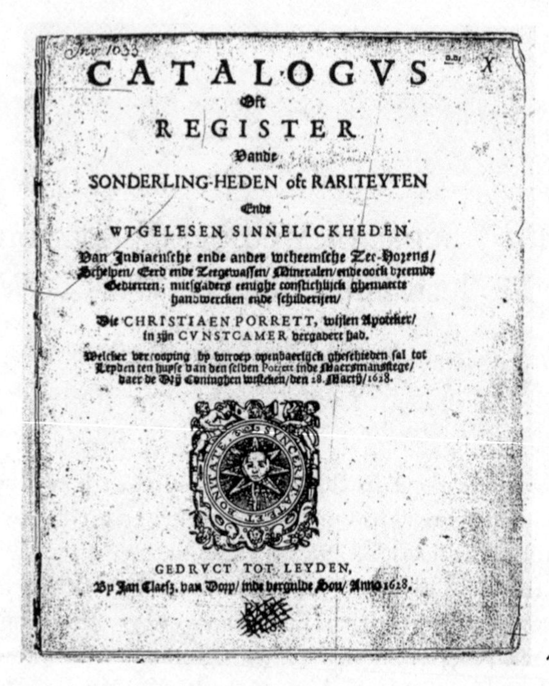
CATALOGVS
Oft
REGISTER
Vande
SONDERLING-HEDEN oft RARITEYTEN
Ende
WTGELESEN SINNELICKHEDEN
Van Indiaensche ende ander wtheemsche Zee-Hoorens/ Schelpen/ Eerd ende Zeegewassen/ Mineralen/ende oock vreemde Gedierten; nutsgaders eenighe constichlijck ghemaeckte handwercken ende schilderijen/
Die CHRISTIAEN PORRETT, wijlen Apoteker/ in zijn CVNSTCAMER vergadert had.
Welcker vercooping by wtroep openbaerlijck gheschieden sal tot Leyden ten huyse van den selven Porrett inde Maersmansstege/ daer de Drij Coninghen wtsteken/den 18. Marty/1628.

GEDRVCT TOT LEYDEN,
By Jan Claesz. van Dorp/ inde vergulde Son/ Anno 1628.

Fig. 10.1. Title page of auction catalog of Christiaen Porret's *Cunstcamer* (Leiden, 1628).

While each of the men listed here plays a role in what follows, Christiaen Porret will assume the lead. His collection is largely unstudied, and what is known about him and the rather spectacular objects he amassed affords a unique view of the practices associated with collecting and with study (knowledge making) in the early seventeenth century. A year after Porret's death in 1627, the contents of his collection were sold by public auction at his home in Leiden (fig. 10.1). The catalog, a printed pamphlet that is our only record of his collection, advertised the goods on offer as "the Exceptional Items or Curiosities and Rare Sensualities [*Sinnelickheden*] From Indian and other foreign locales conches/ shells/ terrestrial and maritime creatures/ minerals/ and also strange animals; and some artfully made handicrafts and paintings Which Christiaen Porrett [*sic*], late Pharmacist /assembled in his Cabinet [*Cunstcamer*]."[2] Like the contents of the collection, the catalog's title may offer potent evidence useful for reconstructing early modern Dutch epistemology. The objects he owned, many of them foreign and most of them unusual in some regard, are ultimately bound together by the force of ownership: Porret is, literally, the organizing principle of this accumulation of curiosities. The fact that

Porret's collection was assembled in his "cabinet" (*cunstcamer*) is key to the following account of Dutch medical collections and their uses, which aims among other things to assess what items such as those collected by Porret—"exceptional," "curious," "rare," and "sensual" objects—have to do with medical knowledge in early modern Holland.

Knowledge and Its Making in Holland

How did the Dutch know? Various recent scholars have characterized early modern Dutch epistemology as bound up with vision and visual representation. Svetlana Alpers's *The Art of Describing* offered a groundbreaking analysis of the extent to which the primacy of visual experience in scientific inquiry went hand in hand with indigenous artistic practices. The Baconian imperative to observe, describe, and chart the natural world and to do so while and by disassociating oneself from inherited precepts resonates with the dominant mode of representation in the Dutch domain: naturalistic and, according to Alpers, descriptive, rather than narrative, picturing.[3] More recently, David Freedberg has called attention in his "Science, Commerce, and Art" to what the subtitle of the essay names "Neglected Topics at the Junction of History and Art History."[4] According to Freedberg, Dutch culture of the Golden Age featured triangular relations among art, science, and commerce; knowledge production is necessarily associated with global trade of the sort at which the Dutch excelled, and science is inextricably linked to representation on the one hand and the market on the other.

Other studies, less strictly focused on the Netherlands and less art historical in orientation, have compounded this view of early modern knowledge production. Pamela Smith and Paula Findlen have demonstrated how closely interlinked the domains of commerce, science, and the arts were and, in the Dutch case among others, how firmly socially and economically grounded natural inquiry was. They have underscored the bonds between patronage and commerce and the impact of both on cultural production concerned with the investigation of nature. "Individuals who claimed to imitate nature, such as many of the artist-artisans, medical practitioners, and other investigators of nature . . . helped lay the foundations of the new philosophy, which eventually would come to be called 'science.'" This knowledge was derived from hands-on and empirical investigation. "This new natural philosophy, pursued with increasing enthusiasm in the late sixteenth and seventeenth centuries, emphasized practice, the active collection of experience, and observation of nature."[5] One of the arenas in which Smith and Findlen locate enthusiasm for this new natural philosophy is medicine. Like Harold Cook, who has asserted that a key role was played by natural historians, individuals interested in natural things (*res naturae*), and, in many instances, medical professionals, Smith and Findlen point out that medical practitioners were prominently involved in "formulating, articulating, and disseminating the new philosophy."[6] Cook, whose research focuses on seventeenth-century Dutch practices, suggests that "the most lively work done in the period" of the Scientific Revolution was "related to medicine and natural history."[7]

Much lively work of precisely this sort took place around the turn of the seventeenth century in Holland. Enkhuizen, a medium-sized port town in which the municipal doctor Bernardus Paludanus lived, worked, and welcomed visitors to his collection, provides a point of entry. Enkhuizen served as the primary Zuider Zee port from the time it distinguished itself as the first Netherlandish city to throw off Spanish rule in the 1570s and would later become one of the six towns in which the Dutch East India Company established headquarters; in the early seventeenth century it harvested the profits of both herring fishing in the North Sea and global trade. Paludanus had traveled extensively before settling in Enkhuizen—in eastern Europe, the Middle East, Egypt, Italy, and the German territories. During these travels, he acquired a medical education (he received his doctorate in philosophy and medicine in Padua), hands-on experience of some of the most celebrated European collections of the time, and collectibles.[8]

Paludanus and his fellow Dutch collectors translated the practice of medical collection to the shores of the Netherlands. We know that Paludanus met such prominent naturalist-collectors as the Neapolitan pharmacist Ferrante Imperato (1550–1631) and, in Bologna, the professor Ulisse Aldrovandi (1522–1605) during his voyages, for example.[9] Likewise, the foreign *naturalia* Dutch collections contained included Asian, African, and American items, in step with the advances of Dutch trade. The Enkhuizen doctor's cabinet contained many thousands of objects—dried plants and seeds and resins and, as one German visitor recalled, "all manner of beautiful and remarkable rarities and unusual things from China, India, America, Africa, Asia, Peru, Egypt, Moluccas, Spain, Canary Islands, Turkey, Greece, etc."[10] Around the turn of the seventeenth century, at a time when he maintained close contact and collaborated with the "Dutch Magellan," the merchant-voyager and author Jan Huygen van Linschoten (ca. 1562–1611), Paludanus began to collect ethnographic items in large numbers as well. Paludanus's collection featured natural things and foreign items (*res naturae* and *res exoticae*) commingled in impressive numbers. The Dutch jurist Hugo de Groot (Hugo Grotius, 1583–1645) was particularly inspired by Paludanus's extensive possessions, which he described as "the treasury of the globe, collection of the whole, ark of the universe, sacred sanctuary of nature, and temple of the world."[11] A microcosmic assemblage, Paludanus's collection offered its visitors the experience of the totality of nature. It consisted mostly of *naturalia,* supplemented by ethnographic materials, many of them new to the European markets.[12]

More often than not, studies of Paludanus's collection mention his profession only in passing. This is in keeping with a more general trend in the literature on early modern collecting, which construes collecting on the part of pharmacists and doctors as a hobby, while in the case of princes and rulers it is thought to have constituted a potent means for sociopolitical maneuvering and the representation of power.[13] As Paula Findlen has demonstrated, the study of substances used in the preparation of medicines (*materia medica*) was a critical link between the medical profession and collections. In the context of a widespread curricular reform that took place during the sixteenth century and affected university medical training directly, the study of *materia medica* rose to new prominence, and instruction in simples became de rigueur at universities throughout Europe.[14] The most renowned sixteenth-century teacher of medicinal preparations, Luca Ghini (1490–1556), was hailed in his own time as a "prince of the science of simples."[15] Simples were taught after midcentury at Montpellier, where free, public lectures were offered to barber-surgeons and apothecaries as well; and, at Leiden, simples were taught to medical students in the university garden as well as in professors' homes.[16] Whether Paludanus, for example, actually used the specimens gathered in his encyclopedic collection for medical purposes is unknown; the scope of the collection and its renown indicate that it was far from simply a practical resource. Paludanus exemplifies the proximity of early modern medical professionals to the domain of natural goods, procured increasingly via trade channels such as ran through Enkhuizen.

Thousands of foreign travelers visited Paludanus's collection, and his guestbook provides a wealth of information about the republic of virtuosi who made excursions for the sake of natural knowledge.[17] His collection was also a professional resource in the sense that when, in the early 1590s, the trustees of Leiden University approached Paludanus and offered him the position of director of the university garden, they stipulated that he was to bring with him to Leiden his collection of dried specimens for use in the instruction of medical students.[18] In the event, Paludanus refused the offer and remained in Enkhuizen.[19] The next candidate considered for the position of director of the garden, the prominent Delft pharmacist Dirck Cluyt, was also renowned for his collections.[20] Because he lacked an academic degree, Cluyt was not offered the position of director, but later, in 1594, he moved to Leiden as prefect of the garden. At the time of his death in 1598, he owned roughly four thousand dried specimens and over one thousand watercolors of plants. Cluyt, Paludanus, and the pharmacist Porret represent a systematically underestimated factor in histories of early modern knowledge. Their commitment to amassing sizable collections of *naturalia* and rare or curious objects is doubly difficult to account for because, on the one hand, it is obscured by the emphasis in histories of the period on other forms of scientific pursuit and, on the other hand, their professional status was in flux, if not in question. During the early modern period, the status of pharmacists and doctors within the medical profession shifted; the late sixteenth century in particular witnessed a trend toward professionalization. To some extent, as illustrated in the requirements established by the trustees of Leiden University for the position of director of the university garden, legitimacy was leveraged on the stuffs of nature; Paludanus's and Cluyt's qualifications for the position were quantifiable and transportable in the forms of fossils, minerals, and other specimens to be used in teaching. Paludanus turned down the offer of employment at the university and Cluyt was deemed not hirable because he lacked an academic degree, but in both cases their collections had helped to qualify them for the position and they each subsequently continued to acquire renown for their collections. We will return later to the issue of social legitimation through collecting.

Ivory Towers

On 28 March 1628, within a year of his death, Porret's collection was sold at his home on the Maersmansteeg in Leiden. What became of the amazing range of objects listed under 719 headings in the printed catalog is not known. Like the phrases animating the title page, the entries vacillate between categories in bewildering ways. Porret's collection contained exceptional, curious, rare, and foreign items that ranged from shells and sea creatures to animals and minerals and to works of art as well. The catalog opens with itemized listings of vessels made of semiprecious stone, an ivory lathe-work tower of enclosed spheres, a spiral staircase in ivory, a Persian cloth in the form of a turban, a sketch of Prince Maurits, and an oblong agate; and it closes with a long series of entries describing watercolor renderings of animals, plants, and flowers. While it is remarkable how many semiprecious stones, natural history watercolors (upward of seven hundred), and exotic items Porret owned, it is even more surprising that his name figures only dimly in histories of collections such as his own.

Within the European context of collections assembled in the sixteenth and seventeenth centuries at courts from Prague and Vienna to Brussels and in ducal residences in between, as on a smaller scale privately, the combination in Porret's collection of natural items, works of art and handicraft, ethnographic specimens, and even optical devices is entirely congruent with more general developments. Early modern European microcosmic collections consisted of famously heterogeneous compilations of goods. The polymath Francis Bacon recommended, in 1594, that the learned gentleman maintain "a goodly, huge cabinet, wherein whatsoever the hand of man by exquisite art or engine has made rare in stuff, form or motion; whatsoever singularity, chance, and the shuffle of things hath produced . . . shall be sorted and included."[21] Many a learned gentleman did. Taste ran to instruments, ethnographical items and imports from the New World, and antiquities, as well as to narwhal horns and bezoar stones. Bacon's inclusive prescription helps to explain the coexistence in such collections of dwarfs and hirsutes, for example, with artifacts of nature such as malformed antlers or "painted" stones.

Early modern collecting and collections of the kind that fall under the joint rubric of *Kunst-* and *Wunderkammern* have been studied in recent years by historians of art and science alike. Analyses of European collections of the sixteenth and seventeenth centuries are often driven by polarities. This is particularly true of studies of collecting that consider them emblematic of epistemological and social interests. Erwin Panofsky's 1954 book *Galileo as a Critic of the Arts* treats the relationship between art and science as a matter of taste and does so with direct reference to early modern collecting practices. Panofsky's book is a fascinating study of sensibility—of Galileo's poetic sensibility and the ways in which it informed his scientific disposition. In formulating an opposition between two poetic modes (and in turn two kinds of collections) Galileo lays out his aesthetic proclivities. Panofsky cites an extended comparison Galileo drew between the poetry of Ariosto, whom he preferred, and Tasso; the two modes of writing are compared to two different kinds of collecting. When he read Ariosto, Galileo wrote, he beheld "opening up before [him], a treasure room, a festive hall, a regal gallery adorned with a hundred classical statues by the most renowned masters," whereas Tasso's poetry called to mind "the study of some little man with a taste for curios who has taken delight in fitting it out with things that have something strange about them . . . but are, as a matter of fact, nothing but bric-a-brac—a petrified crayfish; a dried-up chameleon; a fly and a spider embedded in a piece of amber; some of those little clay figures which are said to be found in the ancient tombs of Egypt."

The terms of Galileo's spatial metaphor make clear that different sorts of collection were associated with different modes of thought and expression. As Panofsky noted, "Galileo portrays to a nicety, and with evident gusto, one of those jumbled *Kunst- und Wunderkammern* so typical of the Mannerist age." Compared to "a formal gallery full of Roman marbles and Raphaels," it comes off as lacking in conceptual order or stability and fails to surpass the quirky pleasures of the little man (*ometto*) at its center.[22] Panofsky suggests that this *paragone* serves as a crucial iteration of the astronomer's "aesthetic attitude" and, in turn, attributes Galileo's scientific proclivities and specifically his distaste for Kepler's models of planetary motion to his affinity for coherent patterns of symbolic thought—as reflected in Ariosto's poetry and collecting "high art" on a grand scale. Galileo's evocation of a cabinet of curiosity renders it intellectually flaccid, leaving the *ometto* in the dust, and in this regard it does not reflect the sort of epistemological undergirding characteristic of the Dutch collections under discussion. Nonetheless, the implicit connections Galileo makes between an individual (the *ometto,* for example) and his collection is underlined by Panofsky. On this model, collections are fossils of philosophical or poetic conceptions.

Panofsky claimed that the passage partially cited above is "fully appreciable only by art historians."[23] Recent literature on collecting and specifically on cabinets of curiosity and *Wunderkammern,* however, belies the notion that any single historical discipline has more traction on the subject than any other. Most recent studies of collecting and of the philosophical category of wonder have been written across historical disciplines.[24] Distinctions between sort and scope of collections are no longer traced along national or regional lines; instead, differences have come to be charted in social and professional terms. H. D. Schepelern has proposed that collecting principles were by and large uniform throughout Europe but that differences pertain between the aims of natural historians and philosophers, on the one hand, and royal or noble collectors, on the other.[25] Giuseppe Olmi, in his studies of early modern Italian collections, has distinguished the private from the princely: "social and economic status of the collectors, and, more importantly, their intellectual and professional interests," are the coordinates according to which the two kinds of collections are organized. Olmi specified that, in his view, the contents and the arrangement of collections assembled by (medical) professionals were "purely functional rather than symbolic."[26] Similarly, in her discussion of "Museums of Medicine," Findlen treats apothecaries' collections as a distinct genre: ownership of such a collection by a pharmacist was unequivocally bound up with the "study of nature as medically necessary knowledge."[27] Pharmacists' collections were, according to Findlen, fairly straightforward extensions of their professional, medical interests. Apothecaries "collected specimens as a natural part of their professional activities; they were the ingredients for the medicines sold in pharmacies."[28] Possession of a collection was worth something in social currency, but even the cultural or social capital at stake was tied directly to professional practice: "Collecting increased the status of men such as [Francesco] Calzolari and Imperato," Findlen writes, "by publicizing their possession of the most exotic ingredients that nature could supply."[29]

On this view, natural historians and philosophers enacted in their collections a commitment to obtaining scads of *naturalia* that confirmed their active familiarity with and control over elements of the natural world. The relationship between the medical profession and *Wunder-* or *Kunstkammern* in the early modern period was built of a common interest in the natural world—in natural philosophy and natural history. Lorraine Daston and Katharine Park have asserted that "the emergence of collecting as an activity not just of patricians and princes, as in the High and later Middle Ages, but of scholars and medical men as well" was "closely connected with [the] new surge of interest in natural wonders."[30] Princely collections were bound by symbolic order; their spectacle ultimately reflected the power of their owners. Daston and Park also adduced social identity as a primary defining factor where they cited a "spectrum ranging from the princely collection . . . to the professional collection," with scholars, physicians, and lawyers actively collecting under the professional rubric, and medical professionals most assiduously collecting *naturalia*. The spectacle of learning, though, that such collections produced was distinct from the spectacle of power staged in princely collections.[31]

Generally speaking, it might be assumed that a pharmacist's collection assembled around the turn of the seventeenth century would feature items relating to medicinal preparations—plants, spices, resins, and minerals. Porret's collection did, in good number. In addition to *naturalia,* however, it contained numerous man-made or artfully natural items, ethnographic objects, and *scientifica* as well. The initial entries in the catalog for Porret's collection list "two serpentine containers/ that serve as cups or mugs"; "two crystal glasses/ with white stripes"; "a platter of serpentine stone"; "an ivory sphere or globe/ with various balls/ that turn within each other/ on a pedestal/ or foot of ebony"; and "a spiral staircase made of ivory." Here and throughout the catalog, cast animals painted with lead glaze are cited: "a cast frog, painted with lead," and "a salamander painted in lead" (no. 29) are just two of the artifacts reminiscent of those produced in Paris by the natural historian and ceramicist Bernard Palissy (1509–90).[32] The first page of the catalog also cites a round piece of quartz and a shell and "two mother-of-pearl fishing rods from the Straits of Magellan." The variety of objects is remarkable. The catalog also lists, for example, "a sea plant like cauliflower" (presumably a coral) and hundreds of shells in all sizes and shapes and colors, including at least one "mother of pearl shell, carved and painted" (no. 168) (see plate 4). A "covered nut from the Indies" (no. 42) is found in the company of a "covered head, from a fruit from the Indies"; either of these may have been a coconut with elaborate decoration. The list goes on, citing a carved wooden crucifix in an ebony case (no. 61); a blue sapphire in lead (no. 62); a large piece of white coral, painted red and gilded (no. 69); a couple of beaks of birds from the Indies; and a number of rather conventional pictures, among them a pair of painted landscapes in the round and images of contemporary rulers. Porret also owned a "bird's nest in a red drawer, with five or six little birds very beautifully constructed of feathers in all colors" (no. 133); "a small box that screws shut, artfully carved, containing wooden toothpicks" (no. 130); numerous groups of "old medals or coins" (no. 99); and a number of foreign pieces of cloth and clothing. Not to mention a peach, a quince, a pear, and a cucumber, each sculpted in wax and containing "two Venetian gloves"; numerous natural stones and painted stones; green eggs of the emu; Indian and Chinese inks; gems and fossils and herbs, both dried and painted; lacquer work; sulfur; a magnifying glass, a kaleidoscope, and other optical devices; Hungarian and Turkish shoes; whistles devised to attract various animals; a blowfish; a large crocodile and a small crocodile; and sheets and sheets of watercolors and drawers and drawers filled with resins, stones, minerals, and fruits.

How unusual was Porret's collection? It is similar to Paludanus's in magnitude, though not entirely in scope: Porret owned works of art (paintings and watercolors and elaborate sculptural items) and Paludanus did not. Porret ventured to collect artful and ingenious items of handicraft; coins and medals; representational works; optical devices. The extent to which his collection comprised a professional resource is not self-evident from the catalog. One measure of its medical or pharmacological role and usefulness may, however, be available through comparison with other lo-

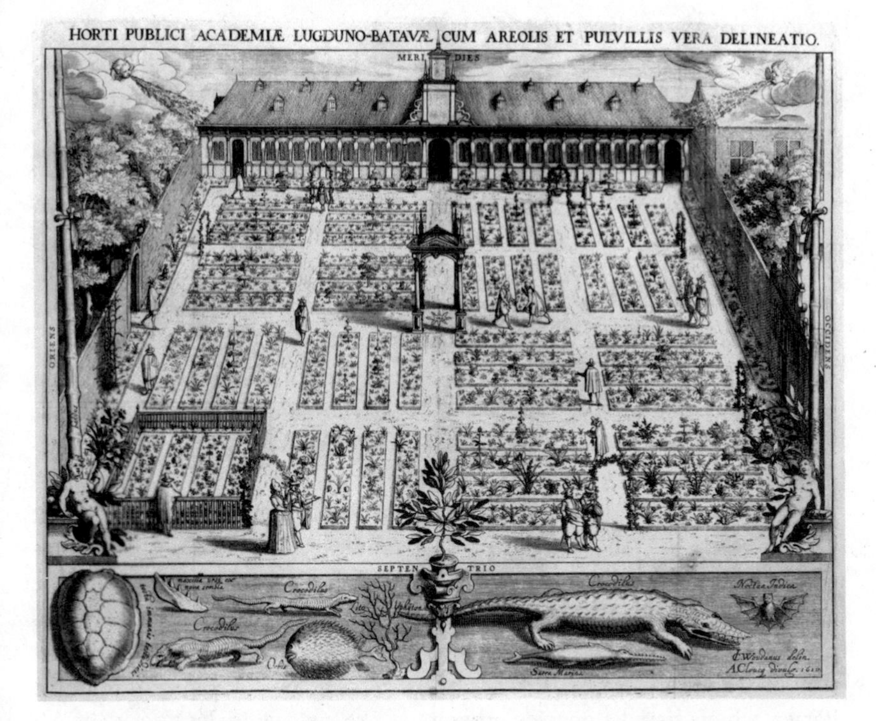

Fig. 10.2. William Swanenburgh after Jan Corneliszn Woudanus, Leiden Garden, engraving, 1610, 33 × 40 cm. Courtesy of the Rijksmuseum, Amsterdam.

cal instances of medical collection. The town of Leiden boasted one of the great public collections of *naturalia* and other curiosities—the university garden—and its extended holdings were, by the first decade of the seventeenth century, popular tourist destinations (fig. 10.2).[33] The anatomical theater too housed a small collection of prints, paintings, and various curiosities; this collection was significantly expanded later in the seventeenth century. In the garden, a long gallery (*ambulacrum*) was built in 1599; it was the brainchild of Pieter Pauw, professor of medicine (the garden opened in 1594). Originally intended to shelter students and visitors from rain and to provide protection for plants during the winter, by the second decade of the century, by which time its floor had been paved, the Leiden *ambulacrum* housed a sort of mini-*Wunderkammer*, with an emphasis on *naturalia*. In 1614 the city historian Jan Janszn Orlers wrote that it was "decorated and hung with many and various maps and geographical depictions, as with some foreign animals and plants, brought here from both of the Indies and other places."[34] The earliest inventory of the contents of the gallery, which refers to the contents as "curiosities," records a number of "foreign animals and plants," some of which may have arrived in the Netherlands on the first Dutch ships to return from the East Indies. They include bamboo stalks (*Arundo indica*) presented to the garden by the directors of the Dutch East India Company, boxes of resins and extracts, and various fruits or nuts. Animals were more numerous and ranged from crocodiles, penguins, and blowfish to parts of animals: the foot of a cassowary bird, a walrus penis, various parts of a bear, and the "beak of a strange bird." Ethnographic items are also listed, among them Pygmy vestments, two Indian hammocks, an Indian skirt, and an Indian ink pot ("Indian" as in either the West or the East Indies).[35]

The collection housed in the *ambulacrum* was coextensive with the university garden, which also contained foreign, rare, and valuable specimens. What sorts of connections existed between collections such as Porret's and "living collections," or gardens? Porret too owned at least one garden, as did his fellow collectors Paludanus and Cluyt and others in foreign lands. One of the things for which Porret came to be known during his lifetime was, indeed, his garden—among other things, a more or less requisite resource for the contemporary pharmacist. In a catalog of the Leiden garden published by Pieter Pauw in 1601, a dedicatory poem refers to Porret, who is listed alongside other great "keepers of gardens" such as Aldrovandi, Pietro Bembo (1470–1547), Felix Platter (1536–1614), and Caspar Bauhin (1560–1624).[36] In 1621, in an extended paean to his own home and garden in Zeeland, the minister Petrus Hondius (1578–1621) made a rather lengthy digression to praise Porret's talents and garden. Hondius called Porret a "famous pharmacist, simplicist, and herbalist" and spoke of his generosity and diligence, invoking his contemporaries Carolus Clusius and Dirck Cluyt in the process.[37] One of the primary links between Porret the pharmacist, Cluyt the pharmacist and university garden prefect, and Clusius the director of the Leiden University garden was their shared interest in horticulture, which is to say in cultivating living collections.

Social Legitimation through Nature

The period during which Porret, Paludanus, Cluyt, and their compatriots amassed their collections was a time of flux in the medical professions. The knowledge of simples among medical professionals was seen, across Europe, to have descended to an all-time low by the early decades of the sixteenth century. In the introduction to his groundbreaking herbal, the *Herbarum Vivae Eicones ad Naturae Imitationem* (Living Images of Plants in Imitation of Nature, 1530–36), Otto Brunfels (1464–1534) told a pointed story about the decline of practical knowledge among medical professionals that was intended to highlight a weakness his publications might ameliorate. Citing Erasmus, Brunfels recounted an instructive prank a certain Basel doctor (Guilielmus Copus, d. 1532?) pulled on the medical faculty at the University of Paris. At a dinner with the Paris professors Copus extracted an herb from the salad and challenged them to name it. Dumbfounded by its appearance, the learned professors concluded that it must be a rare and foreign vegetable. A kitchen maid was called to the table and declared the herb to be common parsley.[38] Professors of medicine and professional naturalists levied accusations of ignorance in matters horticultural and pharmaceutical against unlearned doctors, pharmacists, and other practitioners of the healing arts more often than the tale of Copus might suggest.[39] Accusations of misreading dispensatory manuals or texts on the *materia medica* were directed at pharmacists with some regularity. Mocked in contemporary texts, pharmacists were also in many cases subjected to increasingly stringent controls, often enforced by faculties of medicine who were authorized to license apothecaries and herbalists.[40]

In the context of these disputes about legitimacy and medical knowledge, class divisions between university-trained medical professionals and "unlearned" apothecaries were stressed to the point of outright ridicule. A text on medicinal simples by Antonius Musa Brasavolus published in 1536 offers a case in point.[41] In the course of the book, the narrator (the author, Brasavolus) encounters the aged pharmacist Senex and his helper Herbarius collecting herbs in the hills outside Ferrara. Brasavolus takes Senex to task for "the surprising listlessness of apothecaries" in general and for stubbornly misidentifying medicinal plants in particular; Senex is ruthlessly characterized by the Ferrarese nobleman-author as coarse, ill-mannered, and, initially at least, chauvinistic in his defense of tradition and acquired knowledge over innovative and open-minded study.[42] Social or class dominion was at stake in numerous early modern disputes over the legitimacy of remedies and their contents. As late as 1622, Caspar Bauhin, quoting the Parisian physician Jean Fernel (1497–1558), wrote: "The knowledge, collection, choice, culling, preservation, preparation, correction, and task of mixing of simples all pertain to the pharmacists; yet it is especially necessary for the physician to be expert and skilled in these things. If, in fact, he wishes to maintain and safeguard his dignity and authority among the servants of the art, he should teach *them* these things."[43] In his Flemish herbal (*Cruijdeboeck*), published in 1554 in Antwerp, Rembert Dodoens (1517–85) explained that he had compiled a catalog of the plant

in the practice of natural history. These expectations and these questions have been taken up by Daston and Park in their *Wonders and the Order of Nature*—in particular where they discuss the terms of "preternatural history" as practiced by medical professionals.[47] "Preternatural history" is a term they coin to denote a new form of natural philosophy practiced in the early modern era that encompassed the study and indeed the privileging of "marvelous effects of all sorts" and wonders.[48] Though he could not have written it, the title of the catalog of Porret's collection is as clear a digest of the central terms of emergent "preternatural history" as any other contemporary account. Porret, it seems, participated in the contemporary interest in natural wonders. *Sinnelickheden* are more than the makings of medicine. The emphatic concern with the stuff of nature and sensory engagement with it is crucial to an account of medical collecting. Porret's catalog bristles with natural particulars. We might say that it provides further evidence of Aristotelian investment in accumulating experience dovetailing, in the early modern era, with Baconian refutation of all but the facts of nature to effect a shift in the production of early modern knowledge.

Medical collecting in Holland at the turn of the seventeenth century was more widespread than is generally acknowledged either in accounts of early modern collecting or in histories of science. While some collections were clearly used for pedagogical purposes (the Leiden garden and Cluyt's collection of *simplicia* and watercolors come to mind straightaway), others were less clearly functional in that regard. The category of wonder, as recently explicated by Daston and Park, helps substantially to account for the impact these collections may have had and, indeed, for the ways in which they functioned philosophically. To know was, for these medical professionals, to know nature. And to know was to experience—to engage with the *res naturae* in all their wondrous particularity.

By 1621, Porret, then sixty-seven years old, was unable to visit his garden as frequently as he once had. Petrus Hondius wrote of him, in a dedicatory poem published in 1621: "Your old age prevents you more and more from walking two and three times a day up and back to your garden outside the city."[49] How important the garden had been to him as a resource is clear. Perhaps by this time Porret's collection served him not merely as a source of wonder and means of access to the natural particulars that were so fundamental to the discipline of natural history but also as a source of consolation. Whether or not this was the case, it offered a range of specimens and objects not unlike those for which gardens were built and in so doing constituted a crucial resource for a medical professional interested equally in use and wonder.

CHAPTER ELEVEN

In Search of True Knowledge: Ole Worm (1588–1654) and the New Philosophy

OLE PETER GRELL

The late sixteenth and early seventeenth centuries were a period of great intellectual ferment, particularly in natural philosophy and medicine. Paracelsus and others challenged the learning of the universities, proclaiming that observation and experience must form the foundation of natural knowledge. The works of Andreas Vesalius and William Harvey, while attempting to renovate the medical and philosophical authority of the ancients, ended up overturning them. Nature and sensory observation were emerging as new means of authorizing and legitimating knowledge. Whereas the old story of the "Scientific Revolution" presented this process as self-evident—a general, progressive enlightenment taking place as the result of new discoveries—the historical reality of the process by which new knowledge was gathered, disseminated, and accepted or rejected was far more complex.[1]

In order to understand this process we need to establish how early modern knowledge was gathered and, more importantly, how it was validated. Fortunately, the Danish natural philosopher and physician Ole Worm (fig. 11.1) provides us with valuable insight into these processes, especially through his voluminous correspondence (close to 1,800 letters written between 1616 and 1654), giving us a deeper understanding of early modern knowledge making. Ole Worm's works (twenty books) and correspondence form a unique microhistory of knowledge-making practices in early modern Europe. In particular, his evaluation of Paracelsian controversies and of Harvey's theory of the circulation of the blood points up the unexpected results of his commitment to observation and experiment: this champion of empiricism only reluctantly accepted one of the greatest discoveries in medicine in his lifetime.

Few scholars equaled Ole Worm in the gathering and dissemination

Fig. 11.1. Ole Worm (1588–1654), engraving by Simon de Pas showing Worm at the age of thirty-eight. The poem, written by Worm's colleague Hans Rasmussen Brochmand, praises Worm as a physician and a natural philosopher.

of knowledge in Scandinavia during the first half of the seventeenth century. Worm, by training a physician, did not limit his interests to natural philosophy and medicine but was deeply interested not only in medically related topics, such as botany, natural history, alchemy, astrology, and mineralogy, but also in history, archaeology, philology (especially ancient languages, including the runes he sought to decipher), numismatics, chronology, and collecting. His interest in the world was encyclopedic, a characteristic the Lutheran Worm shared with a number of other scholars of the age, as diverse as the Calvinist Johann Heinrich Alsted and the Jesuit Athanasius Kircher.

Ole Worm's modes of acquiring knowledge were typical for the age: he depended on travel, books, face-to-face discussions, correspondence, collecting rare objects, and finally empiricism and experiment, all of them closely interrelated and interdependent. Ole Worm was barely thirteen years old when he began what in effect became a twelve-year *peregrinatio academica.* Even in an age characterized by increasing travel, Worm's *peregrinatio* was of unusual duration, offering him the chance to spend considerable time at most of the prominent European urban and academic centers. Worm set out in 1601 at the unusually tender age of thirteen on the first of his educational journeys. He originally intended to study theology at the then Lutheran University of Marburg. However, shortly after Worm's arrival, the university became a Reformed institution, having come under the authority of the Calvinist Duke Moritz of Hesse-Kassel.

Worm followed his Lutheran teachers into exile in Giessen, where he continued his theological studies.[2]

In 1607 Worm moved to Strasbourg to study at the Lutheran academy, where he abandoned theology altogether. Now interested in medicine, he took up residence with one of the medical teachers, Dr. Johann Rudolf Salzmann (1573–1656), a keen anatomist and botanist. In August 1608 on his way to Basel to further his medical studies, Worm received a letter from Salzmann informing him that the "medical exercises had temporarily stopped" in Strasbourg, but that he intended to dissect a calf in order to keep his skills honed and he hoped to obtain a corpse for dissection next winter.[3] If we can believe Worm's autobiography, it was Salzmann's experimental and practical approach to medicine that encouraged Worm to further his anatomical and botanical studies at the University of Basel under the tutelage of the famous professors Felix Platter, Caspar Bauhin, and Jacob Zwinger. When in Basel, Worm did not limit his medical studies to the university curriculum offerings but actively participated in Bauhin's private botanical exercises and anatomical dissections, as well as Zwinger's practical bedside teaching.

After a little over a year in Basel, Worm continued his training in Padua, where he matriculated in October 1608. During his eighteen-month stay, Worm studied anatomy under Fabricius Aquapendente, and he claimed that his knowledge of surgery was much improved by the teaching of Julius Casserius Placentinus. Considering Worm's later interests in botany, it is noteworthy he does not refer at all to the famous botanical garden in Padua already created in 1545. Before moving on to France, Worm spent the summer of 1609 touring Italy, traveling as far south as Naples.[4] Worm wished to go to France to study iatrochemistry, which by spring 1609 had become a primary interest, as he informed his Basel teacher Jacob Zwinger, who may have been the first to direct Worm's interests toward Paracelsianism. He specifically asked for letters of introduction to the famous Parisian Paracelsians Joseph du Chesne and Theodore de Mayerne.[5] However, when Worm finally arrived in Paris in early March, Joseph du Chesne was already dead and the Huguenot Theodore de Mayerne had left France for England. Instead, Worm appears to have befriended the anti-Paracelsian physician Jean Riolan (the Younger) and the philologist Isaac Casaubon while also practicing medicine in the French capital. Worm's stay came to a sudden end with the murder of King Henry IV on 14 May 1610, and Worm left Paris for the safety of the United Provinces, where he spent time at the increasingly important University of Leiden.[6]

By June 1610 Worm was back in Denmark visiting the nobleman Holger

Rosenkrantz, who became a close friend and patron. In September that year Worm matriculated at the University of Copenhagen, but this proved a brief stay of one term only. Worm's interests in Paracelsianism remained strong, and in early 1611 he was already on his way to the University of Marburg, where the Paracelsian Johannes Hartmann occupied the first European chair in chemistry. On his way to Marburg he stopped in Hamburg to visit the well-known iatrochemist Conrad Khunrath.[7] According to his autobiography Worm spent his time in Marburg on iatrochemical studies and experiments under Hartmann's supervision; however, Worm found his studies under Hartmann expensive, as the Marburg professor was willing to part with important alchemical information and insight only in return for hard cash.[8]

Worm probably long intended to obtain his MD from the University of Basel, as he wrote to Caspar Bauhin in October 1611, for Worm had studied longest here and this institution maintained the closest relations to the medical faculty in Copenhagen.[9] Although his autobiography states that he intended to obtain the degree from the Marburg academy, this should probably be read as an appreciation of that university rather than real intent. Despite the plague that caused the University of Marburg to close over the summer, Worm could easily have returned to Marburg from nearby Kassel, where he was staying with his friend the alchemist physician Arnold Gillenius, who had become court physician to Duke Moritz of Hesse-Kassel and was heavily involved in the duke's alchemical laboratory. Instead, Worm made sure his promoters within the medical faculty in Basel occupied important positions that might further his interests before he defended his doctoral dissertation, *Selecta Controversiarum Medicarum Centuria*, in December 1611.[10] Having obtained his MD, he traveled down the Rhine from Basel to Amsterdam with the intent of crossing over to London, where he arrived in February 1612 and, within three months, became personal physician to Robert Rich, the later Earl of Warwick. This evidently gave Worm an opportunity to befriend the Huguenot Paracelsian Theodore de Mayerne, now physician to James I of England, whom he had been unable to meet in Paris.[11] After a year and a half in London, Worm returned to Denmark for good in July 1613 to take up a professorship in pedagogy at the University of Copenhagen.

At the time of his appointment to this first professorship, Worm had sampled the best medical education available at the leading universities in Europe and had been educated by some of the most prominent medical figures of the age, often participating in their private colloquia and classes. Through his varied experiences Worm early on developed a hands-on, observationally based experimental approach to natural philosophy and medicine, happily mixing a Hippocratic approach with aspects of Paracelsianism, a sound skepticism with an open-minded inquisitiveness. Furthermore, Worm had practiced medicine in many of the leading European cities, mastering several languages, while simultaneously finding time to benefit from the company of some of the leading intellectual figures of the age. By the time of his return to Copenhagen, Worm had established a comprehensive network of friends and contacts across Europe that would prove particularly important for a man who was to spend the rest of his life teaching at the University of Copenhagen. Without it, Worm's entry into the international republic of letters would have been nearly impossible, as would his attempts to keep abreast of the discoveries and challenges to established knowledge in the early seventeenth century. However, from the outset, the driving force behind Worm's thirst for knowledge lay in the hope of obtaining true or ultimate knowledge and understanding of the natural world as God had created it.

But how did Worm go about acquiring new knowledge while pursuing his academic career in Copenhagen, moving on to a professorship in Greek in 1615 and finally obtaining a chair in medicine in 1621? Worm's engagement with Rosicrucian and Paracelsian ideas and with Harvey's circulation of the blood illustrates how he learned about and evaluated new ideas and discoveries from his remote outpost. Worm's interest in iatrochemistry and Paracelsianism remained strong, as can be seen from his early letters and his 1611 doctoral dissertation at the University of Basel. By 1611 Worm had become interested in the Paracelsian-inspired Rosicrucian phenomenon, which was to sweep across northern Europe in the years leading up to the Thirty Years' War. That year Worm had been shown a manuscript copy of the *Fama Fraternitatis*, one of the Rosicrucian manifestos circulating in manuscript in Germany some years before their publication. The "famous iatrochemist" who showed it to him was undoubtedly Johannes Hartmann, professor of chemistry at the University of Marburg. With hindsight, Worm claimed that he initially considered Rosicrucianism to be pure fantasy and deeply obscure but that Hartmann had convinced him of its value through his weighty arguments, and Worm remained fascinated by and attracted to Rosicrucianism until at least 1618.[12]

Writing in August 1616 to his brother-in-law, Jacob Finke, who was studying in Giessen, Worm requested him to pass on any reliable information that might come his way about the Rosicrucian Brethren. Worm wanted to know who they were, where they were located, and whether or not the local prince, as rumored, belonged to the order.[13] In fact, he wanted to know anything that was *not* known from their published writings, which were already in his possession.[14] In the spring of 1617 Worm wrote to his friend Niels Foss, then in Strasbourg, asking him if he had heard anything definite about the "Rosicrucian Society." Worm realized that people differed fundamentally in their judgment of this phenomenon, but the publication of Georg Molther's answer to the Rosicrucians published at the last Frankfurt Book Fair had convinced Worm that the movement could not be "pure fabrication."[15]

By January 1618 Worm was still trying to obtain information about the Rosicrucians from Jacob Finke in Giessen, pointing out that "strange things" were circulating in print and rumor about them in Copenhagen. Worm, however, remained positively inclined toward Rosicrucianism,[16] and he was prepared to discuss the matter with at least one colleague within the theological faculty. He had lent Professor Jesper Brochmand, who later became a forceful advocate of Lutheran uniformity in Denmark, some of his Rosicrucian pamphlets. Brochmand was clearly baffled by this material, for, on returning it, he wrote: "But to explain the new things, about which you have enigmatically written to me, an Oedipus would be needed."[17] One wonders whether Worm, already nurturing some religious reservations about Rosicrucianism, was in fact trying to elicit from his colleague in theology a verdict about the movement.

Worm's letter of March 1618 to his friend Laurits Scavenius, then staying in Strasbourg, certainly reveals growing doubt about the Rosicrucians. Evidently, Scavenius had informed Worm about a busybody (*ardelio*) who had declared himself a Rosicrucian. This excited Worm, who wanted to know more about him or others who claimed to belong to this "sect," as Worm put it. Clearly, Worm now saw the potential dangers of religious heterodoxy. Writing to Scavenius in Strasbourg a couple of months later, he described the Rosicrucians as lunatics who had promised much but delivered nothing.[18]

In his oration to graduating students on May 1619, Worm described the Rosicrucians as a dangerous sect, and by August 1620 his rejection of Rosicrucianism had become unequivocal. His student Anders Jacobsen Langebæk, studying in Jena, informed him of a recent Rosicrucian publication already possessed by Worm. From this tract Worm concluded that Rosicrucianism was nothing but an amalgamation of sects, adding that the claims of this and other Rosicrucian tracts to be Lutheran were spurious. Instead, he considered the Rosicrucians to represent a dangerous mixture of fanaticism, Anabaptism, and Paracelsianism.[19] Despite Worm's rejection of Rosicrucianism in 1620, he retained some interest, at least privately, in it. As late as 1623 he requested his student Hans Andersen Skovgaard in Wittenberg to obtain books for him written by known Rosicrucians.[20]

While in the midst of rejecting Rosicrucianism, Worm remained interested in iatrochemical and Paracelsian medicine and in March 1618 recommended that Niels Foss travel via Orléans and meet the iatrochemist De Trogni. When Foss reached Paris, Worm wanted him to meet Jean Beguin, whose book *Tyrocinium Chymicum* (1610) Worm described as an excellent beginner's introduction to chemistry. Worm also encouraged Foss to meet Etienne de Clave, "a refined man," whom Worm had met in Paris. De Clave was, according to Worm, "exceedingly experienced in chemistry and an excellent practician," from whom Foss "would not depart without having learned a lot." Worm may well have met the Paracelsian Jean Beguin too during his brief stay in Paris in spring 1610, since Beguin was a protégé of de Mayerne, whom Worm sought to meet while in Paris.[21]

In 1623 Worm's student Hans Andersen Skovgaard was studying medicine in Wittenberg, where he was instructed in iatrochemistry by Daniel Sennert, who offered private laboratory exercises for a small group of students paying ten thalers each. Skovgaard was convinced of Sennert's honesty because he promised the students, not the discovery of great secrets, but only knowledge of practical processes. He informed Worm that Sennert and Hartmann were considered the only prominent teachers of chemistry, and Skovgaard forwarded a copy of Sennert's *Institutiones Medicinae.* Worm was delighted with Skovgaard's iatrochemical undertakings and encouraged him to participate actively in the experiments even if he should get himself dirty. Likewise, he was positively impressed with the price Sennert charged for his instruction, which compared well with the excesses Worm had personally experienced when he studied under Hartmann. He encouraged Skovgaard to continue to keep him informed about the experiments he undertook under Sennert and asked how Sennert cured a specific disease. It would appear Worm had little or no knowledge of Sennert before his student Skovgaard arrived in Wittenberg.[22]

In 1638 Worm's nephew and student Henrik Fuiren was studying medicine in Paris, and he informed his uncle in July that he had been taught chemistry by the Scotsman "Dr. Davidson," a reference to the Paracelsian iatrochemist and later first professor of chemistry at the Jardin Royal des Plantes in Paris, William Davidson.[23] Fuiren had also sought instruction in botany from Guy de la Brosse (ca. 1586–1641), founder of the Jardin Royal des Plantes. Fuiren had been impressed by de la Brosse's *De la nature, vertu et utilité des plantes* (1628), but he had been put off by de la Brosse's surly, unfriendly manner. Another Worm student, Niels Bertelsen Wichman, was more positive, having been shown plants by the garden's "learned director, the chemist de la Brosse."[24] De la Brosse was a keen iatrochemist and admirer of both Paracelsus and Peter Severinus the Dane, as can be seen from his work of 1628, which dedicates one out of five books to a discussion of chemistry and Paracelsianism.[25] Worm wrote to Wichman and asked him to obtain de la Brosse's work on plants, which he was keen to read, and suggested it be brought back to Copenhagen with another Danish student about to return home. Wichman obliged, also informing his mentor about de la Brosse's invention of mechanical statues that could indicate both the time and the zodiac and that were exhibited in the middle of the Jardin Royal des Plantes. Wichman promised to try to obtain a drawing of them for Worm. Worm probably received de la Brosse's book in the summer of 1639; in May 1640 he wrote to his nephew Thomas Bartholin in Paris: "There is in Paris a certain de la Brosse who is an excellent botanist and chemist; I own his work on plants, where he promises to publish a book on both the natural and the methodological art of medicine." Worm instructed Bartholin to send him this work as soon as it was published. At the end of August 1640 Bartholin informed Worm that the work had not yet been printed in Paris.[26]

Ole Worm managed to keep himself continuously informed about new developments within iatrochemistry and Paracelsianism. By January 1646, having become aware of Jean Baptiste van Helmont (1579–1644) through his many Leiden contacts, Worm wrote to Thomas Bartholin in Paris, referring to Van Helmont as "a peculiar paradox-monger" who claimed that a physician did not deserve his name if he was unable to heal a fever patient within four days. Worm had been informed that Van Helmont lived near Brussels and requested that Bartholin visit him if he traveled in that direction to discover whether there was any truth behind his "bombastic statements." From another letter written the following year from Leiden by another nephew, Rasmus Bartholin, it seems that Worm had already read some of Van Helmont's publications. Rasmus informed his uncle that Van Helmont's collected works were being printed by Elzevier (*Ortus Medicinae*, 1648); the edition would contain all Van Helmont's published works and much new material by his son Franciscus Mercurius van Helmont. Worm continued to take an interest in Van Helmont's ideas until his death. In 1650 he encouraged Rasmus Bartholin to visit the Helmontian physician Walter Charleton and his associates while in England.[27]

Among the manuscripts with which Thomas Bartholin returned from his *peregrinatio academico* was Hieronimus Bardi's *Theatrum Naturae Iatrochymicae Rationalis* (published in Rome in 1654), which Worm read with great interest. Bardi, a protégé of Athanasius Kircher, promised to develop a method for extracting the quintessence of plants, thereby making it possible to reproduce the plants in a gaseous state in a vessel, where they would become visible with real colors, leaves, and flowers.[28] This phenomenon, known as palingenesis, had been widely discussed since the sixteenth century among Paracelsians, who believed plants could be calcinated to ashes, and then, when heated in a hermetically sealed glass container, the plant's original shape would appear in a gaseous form in the vessel. Palingenesis was promoted by Paracelsus himself and a number of his prominent followers such as Joseph du Chesne and Athanasius Kircher, to mention only two. It was an aspect of Paracelsianism closely linked to the search for the Philosophers' Stone as the ultimate true knowledge, leading to eternal life, and intermixed with apocalyptic expectations.[29] Worm, long interested in palingenesis, discussed it with his students, and in April 1648 his nephew Rasmus Bartholin wrote to him from Leiden asking if he had yet received "the herb, mentioned by Kircher, which grew in a bottle with water." According to young Rasmus, the Paracelsian physician Joel Langlot, who had accompanied him on his journey to England and recently become court physician to the Duke of Schleswig, knew this secret, and had, on Rasmus's insistence, promised to share it with Worm.[30]

Worm responded by pointing out that he was not interested in the herb referred to by Kircher in his *Ars Magna Lucis et Umbrae* (1645), because it "had emerged not by art but by accident," as he put it. Instead, he was looking for the method whereby plants could be resurrected in spiritual form in all their colors from their ashes or salt by the use of gentle heat, that is, palingenesis. He asked Rasmus if he had actually seen such plants being resurrected and reported that he had heard that Estienne de Clave in Paris had performed palingenesis. Meanwhile, he was trying to obtain one of de Clave's plants for "whatever price the inventor might ask" through one of his Parisian friends.[31]

Worm, however, did not wait for Langlot to contact him, but wrote immediately to the physician in Gottorp Castle congratulating him on his new position and informing him that his nephew had told him that Langlot possessed the "secret, which Quercetanus [du Chesne] referred to in his book on hermetic medicine, chap. 23, whereby gaseous plants through heat are revitalized by philosophical ash with real colors and the shape of all parts." Worm reminded Langlot of the promise made to Rasmus Bartholin and asked him to forward a specimen for his museum. This would serve Worm well, not least because he was involved in a running controversy with one of his learned friends who denied the possibility of palingenesis. Worm had already received "some methods of preparation" from friends, but none had succeeded for those who had tried them, possibly, as Worm added, "because the correct implementation had not been adopted."[32]

Worm returned to the subject of palingenesis five years later in a letter to Langlot. By then he had clearly received Langlot's formula for palingenesis, which he claimed differed significantly from all the other formulas he had been given:

> Some years back I received the following recipe from a friend: The whole plant in flower with its roots is dissolved into three elements which are cleared of impurities, reunited, and placed in a hermetically sealed vessel, until it exists in the form of salt, which is then exposed to light heat; through the hot fumes the whole plant is reproduced out of the extract of the whole thing; this strikes me as more likely than if this is attempted through the seeds alone. But I am not primarily concerned with the information as such, but only the act itself, in order to acquire proper knowledge of the possibility of this art; because, as you are aware, there are a considerable number of people who reject the possibility.

Worm added that if Langlot wanted to carry out the experiment, he was happy to send him information about the procedure.[33]

Langlot was evidently engaged in experiments about palingenesis, as can be seen from Worm's letter of September 1653 in which he praised Langlot's undertakings:

> It pleases me that you are concentrating on the capital secret of reawakening plants in a gaseous, bodily form. If there is anything within chemistry which deserves attention, this is truly it. But you have ample opportunity to try everything under the protection of the noble duke; you will not miss what you need, neither will your work be unprofitable; therefore, continue firmly with your purpose; I am hoping for a fortunate outcome.

Evidently, Worm was envious of Langlot, who could conduct such experiments with no shortage of funds and under princely protection. Worm's deep concern with palingenesis is further illustrated by his recommending to Langlot no fewer than three formulas while also suggesting appropriate plants for the experiments.[34]

Worm's lifelong interest in Paracelsianism and iatrochemistry is also evident in the advice he gave his son Willum during his study tour abroad. He praised Willum, then in Leiden, for having enrolled in a "chemical college," while recommending him "to learn Glauber's method."[35] Clearly, Worm knew of the German Paracelsian chemist Johann Rudolph Glauber, one of the leading practical chemists of the seventeenth century. Like Paracelsus, Glauber was a self-taught controversialist, strongly immersed in the artisanal, practical aspects of iatrochemistry. Worm evidently knew Glauber had gained his reputation in Amsterdam.[36] In July 1654 Willum sent Worm a copy of Glauber's book *Menstruum universale oder Mercurius philosophorum genannt* (1653).

Worm's early interest in Paracelsianism and iatrochemistry continued throughout his life and did not diminish with his rejection of Rosicrucianism in March 1618. As we have seen, Worm not only relied on books and correspondence with academic colleagues for news about Paracelsian and iatrochemical undertakings but also depended in particular on the observations of, and intelligence from, his students (many of them also his relatives) traveling abroad and attending leading universities of the age. Like their teacher in Copenhagen, their interests were not limited to iatrochemistry but included a host of subjects, prominent among which was anatomy.

Despite his diverse interests, Worm always considered himself first and foremost an anatomist. His letters constantly emphasize the importance of human anatomy and regular dissections of animals and humans. He would have been delighted to know that the small supernumerary bones in the sutures of the skull—his only anatomical discovery—are today known in English as Wormian. With his interest in opening up bodies, both animal and human, Worm was very much at the cutting edge of the early seventeenth century's search for better anatomical knowledge.

Two major anatomical discoveries, both published in the 1620s, were to preoccupy Ole Worm and many of his academic colleagues across Europe for a considerable time: Gasparo Aselli's discovery of the chyliferous vessels of the intestine, which Aselli labeled lacteals (white veins), and William Harvey's discovery of the circulation of the blood. The former was significant and vigorously debated; the latter was truly revolutionary and remained deeply controversial for decades. Harvey, who had already made his discovery around 1618, was aware of its controversial nature and accordingly refrained for a decade from publishing his results, which contradicted the reigning medical orthodoxy of the previous fifteen hundred years on the functioning of the body and the movement of the blood. Finally, in 1628 he published his results in a short tract entitled *Exercitatio Anatomica de Motu Cordis et Sanguinis in Animalibus* (Anatomical Exercises on the Motion of the Heart and Blood in Animals). By early April 1631 news of Harvey's discovery had reached Ole Worm in Copenhagen. Writing to his student Jacob Svabe, who was then continuing his medical studies at the University of Leiden, Worm made the following inquiry:

> Because you write that you have seen three dissections of the human body this winter, I would like to know whether your anatomists have demonstrated Aselli's lacteals in the mesentery? Likewise, what do they think of the circulation of the blood in the body according to the opinion of this Englishman? This winter I have for the first time observed it myself in a dog, following the author's instruction. The matter is certainly of great importance and benefit.[37]

Worm's letter was a response to Jabob Svabe's long and informative letter of February that year. Svabe, who had originally matriculated at the University of Copenhagen in September 1627, was clearly close to his teacher Worm. It was undoubtedly on Worm's advice that Svabe went on to matriculate in medicine at Leiden in May 1630. Worm repeatedly advised Svabe to focus on anatomy; it was therefore with some pride that Svabe reported back on three public human dissections by Professors Otto Heurnius and Adrian van Valkenburg that he had witnessed over the previous couple of months. Svabe added an observation clearly meant to please his mentor in Copenhagen:

> Truly, it cannot be denied that the immediate view of an ingenuously dissected corpse is much to be preferred to the beautiful and detailed drawings that a considerable number of people consider to be the anatomical training's alpha and omega. However, they should not be dismissed, since they inspire the eye and the mind through a shadow figure.[38]

Despite his advocacy of practical anatomy, Worm seems to have been unable to conduct any human dissections in Copenhagen. When he referred to his own involvement in human dissections, Worm always returned to his experiences in Basel and Padua from 1607 to 1609. The University of Copenhagen did not acquire an anatomical theater until 1644, with the first public dissection occurring only in the following year. This might explain Svabe's qualified acknowledgment of the usefulness of drawings. Worm dissected animals, however, and performed vivisections, thus emulating Aselli's vivisection of a dog.

Considering Worm's constant endeavor to obtain new books and his many excellent contacts with major booksellers and printers across northern Europe and at the Frankfurt Book Fair, it is surprising he had not yet obtained a copy of Harvey's work three years after publication; it is even more surprising because he had been able to obtain Gasparo Aselli's *De Lacteis Venis,* probably in the edition published in Basel in the same year as Harvey's book. Instead, Worm waited until Jacob Svabe had returned from his studies in Leiden to borrow Svabe's copy together with Adrian Speigel's *De Humane Corporis Fabrica* (1632), which supported Harvey's views. Worm thus did not read Harvey's book until the summer of 1632 at the earliest, more than a year after he had first made his inquiries.[39]

Svabe, however, responded with enthusiasm to Worm's original inquiry in June 1631. Surprisingly, Svabe, who by then had been studying medicine in Leiden for over a year, had heard nothing of Harvey's ideas before Worm's letter "informed him about this new and unheard of conception . . . concerning the return-run of the blood." Svabe considered the matter on his own, "ardently absorbed," as he put it, for a week before consulting his more experienced fellow student and friend Herman Conring, who not only possessed a copy of *De Motu Cordis,* which he showed Svabe, but also expressed himself so "excellently and intelligently about the circulation of the blood" that he struck Svabe as being in agreement with Harvey. When Conring realized Svabe was on the verge of being converted to Harvey's "heretical view," Conring quickly backtracked, adding that, even if Harvey's ideas at first glance looked very attractive, his inability to prove them through observation and anatomical demonstration made them less convincing. Svabe, obviously deeply engrossed in the issue, consulted the Leiden anatomists Otto Heurnius and Adrian van Valkenburg, who both agreed with Conring, with the added proviso "that, concerning questions that change the old and accept the new, we should rather show some care and hesitation than be daring and improvident."[40] In his response, Worm declared himself in agreement with Conring and the Leiden anatomists, pointing out that in medicine, "especially where it is concerned with the connections between the different parts of the human body, we need an eye on every finger in order that we believe only what we observe."[41]

Unlike Herman Conring, who became a follower of Harvey and the first to defend the theory of the circulation of the blood in Germany as early as 1632 on taking up the chair of medicine in Helmstedt, Germany, Worm remained skeptical, not least because he insisted on the necessity of observing the unobservable. Even so, the question of circulation remained central to Worm's interests. In his annual thesis for 1632, which Worm as a medical professor was obliged to present to the medical faculty with a student as respondent, the issue of circulation loomed large. It was between Svabe's return to Denmark in June 1632 and the presentation of Worm's thesis in November that Worm must have borrowed and read Svabe's copy of *De Motu Cordis*. He quotes Harvey's work at length in the second and fifth controversy of his thesis. Dealing first with whether the blood is formed in the parenchyma or in the veins of the liver, the thesis then discusses whether the function of the heart is to pump the blood synchronously with the pulse from the arteries to the veins so that it can circulate through the body. Worm raised a number of objections to Harvey's work, mainly in terms of the missing observational proof, but admitted that the short form of his thesis did not permit an in-depth analysis of Harvey's views.[42] By the end of 1632, having finally read Harvey, Worm clearly felt able to reject his account of the circulation.

Aselli's discovery of the lacteals, whose vivisection on a dog Worm had been able to copy and see for himself, was another matter, as can be seen from his letter of February 1634 to his student and nephew Henrik Fuiren, who was by then pursuing his medical studies in Leiden. Fuiren had been present when Worm had copied Aselli's experiment, and Worm now wanted to know whether the Leiden anatomists had observed anything relating to the lacteals in the mesentery. Worm was very disappointed with Fuiren's response, pointing out:

> I am deeply surprised that your anatomists treat their subjects with such carelessness, that they dare, contrary to the testimony of every spontaneous observation, to deny or hide those vessels which are so necessary and useful, and which have been observed and demonstrated not only by so many and precise anatomists but also by students and beginners. Such a thing was never done formerly by Pauw; nor by others, who established the fame of this university.[43]

Thus, Worm was convinced by Aselli's discoveries, which he could replicate and observe (even if he sought further confirmation from others), but he rejected Harvey's discoveries about the circulation of the blood, which ultimately rested on calculation. Worm would appear to have taken no further interest in Aselli's and Harvey's ideas until another of his nephews and students, Thomas Bartholin, matriculated in Leiden in 1638. In January 1639, by which time Bartholin had already provided his uncle with much information about his anatomical undertakings in Leiden, especially the dissections he was regularly participating in at the local hospital, Worm wrote to his nephew. He praised him for his anatomical undertakings and observations, reminding Thomas to keep him informed about any new anatomical discoveries. He finished by emphasizing how important it was for him to know what the Leiden anatomists thought about Aselli's lacteals and whether they themselves had successfully observed them.[44]

In two letters from February and April 1639 Thomas Bartholin provided Worm with detailed information about the views of the Leiden anatomists on Aselli's lacteals. In February Bartholin had been part of a group, supervised by Van Valkenburg and Johannes Walaeus, who had performed an animal vivisection to demonstrate the presence of the lacteals. The dissection had proved inconclusive, because the lacteals had disappeared from view before they could be properly observed, and only Johannes Walaeus had been convinced. In the first half of April Bartholin and his fellow students took the initiative to dissect a dog in order to demonstrate the lacteals, which this time were clearly observed. However, among those present, Professor Adolph Vorstius, "a defender of the old," as Bartholin put it, still rejected the existence of the lacteals despite the fact that they had been clearly shown, claiming it was all "a figment of the imagination." Bartholin also reported that the celebrated Amsterdam anatomist Nicholaus Tulp had told him that he had been able to observe the lacteals in a recent dissection of a hanged man dead less than five days.[45]

Not surprisingly, Worm was deeply disappointed by Vorstius's reaction, wondering how "such a learned and famous man" could doubt "his own senses and eyes." Worm accepted that one might still have some queries about the exact role of the lacteals, but only the blind could doubt their existence. Worm claimed that Aselli had more than satisfied the inquisitive mind until a better and more certain knowledge could be obtained in the future. Further careful research into the role of the lacteals would add new and precise observations to the excellent foundations laid by Aselli.[46]

During the first months of 1640 Thomas Bartholin was approached by Leiden booksellers who wanted to issue a new "well-printed" and illustrated edition of his father Caspar Bartholin's famous anatomical textbook, first published in Wittenberg in 1611. The Leiden professor and mentor of Thomas Bartholin, Johannes Walaeus, was closely involved in this project from the start, recommending appropriate engravings for the volume and asking Thomas to add "a few words here and there" in the book about the new medical discoveries made over the last thirty years, especially those of Aselli and Harvey. This plan quickly changed, however, not least because Waleaus, after initially rejecting Harvey's circulation of the blood, had come to accept it by late March 1640. Together with other students, Thomas Bartholin had been involved in Walaeus's frequent vivisections on animals, which found supporting evidence for Harvey's view. At the beginning of April he reported to Worm that together with Walaeus he was daily engaged in vivisections on different animals, all relating to the circulation of the blood. They had made many discoveries that supported Harvey, and Walaeus had publicly announced his conversion to Harvey's view during the disputation of a thesis. The other medical professors in Leiden were, according to Bartholin, yet uncertain about whom to support.[47] Initially, Walaeus appears to have wanted Bartholin to summarize in writing his results in support of Harvey, but Bartholin managed to convince his mentor that Walaeus himself should provide this summary in a letter to be appended to the new edition of Caspar Bartholin's anatomy.[48]

From Worm's response of April 1640 it is evident that he had maintained his interest in the circulation of the blood, but that he remained unconvinced by Harvey's account. Worm admitted that Harvey's observations were applauded by some scholars, including Descartes. If Worm had not already seen a copy of Descartes's *Discourse on Method*, published three years earlier, he was at least aware that it contained a section on the heart and the motion of the blood, even if the differences between Descartes and Harvey had escaped him. But, as he emphasized to his young nephew, there were many other scholars who rejected Harvey's observations. He referred specifically to James Primrose's recently published tract against Harvey, *Observationes in Johannis Wallaei Medicinae apud Leydenses Professoris Disputationem Medicam Quam pro Circulatione Sanguinis Harveana Proposuit* (Amsterdam, 1640), asking Bartholin whether he had seen this publication, which he had managed to have forwarded via his contacts in England. Worm emphasized that he was drawing Primrose to Bartholin's attention, not to cause him to reject Harvey's observations, but only to make him aware of what Primrose had found deficient. Worm wrote to Bartholin again in May, telling his young nephew that he would be very pleased if he would send him what Walaeus had published on the circulation of the blood. He added that he still found it impossible to believe in the circulation of the blood and that his doubts long predated his reading of Primrose. He also complained that he had been unable to obtain a copy of the Venetian physician Emilio Parigiano's refutation of Harvey, first published in 1635, but Worm took the opportunity to send Bartholin a copy of his 1632 medical thesis, in which he had raised his objections to Harvey.[49] When Bartholin responded in June on his way to Padua, he enclosed a copy of Walaeus's thesis supporting Harvey and thanked his uncle for having sent him his thesis with its arguments against Harvey. On the request of Walaeus, he had presented him with Worm's thesis. Bartholin added, however, that he himself had already raised objections identical to Worm's some months back when he was directly involved in Walaeus's vivisections. Walaeus had dealt convincingly with them then but had added new "assumptions" of greater weight in his publication that had to be considered.[50]

A year and a half elapsed before Worm and Bartholin returned to the question of the circulation of the blood, when it was occasioned by Bartholin's receiving a visit from the Padua professor of medical theory Fortunio Liceti, who had just finished a manuscript on the circulation of the blood. He rejected Harvey's account and introduced his own double-circulation, apparently designed to make it possible to integrate the idea of circulation into an Aristotelian position. The fact that Bartholin had published Walaeus's letters appended to his father's revised anatomy book two years earlier obviously made him an important person in the international republic of letters and, for Liceti, worth consulting, despite still being a student. Bartholin informed Worm that he had raised some objections to Liceti's account based on common sense, as he put it, and wondered if Worm had anything to add.[51] Worm responded by expressing his delight that "the learned Liceti, with all his excellent acuteness, had examined the yet-undecided controversy about the circulation of the blood." He then went on to express the hope that Liceti would be able to settle the matter and bring an end to the controversy. Worm, however, declared himself as yet unable to reach a decision, not least since the objections he had raised in his public thesis of 1632 had not been answered by either Walaeus or Harvey. From what Worm could gather, Liceti's concept did not raise similar problems, adding, however, that Liceti's "sound reflections" would be much more convincing if supported by precise observations.[52]

This was evidently not the answer Bartholin had hoped for. His letter of 3 August 1643 to Worm expresses considerable exasperation with his uncle and mentor, bluntly stating that he was unable to see how Liceti could defend his account of the circulation because he had no experimental basis for it but depended solely on speculation. Only through vivisection of animals could this matter be resolved, according to Bartholin.[53]

Worm would certainly have agreed with Bartholin about the need for observation and experiment in order to settle this major controversy. However, it seems to have been a book, *Tractatus de Sanguinis Generatione & Motu Naturali,* published in 1643 by one of Harvey's earliest supporters, the Helmstedt professor of medicine Herman Conring, that finally caused Worm to accept Harvey's account of the circulation of the blood. Having just read this book in December 1643, Worm warmly recommended it to Bartholin, pointing out that he would find in it "many surprising things which go against the common assumptions of the anatomists."[54] A couple of years later, Worm had occasion to write to Conring, when Conring's Danish student Frederik Arsinæus, whose father had been one of Worm's close friends, visited him in Copenhagen. Worm informed Conring that he had read his learned work on the circulation of the blood, adding that he had found the book most engaging, "not so much because of the novelty of the content—for this has already preoccupied me in different ways—but because the truth of this new discovery has been confirmed with solid proofs."[55]

Thus, fifteen years after Worm had first become aware of Harvey's account of the circulation of the blood, one can truly say the Worm had turned. That it happened to be the work of Hermann Conring, one of the earliest converts to Harvey's ideas and who had featured so prominently in Worm's earliest inquiries about the circulation, who eventually convinced Worm adds an ironic twist to this story. Unfortunately, Worm does not tell us which were the "proofs" that finally convinced him.

It is noteworthy that Worm seems to have been aware of Harvey's work before it had come to the attention of the medical professors in Leiden. Without his extensive *peregrinatio academica,* which took him to England, it is unlikely that Worm would have known about Harvey so early. News of Harvey may well have come from the same sources in England who provided Worm with the work of James Primrose. Likewise, it was Worm's Leiden contacts, some of whom went back to his student days, who made it possible for Worm to ensure that his own students were well received at that university.

However, more important for the outcome of this story was the role of Worm's own students, several of whom were also his relatives, studying abroad in the 1630s and 1640s and serving as his eyes and ears. It was primarily through them and their letters that Worm was able to keep abreast of new discoveries and new knowledge. Worm inculcated in them the need for anatomical observation and an experimental approach to medicine and natural philosophy. Their letters to their mentor, along with Worm's extensive contacts within the European republic of letters, not to mention booksellers and printers, made it possible for Worm to acquire new knowledge rapidly. As we have seen, he heard about Aselli's discoveries of the lacteals more or less simultaneously with Harvey's discoveries. That he

CHAPTER TWELVE

Stone Gods and Counter-Reformation Knowledges

CARINA L. JOHNSON

In early modern Europe, two forms of knowledge—ideational and material—intersected within the collection. Nascent institutions (museums) and disciplines developed through this rich interplay of concepts and objects, through this dynamic making of knowledge. The collection's scope was both the man-made and the natural: collectors strove to represent authority over cultures throughout the world and to organize nature itself in their collections.[1] Scholars have proposed that the collection formed outside or even as an antidote to the religious tensions and confessional divides of the later sixteenth century. They have argued, for example, that the Holy Roman Emperor Rudolf II amassed his collections in a search for universal knowledge that could bridge confessional fissures.[2] Such studies, however, have underestimated the pervasiveness of religious division in sixteenth-century cultural production. The pressures of Reformation era doctrinal debates drove theologians and other thinkers to define religious orthodoxy more narrowly and to critique some practices and beliefs as false. For both reformed and Catholic thinkers, "idolatry" became the very concept or metonym that defined false belief. Through the rhetorical and material instantiation of idolatry in collections at the end of the sixteenth century, Europeans expressed their shifting understanding of the world and its religious and cultural boundaries. New ideas about idolatry and the display of idols developed in elite court circles, notably the Catholic courts of Philip II of Spain, the ducal Bavarian Wittelsbachs, and Emperor Rudolf II. By reaffirming the presence of idolatry in cultures beyond Europe, Counter-Reformation thinkers could redefine their own culture against those demonically inspired societies.

Within this shift to a new understanding of idolatry and culture, the position confirmed by the Council of Trent was a crucial conceptual turning point. In the last few days of the council, 3 and 4 December 1563, the council passed the decree "On the Invocation, Veneration, and Relics of Saints, and on Sacred Images." In the text, the council affirmed the importance and value of invoking saints and of venerating and honoring images and relics. It thus rejected the idea that such acts were idolatry (*idolatria*) and emphasized that veneration, as practiced in the true church, was clearly distinct from wrongful superstitious acts. Idolatry was a superstition that located the divine within images themselves; such inappropriate practices of veneration were committed "of old by the Gentiles, who placed their hope in idols." On the other extreme, people who believed that either prayer directed to saints or the veneration of relics and images was in vain should be utterly condemned.[3] The last warning was, of course, directed at Protestants, who rejected such practice as erroneous and criticized Catholics as mired in superstition.

Protestant critiques of devotional images had circulated since the initial days of the Reformation. The early reformers had resuscitated charges of idolatry against the cult of the saints, the veneration of images, and other superstitious ceremonies, charges that John Wyclif and Jan Hus had leveled in the late fourteenth and early fifteenth centuries.[4] The persuasiveness of these arguments was crucial to the early success of the Reformation: during the 1520s, outbreaks of iconoclasm focused popular calls for religious reform, forcing magistrates to allow moderate reform in order to quell threats of further destruction and unrest. Protestant attacks on the idolatrous practices and beliefs of the unreformed church reverberated for over forty years, with a substantial midcentury rearticulation by John Calvin. For Calvin, even the mere physical presence of relics, images, and other idolatrous objects was actively contaminating.[5] It was not only dangerous to look at them; it was dangerous to be near them. In 1561–62, new waves of iconoclasm in France demonstrated the ongoing divisiveness of the issue. These acts of material violence led the French delegation to press the assembled Tridentine council for a doctrinal resolution on the status of images and relics.[6] By reasserting the value of images and the cult of the saints in 1563, the Catholic Church affirmed a key doctrinal distinction between itself and other Christians that would persist. Despite further outbursts of intense iconoclasm linked to Calvinist political rebellions in the Netherlands, the Catholic Church stood resolved that their treatment of images constituted true religious practice.[7]

After the conclusion of Trent, the Counter-Reformation church turned to promulgating its doctrinal decisions. Important Catholic theologians positioned their defenses of devotional images against vociferous Protestant charges of idolatry. A particular target was the work of the Magdeburg Centuriati, a group of Lutheran Protestant thinkers who (beginning in 1559) had composed a chronological exposé of the Antichrist's activities and the general heresies accreting to the church through the centuries. Nicholas Harpsfield, publishing in 1566 under the name Alan Copus, formulated an initial post-Tridentine rebuttal of the Magdeburg group in his *Dialogi Sex contra Summi Pontificatus, Monasticae Vitae, Sanctorum, Sacrarum Imaginum Oppugnatores, et Pseudomartyres.* Harpsfield drew upon John of Damascus and the second Council of Nicaea to buttress Tridentine positions on devotional images and the cult of saints. He showed little interest in cultures beyond Europe, except as examples illustrating the active presence of saints or the church's missionary vigor.[8] Although Harpsfield's dense treatise shored up Catholic positions in the short term, the incisive writings of Jesuit theologian Robert Bellarmine soon replaced it as the favored reference for Counter-Reformation writers.[9]

When Bellarmine was called from Louvain to Rome in 1576 to fill a chair in controversial theology, he introduced the plan for his major counterattack, *De Controversiis Fidei Christianae,* against Protestant arguments. Like Harpsfield, Bellarmine looked back across historical time to dismiss attacks against the church. During the past six hundred years, he argued, the devil had begun inciting people to attack inadequately defended articles of faith. In the present day, heretical Protestants continued their demonically inspired attacks on papal authority, grace and free will, and the veneration and invocation of saints, relics, and sacred images.[10] These religious controversies demanded urgent resolution and would become the focus of the *De Controversiis Fidei Christianae,* published in three parts in 1586, 1588, and 1593. The 1588 installment addressed the cult of saints, relics, and the proper use of devotional images.[11] For Bellarmine, the principal troublemaking heretics were no longer the Magdeburg Lutherans but instead Calvin and his followers,[12] whose critique of images had inspired iconoclasm in France and the Low Countries during recent decades. In his rebuttal of Calvin's complete repudiation of images, Bellarmine cited biblical and patristic sources voluminously to defend the Tridentine decree.[13] Although non-Christian (*"ethnici"*) idolaters appeared rarely in his text, Bellarmine employed them to define idolaters as polytheistic in contradistinction to Catholic monotheists. One sign of the orthodoxy of Catholic images, then, was their representation of one god in contrast to idolatrous images, which represented many gods.[14] This tactic of turning from comparisons between Catholics and Protestants to comparisons between Catholics and non-Christian idolaters also appeared in Peter Canisius's early catechisms. Beginning in the 1550s, the Jesuit Canisius had illustrated true religion's veneration of saints and use of images through the counterexample of heathens, who adored false gods. His widely reprinted catechism, like Bellarmine's theological arguments, left the identification of heathens unfixed in space and time.[15]

The debate on idolatry also played itself out in the material world, where Protestant iconomachy wreaked lasting changes. After reforming iconoclasts stripped bare the churches of western and central Europe, those sacred spaces often remained unadorned. In the arena of post-Tridentine Catholic and Protestant polemics, charges of idolatry against Catholicism were not limited to intangible theology or tangible church decoration. They extended to other, more profane objects. For Protestants, Greek and Roman sculptures as well as devotional images were dangerous, since a beautiful statue might well promote idolatry.[16] Catholic responses explained to the pious Catholic worshiper and the artist how they could view sacred and profane images virtuously.[17] Any display of religious or non-Christian figurative art had to grapple with these debates about images and answer the questions: What constituted an appropriate image? Which images were idols?

As new showcases for art and other objects, princely collections or *Kunstkammern* began appearing in the last third of the sixteenth century. When Samuel Quiccheberg produced the first programmatic description of an ideal collection (*Kunstkammer*) in the Holy Roman Empire in 1565, he sidestepped debates about idolatry and confessional issues more generally. Quiccheberg, the court historiographer for the Bavarian Wittelsbachs, promoted a universal organizing structure for the collection.[18] The ideal collection should represent the world in an encyclopedic manner, to demonstrate princely mastery of that world. Emphasizing both textual and material forms of knowledge, Quiccheberg's ideal collection or "universal theater" was to be divided into a library, workshops, and material sections. The theater began with a universal library containing the subjects of theology, law, medicine, history, philosophy (including magic), mathematics (including astrology), philology (including military matters, architecture, and agriculture), sacred and profane poetry, music, and grammar.[19] The ideal theater also included workshops for crafts, a pharmacy, and a smithy, along with an assembled collection of unusually fine examples of these arts. The material collection (*artificialia* and *naturalia*) itself should include images and sculptures depicting sacred and profane subjects. Quiccheberg explicitly included ancient and new statues of gods (*numina*), whether made of stone, wood, clay, or bronze, as a section in his collection. Quiccheberg also incorporated the study of "foreign" or "exotic" peoples ("*exterarum nationum*," "*gentium peregrinarum*") into his encyclopedic plan. He encouraged the collection of information about foreign peoples' mores, culture, and products—both military (arms) and domestic (clothing)—so that the collection could serve both symbolic and practical concerns.[20] Quiccheberg innovated by separating "foreign peoples" from European Christians, even as he recodified the ethnographic or descriptive categories found in travel narratives dating back to thirteenth-century Franciscan descriptions of Mongols. Although Quiccheberg's organizational structures revealed heightened interest in extra-Europeans, they avoided the dangerous issue of non-Christian religious objects, showing no concern or space for these peoples' religious paraphernalia or worship. Within a generation, however, the conceptual framework of the collection became confessionalized.

In 1587, Gabriel Kaltemarckt proposed a Protestant Lutheran answer to the question "What constituted an appropriate image?" He argued that the Saxon Wettin collection should follow a Protestant aesthetic and organizational scheme. Pre-Reformation or Catholic art was idolatrous, and Kaltemarckt placed it alongside other heathen art:

> [J]ust as no people under the sun (except for the orthodox Christian church) has a thorough understanding of God, so too the right use of sculptures and paintings is confined to the true religion. This requires no proof, since it is obvious that the ancient and highly intelligent heathens, // the Greeks and the Romans etc. . . . , the papists . . . , the Turks and the present zealots and iconoclasts all possess neither true religion nor the right use of the visual arts. The heathens and papists misuse the visual arts for the sake of idolatry while the Turks and the zealots have not the least liking for them.
>
> [Das / gleich wie kein Volck under der Sonnen (aus- / serhalb der rechten christlichen Kirchen.) nichts / gründlichs von Gott weis, also ist auch der / rechte gebrauch der Bilder und Gemehle, al- / lein bei rechter Religion zufinden, Und das / darff keines beweises. sintemal offenbar, / das weder die alten hochverstendigen Heiden // Griechen, Römer etc. . . . noch die papisten . . . noch die Türcken und izigen Schwermer / und Bildstürmer, weder ware Religion, / noch rechten gebrauch der Bilder haben. Sintemal / die Heiden und Papisten die Bilder zur Abgötterey / misbrauchen, die Türcken und Schwermer aber, / durchaus keine Bilder leiden noch dulden können.][21]

For Kaltemarckt, the charge of idolatry acted as a tool of exclusion against other Christians as well as non-Christians. He also underscored the deep divisions between reformed parties, repudiating "zealot" Protestants by equating them with Muslims. Only Kaltemarckt and fellow followers of the Augsburg Confession practiced true Christian religion and the correct use of art. Reflecting Kaltemarckt's belief that representations of images or idols were very dangerous, the actual Wettin *Kunstkammer* (inventoried in 1587) focused heavily on artisanal production and scientific instruments.[22] A comparison between the Protestant Wettin collection and contemporaneous Catholic collections of Habsburg and Wittelsbach princes reveals that confessionally specific doctrines on images and idolatry shaped the content and organization of Catholic collections as well.

The most important Catholic princely collections at the end of the sixteenth century followed Quiccheberg's 1565 organizational scheme with fidelity. Their contents diverged from it in one important respect, through the presence of named idols. The collections of Philip II of Spain (inventoried in 1598), the Bavarian Wittelsbachs (inventoried in 1598), and Emperor Rudolf II (inventoried in 1607) all incorporated objects that were marked, displayed, and categorized emphatically as idols rather than images of deities. Such a demarcation not only departed from the Quiccheberg program; it also rejected pre-Tridentine descriptions of some objects. Some of this sculpture and metalwork had previously been housed in Charles V's and Ferdinand I's treasuries, where it had been described as bearing representations of men or animals. Similarly, the midcentury collections of Ferdinand's sons, Maximilian II and Archduke Ferdinand, had not identified extra-European material objects as idolatrous or even as containing depictions of gods. Inventories of those midcentury collections adhered to Quiccheberg's categories, particularly those for extra-European material objects, and ignored religious debates. Indeed, the Protestant-sympathizer Maximilian actively promoted secular themes in all his cultural patronage, trying to avoid reminders of religious division in the Holy Roman Empire.[23] If the princely *Kunstkammer* was intended to conceptualize the world and its history, we must explain the redefinition of objects as idols in Catholic collections at the end of the century. Given Protestant critiques, why not avoid collecting idols entirely, as the collections of the preceding Habsburg generation and of Protestant princes had? Why the Catholic interest and innovation in idols?

By exhibiting idols in collections, Catholics could displace Protestant charges of idolatry onto what Catholics increasingly argued was their proper focus, idolatrous images produced by non-Christian cultures.

This strategy had been employed previously, in the thirteenth century, when Latin Christian authors and artists represented ancient Greeks and Romans, Jews, and Muslims as idolaters.[24] Influential post-Tridentine Catholic authors Bellarmine (in his theology) and Canisius (in his catechism) did not quite take the step of locating the true idolater, but other Counter-Reformation thinkers did so in texts and in collections.

One crucial strand of Counter-Reformation discourse on idolatrous cultures was developed in Iberia and Spanish America. During the first half of the sixteenth century, Bartolomé de Las Casas had explored the relationship between culture and idolatrous religious practice in his writings on the Americas. The evolution of his thoughts concluded with the expansive *Apologética Historia,* completed in 1559. Extending earlier Renaissance ideas about pagan antiquity, Las Casas defended Mexican and Inca culture by equating them with the cultures of classical antiquity. The societies of Greece and Rome were directly comparable to those of Mexico and the Andes, not just in terms of religion and worship but also, and more importantly for Las Casas, in terms of just government, urban development, and other markers of civilization outlined by Aristotle in the *Politics.* By discussing these cultures together, Las Casas tapped into preexisting validations of Greek, Roman, and Egyptian culture and religion. Those arguments justified Christian humanist appreciation of classical texts and ancient Mediterranean cultures; they defended the religious practices of Greeks, Romans, and Egyptians as virtuous. Although Las Casas refrained from making explicit connections with Neoplatonic occult projects of knowledge, he did employ the concept of natural religion. For Las Casas, these pagan religions reflected men's desires for the true religion of Christianity. Because Christ had not yet appeared on earth, these religions were necessarily partial in their understandings of God and thus worshiped him incorrectly.[25] Thus, preconquest virtuous idolatry could easily be replaced with Christianity; "Indian" religion and cultures should be studied in order to facilitate this conversion. Early Franciscan missionaries had embraced the project with much enthusiasm, learning languages and recording the history and culture of the peoples of New Spain.

As Las Casas concluded his thinking on the virtuousness of Indian religion, other texts began to note the presence of ongoing idolatry in the Americas. Even the Franciscans, who had initially heralded the missionary field of the Americas as an opportunity to escape the Protestant heresies of Europe and create the Millennial Kingdom, declared by the 1560s that idolatry persisted in their New World. After the closing of Trent in 1563, Philip II began to show active concern about the practice of true religion in his empire, both in Iberia and in the Americas. In Iberia, he promoted the cult of saints and supported Tridentine reforms. In the Americas, after roughly two decades of following Las Casas's approach to Indian knowledges, Philip II abruptly changed course and took strong measures to curtail all inquiries into Indian religion and culture. In 1577, Philip ordered the seizure of all known texts on the subject, including the monumental works of the Franciscan Bernardino de Sahagún. Philip's ban on knowledge about the Indies explicitly mentioned the "interpretation of idols" as a problematic activity that must be halted through the suppression of information.[26]

By 1598, Philip's collection generally sustained his own ban on Indian religion, culture, and knowledge. Instead of artifacts made in the Indies, he collected objects constructed from "India" materials (balsam, coco, various woods, elephant skins) but made in European forms.[27] The striking exception to this tendency was Philip's Inca treasure, which included "idols." Why were there Inca idols in Philip's collection? The answer lay in the provenance of the Inca treasure. In 1598, material objects from the original 1532 conquest were no longer of interest to Philip. What he treasured, instead, were newer pieces sent to him by Francisco de Toledo, his viceroy in Peru from 1569 to 1581. Toledo had sent twelve "little idols" found in Inca houses, a silver idol of a *xeme,* and symbols of Inca governance.[28] During Toledo's tenure, he had enacted sweeping organizational reforms, reinvigorated the search for idolatry among the Andean Indians, and finally captured the last outpost of non-Christian Inca rule, along with the "rebel" Inca (Tupac Amaru) himself. The "Inca treasures" sent by Toledo, then, symbolized recent victories for Christian faith and Spanish political and religious domination in Peru. They also reinforced notions that Inca subjects were currently and perhaps irrepressibly idolatrous.

By the late sixteenth century, the writings of Jesuit José de Acosta promoted the idea that idols and idolatry dominated all Indian cultures and religious practices. Acosta began his career in the Andes as a counselor to Viceroy Toledo. Upon returning to Iberia in 1587, Acosta developed close connections to Philip II, acting as Philip's representative in negotiations with the papacy. (Acosta's contemporaries found him more loyal to Philip's agendas than those of either the papacy or the Jesuit order.) Acosta's popular texts *De Procuranda Indorum Salute* (completed in 1576 and published in 1588) and *Historia natural y moral de las Indias* (completed in 1588 and published in 1590) were approved for publication by the Crown and likely read and appreciated by Philip II. In the texts, Acosta argued that, in the evaluation of any people, the question of idolatry should be first in importance and influence.[29]

Acosta dismissed Las Casas's positive cultural evaluations of people in the Americas, retaining only Las Casas's strategy of comparing cultural traits found commonly in non-Christian cultures of the present day and antiquity. While he was careful to note that non-Christian cultures were not simply all the same and indistinguishable, Acosta argued for the overarching importance of their commonalities. Societies with a certain level of government organization and religion shared certain cultural practices such as maintaining ceremonies and superstitions and dressing in similar "simple" forms of clothing.[30] Well-developed idolatry, with priests and temples, was usually found in the most organized and "civilized" of these non-Christian cultures.[31] In these assertions, we can see Acosta's model of cultural hierarchy being constructed. Religious error was no longer interpreted as a sign of a society's striving toward God, as Las Casas had understood it. For Acosta and other Tridentine era Catholic thinkers, error was negative, and there was too much of it in the Americas. Even so, concerns about idolatry in the Americas were never isolated from concerns about Protestant heresy: other discussions of idolatry in the New World, like Valentinus Fricius's *Yndianischer Religionstandt der gantzen newen Welt,* were consciously presented in the shadow of European confessional strife.[32]

Acosta's understanding of the forms of idolatry changed from 1576, when he wrote *De Procuranda,* to 1588, when he wrote the *Historia natural y moral.* In *De Procuranda,* the three forms of idolatry were those laid out by John of Damascus and the Book of Wisdom. Each wrongful object of worship had once or still physically existed. The three forms of idolatry could be classified with examples from antiquity: Chaldeans worshiped celestial bodies, signs, and elements; Greeks worshiped the dead as gods; and Egyptians worshiped animals and inanimate things like rocks.[33] More singularly, Acosta argued that idolatry was a disease, inherited through mother's milk and nurtured by culture.[34] Because idolatry was heritable, he continued, converts in the New World could never become full members of the Christian church and must remain neophytes barred from the priesthood.[35] Seven years later, the Third Church Council of Lima (1583), with Acosta as adviser, would embrace this argument and declare that Indians should be assigned permanent neophyte status. By 1588, when Acosta set down the *Historia natural y moral,* the devil rather than culture became the nurturer of idolatry. The three previously cataloged forms of idolatry—worship of nature, of the dead, and of animals—were all found in the Americas, but these three categories were no longer sufficient to describe idolatry there. Acosta introduced a fourth, distinctly Tridentine, form of idolatry: the worship of objects made by human hands. Such pieces of human invention included "statues of wood, stone, or gold, like those of Mercury or Pallas . . . which represent nothing" (estatuas de palo, o de piedra o de oro, como de Mercurio o Palas, que . . . ni es nada, ni fué nada).[36] Although Acosta failed to make any further direct comment on this fourth form of idolatry, it refocused attention on the problem of graven images. Further, by arguing that Indian religious errors lay solely in the imagination and not in mistaking natural signs or past heroes, Acosta magnified the role of the devil and the danger of idolatrous error. Acosta explicated, at some length, the devil's active role in driving Indians to worship falsely, concluding with extensive advice on how to combat such demonic influences.

In the same year that the *Historia natural y moral* was published, Bellarmine also highlighted the devil's role in guiding Jews and heathens to make objects and worship them as idols.[37] Although Bellarmine did not offer a hierarchy of non-Christian cultures in his *De Controversiis,* he shared with Acosta the tendency to conflate such cultures. In Counter-Reformation discourse more generally, succumbing to the devil's temptation was interpreted as a sign of cultural weakness common to non-Christians. Increasingly, Jews, Indians, Egyptians, Phoenicians, Romans, Greeks, and Tartars were all believed to share customs and idolatrous practices. For example, by 1615, Vincenzo Cartari's manual of the deities of Greek, Roman, and Egyptian antiquity was revised to integrate the gods of Mexico and Japan into the text.[38] In such texts, idolatrous religion was located outside the boundaries of sixteenth-century Europe by either time or space. The idea that sincere faith could be found in antiquity[39] and in the far-flung corners of the contemporary world had been rejected. Instead, the idea of progress from an undifferentiated idolatrous state to one of true religion and proper veneration of images was found throughout Counter-Reformation polemics. (With their fear of images, Protestant and Muslim "heretics" had fallen off the path of progress.) The material presence of idols in Counter-Reformation collections, then, must also be read through this shift to cultural hierarchy and conflation. True idols signaled the distance between Catholic European religion and that of demonically inspired cultures.

Unlike Philip II, the Wittelsbachs and Rudolf II held no direct sovereignty over peoples in the world outside Europe. Lacking an empire or idolatrous subjects in the Americas, these rulers struggled instead with unruly, heretical Protestant subjects in central Europe. Their collection inventories more closely followed Quiccheberg's model of a microcosmic world over which a prince demonstrated his symbolic mastery, yet these collections encompassed not only arms and clothing but also devotional objects and idols from the Americas and elsewhere. The Wittelsbach dukes Albrecht V and Wilhelm V embarked on an ambitious program of art patronage and Catholic religious revival in their territories during the last decades of the sixteenth century. They were extremely committed to the project of the Counter-Reformation; the expansion of their own authority was invested in their promotion of Catholicism in Bavaria. They supported religious instruction, and their symbolic efforts focused on monumental building projects (particularly major churches) and a *Kunstkammer* designed for relatively broad viewing.[40]

A portion of that Bavarian *Kunstkammer* or collection was inventoried in 1598 by court historiographer Johann Baptista Fickler.[41] In the inventory, Fickler sorted objects into categories of materials and forms of daily practice. He placed certain kinds of non-Christian religious objects and equipment together: Jewish instruments of circumcision and worship, for example, were placed next to both East and West Indian religious objects.[42] Several Indian priests' masks and head coverings, a Muslim "tract" (probably the Koran), and idols "that the unbelievers pray to and revere . . . [that] looked more like a devil than a man" (daselbst von denn unglaubigen angebettet und geehret worden . . . sollich bildt sihet mehr ainem Teüfl als Menschen gleich) were all present, displaying their devilish aspects.[43] The presence of these idolatrous objects from false religions was not accidental: the Wittelsbachs actively sought such pieces for their collection.[44] Viewers of the collection could see that the non-Christian cultures represented on the display tables included those from the past (Greek, Egyptian, Roman) and present (Turkish and Indian). Viewers could also look up at the walls of the collection to see images of famous or virtuous men and women.

In the Wittelsbach collection, the repudiation of idolatrous relics had a counterpoint in the renewed veneration of true relics. The Wittelsbachs energetically collected relics and reliquaries from throughout the Protestant areas of the Holy Roman Empire, establishing Bavaria as a haven for Catholicism in contrast to the crypto-Protestantism and open Protestantism found in the emperor's hereditary lands. The Wittelsbachs even made a concerted effort to gain custodianship of the imperial relics, which were displayed at coronations. They argued that the relics were neglected in Protestant Nuremberg and should be guarded by true Christians rather than the heretical Protestants.[45] Christian relics were not placed in the same part of the collection as the non-Christian religious objects: the two categories were physically segregated from one another. The collection overall, then, represented idolatry and pagan religion as distinct from the Catholic faith's relics and cult of saints.[46] Through the separate locations of true and false religious objects, the viewer would easily absorb the Counter-Reformation message that Catholic and non-Christian images and art were categorically different. The Wittelsbachs, anxious to demonstrate their leading role in the religious revival of the Counter-Reformation, collected idols as examples of the devil's work, as part of a project to symbolically represent and advance the church's triumph over the devil.

Although we cannot know how much of the collection's organizational scheme was originally Fickler's (he worked with the collection after 1592, but no preceding inventory exists), other evidence suggests that Fickler was keenly alert to the religious dangers of possessing idols. Fickler's published writings reveal a deep engagement with the issues of images, relics, and the veneration of saints. A lay legal scholar, Fickler devoted his life and work to the Counter-Reformation cause: responding to the insult that he was only a "papist jurist," Fickler replied that he was continually thankful that he was one, since this allowed him to work for the good of the church.[47] Fickler began his service to the Counter-Reformation church as a member of the archbishop of Salzburg's delegation at the third session of the Council of Trent, the session that produced the decree on saints and images. After studying law in Italy and solidifying friendships with Counter-Reformation figures active in the Holy Roman Empire,[48] Fickler established his position as a legal scholar with *Theologia Juridica, Seu Ius Civile Theologicum . . .*, that called straying Christians back to the true Catholic Church.[49] He subsequently published aggressive vernacular rebuttals of Protestant doctrinal attacks on the veneration of the Virgin Mary and the saints, and on the Catholic use of images. Fickler's determination to persuade his opponents, or at least a reading public, was dogged: the printed debates with Protestant authors ran through multiple rejoinders on both sides.[50] After moving from the ecclesiastical Salzburg court to the ducal Bavarian court, Fickler continued the parsing of true religion through cultural comparison in his polemical texts, clarifying the differences between Jewish and Catholic uses of images and between the pagan worship of goddesses and the Catholic veneration of the Virgin Mary. By the 1590s, Fickler's rebuttals contained conflated rejections of the "heathen, Jewish, Muslim, and other unbelievers, as well as the Protestants" (Heyden / Juden / Türcken / oder andern Unglaubigen / als bey den Evangelischen). Their religious texts (Talmud, Koran, vernacular Bibles) were all equivalent, and all were the tools of the devil.[51] Fickler's pamphlet rhetoric paralleled his inventory of the Wittelsbach collection, in which idolatrous objects were interchangeable but also infused with the demonic.

The inventory of Rudolf II's collection, dating from 1607, reveals a similar interest in collected but conflated idolatrous religious paraphernalia as distinct from Catholic objects. This inventory separated Catholic religious images and the rest of the collection both conceptually and materially by excluding Catholic religious images from the inventory altogether. Rudolf's relationship to the Counter-Reformation was more vexed than that of the Wittelsbachs. When he succeeded his Protestant-sympathizer father as Holy Roman Emperor in 1576, Rudolf's religious patronage extended to both Protestants and Catholics. Yet in 1599, Rudolf allowed a consequential Counter-Reformation presence to establish itself at his court in Prague.[52] Rudolf's cultural patronage and his collection initially had focused on the search for overarching cosmological order or occult knowledge, avoiding doctrinal debates between the confessions. After 1599, Rudolf did not abandon his occult investigations, but his support of Counter-Reformation thinkers grew stronger during the remainder of his rule.

In Rudolf's collection inventories, Quiccheberg's categories can be found as well as non-Christian religious objects and idols. Such images of false gods were almost invariably identified as "Egyptian." While this description might represent their cultural origin, in other places in the inventory it seems to signal their idolatrous nature or the conflation of pagan cultures: "Egyptian idols" were scattered in among Indian religious objects.[53] Such an attribution connected idols and idolatrous beliefs to Egypt's supposed role in bringing idolatry to Greece and Rome, and thus to all the ancient Mediterranean and Europe (along with writing, art, and, according to some, the calendar).[54] It also evoked a popular late medieval biblical legend symbolizing the triumphal arrival of Christianity and the end of superstition. In the apocryphal story "The Fall of the Egyptian Idol," the Holy Family fled to Egypt, where idols cracked and toppled in the presence of the Christ child.[55] Iconographically, the toppling idol often represented general pagan culture confronted with Christianity. Identifying idols and knowledges as "Egyptian" or pagan did not belie multiple meanings: both Counter-Reformation triumphal church and Neoplatonic occult significations could be contained in such objects and could be of interest to Rudolf. By 1607, Rudolf was sure that he was damned even as he supported Counter-Reformation polemicists. He refused the Eucharist and increasingly explored pagan and occult knowledges to the neglect of his political rule.[56] Although Rudolf's religious purposes were perhaps opposed to those of the Wittelsbachs, the same stark contrast between idolatrous cultures and the true Catholic Church could be valuable for his own explorations of hidden universal knowledge.

The Counter-Reformation Wittelsbachs and Rudolf II collected contemporary extra-European idols as evidence of the distance between their own Catholic religion and culture and those of non-Christians. Other important Catholic rulers followed a similar trajectory of display, notably the Counter-Reformation popes. Rome in the pre-Reformation era had witnessed Neoplatonic enthusiasm for classical knowledge and art. In the initial days of the Counter-Reformation, however, Pius V (1566–72) limited access to classical sculptures that he feared were dangerously idolatrous.[57] From 1572 to 1590 Gregory XIII and Sixtus V physically transformed Counter-Reformation Rome into a symbol of triumphal Catholic vigor. This new Rome touted its Christian disjuncture with antiquity through preaching and extolled its Catholic victory over the "Empire of Satan" that had been idolatrous pagan religion and culture. In the material world, these popes also erected obelisks from antiquity as monuments to that triumph, first carefully exorcising them to expel all impiety and idolatry.[58] The important distinction to be made, heard, and seen was between the Counter-Reformation church and the secret knowledges and demonic idols of pagans.

By 1600, Catholic authorities had reached a new consensus: Protestant criticisms of images could be refuted. Pagan idols no longer placed the Catholic viewer at risk of misunderstanding the nature of images. Instead, demonic idols illustrated the distance between non-Christians and Catholics. The 1563 Tridentine decree encouraged Catholics to do more than flatly deny Protestant accusations of idolatry or produce Counter-Reformation art against Protestant iconoclasm. Counter-Reformation thinkers like Acosta and Fickler displaced Protestant charges by redefining idolatrous cultures as contemporaneous, inferior, and safely located beyond Europe. In collections, Catholic Europeans could look upon and ponder the false paths of demonically inspired superstitious faith and worship, as well as the folly of Protestants.

These Counter-Reformation classifications, developed in both collection and text, would take on new vigor in the seventeenth century alongside European paradigms of cultural hierarchy. The separation of true and false believers through a well-defined idolatry became a valuable rhetorical tool in the popularization of Tridentine doctrines. As the Counter-Reformation began its successful re-Catholicization of Central Europe, catechisms that promoted the denigration of devil-inspired religions and cultures circulated even more widely.[59] Bellarmine's own popular catechisms (two formats produced in 1597 and 1598 and translated into multiple vernaculars) promoted the distance between true Christians and false sects. In so doing, he conflated "the false religions of pagans, Muslims, Jews, and heretics" (von allen falschen Secten der Heyden/ Türcken/ Juden/ und Ketzer). This catechism also reminded the reader that false gods could appear in natural forms as celestial objects, animals, and dead men but also, more idolatrously, as images made of gold, silver, wood, or stone that were inspired by the "hellish devil."[60] From the Council of Trent's closing to the end of the century, extra-European religions and cultures became woven together both conceptually and materially into a category of falsity and denigrated difference located in the world outside Europe, where the busy devil awaited the missionaries of the newly triumphal Catholic Church.

CHAPTER THIRTEEN

Temple and Tabernacle: The Place of Religion in Early Modern England

JONATHAN SHEEHAN

In antiquity, the sacred grove of Diana Nemorensis was the site of a bloody drama. In this grove, "a small, crater-like hollow on the mountainside," lurked a fearsome figure, the King of the Wood, whose task was to serve as both priest and sacrificial victim to the goddess Diana. The king was a man terrified and terrifying: having killed his predecessor and assumed his crown, he awaited the arrival of his murderous successor, warily keeping his watch day and night. "Surely no crowned head ever lay uneasier, or was visited by more evil dreams, than his," commented Sir James Frazer, who made the sanctuary at Nemi into the mythical center of his enormous *Golden Bough*.[1] The shrine of Diana provided Frazer with a microcosm of the religious legacy of the ancient world: it was the kernel around which he condensed the mythological universe of antiquity. In this distant time, the entire world was, like Diana's grove, a space marked by the sacred, roamed by gods, demons, and their priests. For us moderns, by contrast, the sacred has slowly retreated:

> For ages the army of spirits, once so near, has been receding farther and farther from us, banished by the magic wand of science from hearth and home, from ruined cell and ivied tower . . . from the riven murky cloud that belches forth the lightning. . . . The spirits are gone even from their last stronghold in the sky, whose blue arch no longer passes, except with children, for the screen that hides from mortal eyes the glories of the celestial world.[2]

From a world fully occupied by the spiritual, modern science has pushed the sacred farther and farther away: from homes and hearths, to the arch of heaven, and finally out of our universe altogether. The sacred grove of Nemi has become just another Italian village.

Although, in its elegiac tone, the *Golden Bough* hardly stands as a monument to scientific sobriety, Frazer's hymn to primitive sacrality conceals a principle shared by nearly all anthropologists of religion, namely, that the sacred *needs* a place to inhabit distinct and separate from the space of the profane. True religion exists only in its own place: "for religious man," Mircea Eliade declared in 1957, "space is not homogeneous." Instead, religion divides undifferentiated space, offering humans "an absolute fixed point" that, in Eliade's mind, allows for the birth of a world impossible "in the chaos of the . . . relativity of profane space."[3] For Rudolf Otto, place grounded the numinous experience of religion as terror: "'holy' or 'sacred' places . . . [are] spots of aweful veneration." When Jacob exclaimed, "How dreadful is this place! This is none other than the house of Elohim!" (Gen. 28:17), he articulated the essential connection between the existential dread of religion and the place where it dwells.[4] More prosaically, Victor Turner's insistence that "liminality" is the core of religious ritual embraces the spatial demands of the sacred. The *rite de passage,* in Turner's famous essay, staged the tribal chief's physical and symbolic movement from abject outsider into the center of religious and political power, into what Turner called "the tribal territory itself."[5] For the modern human sciences, religion cannot function without a space for its performance.

This sense of religion's place was first conceived in the early modern period by theologians and scholars of religion fascinated by the nature of true religion, its rituals, and the place of their performance. These analysts were largely Protestant, largely English, and clustered in the seventeenth century, and for them, no places were as symbolically freighted or systematically investigated as the ancient Jewish Tabernacle—the portable sanctuary and heart of itinerant Israel—and the Temple of Solomon. These two places of worship were, after all, the only ones explicitly detailed in scripture and expressly designed by the Supreme Architect himself. The relationship between them and the rites prescribed to the Jews after Moses's descent from Mount Sinai was the subject of obsessive inquiry among early modern English scholars of the Bible. This inquiry was not, it must be said, driven by the claims of sober science. Rather, it was energized by the Christian confessional controversy that so shaped seventeenth-century scholarship. Knowledge and polemics have seldom been closer than here, where Christians began to grapple with the ceremonial nature of their own religious practices. And yet it was this very polemical background that made the investigation of these holy places and the rituals performed in them so enormously compelling and ultimately fruitful. Knowledge about religion in its abstract form—and the development of an idea of sacred space that could apply equally to any religion surely represented a monumental act of abstraction—was not birthed by any serene goddess of wisdom. Nor did the conceptual architecture that later made possible the anthropological investigations of comparative religion leap out of the head of scientific inquiry. Instead, both took their origins from a more visceral place, bursting from the belly of seventeenth-century theological dispute. The discovery that *all* religion needs a place for its performance was a child of religious controversy.

Finally, a relevant caveat: the investigation of religion in the early modern period places a certain burden on those modern readers who prefer their theory undiluted. A philosophically abstract analysis of religion was, for the most part, neither conceivable nor interesting to early modern scholars. Indeed, it was quite dangerous in a confessional context whose poisonous atmosphere made direct investigation of Christianity a frightening endeavor. Instead, this religious analytic was *made known:* less described than unfolded inside and against a host of materials that today we call antiquarian. If scientific knowledge was made in the laboratory and not just in the mind of scientists, religious knowledge was made in the laboratory of antiquarian scholarship. It was here that scholars could and did engage with the space of the sacred: using the raw materials of antiquity, they made new knowledge about the world of religion. And so it is into this laboratory that we must plunge.

Figures of the Temple: Typology and Antiquarianism

Judging by the number of reconstructions, it is clear how strongly the Jewish Temple and Tabernacle attracted the early modern scholarly imagination. In printed Bibles, in geographies and descriptions of the Holy Land, in analyses of Jewish ceremonies: the image of the Temple was ubiquitous. In the earliest days of print, images tended toward the schematic.[6] But by the end of the sixteenth century and especially during the Reformation, the visual field became extremely complex. Already in 1540, the Estienne Bible was decorated with images of both the Temple and Tabernacle designed by the Protestant-leaning Franciscus Vatablus for his commentary on Ezekiel.[7] But the Ezekiel temple found its true visual home in the works of the Jesuit Juan Bautista Villalpando, whose 1596 *In Ezechielem Explanationes* offered a panoply of stunning images.[8] Assuming that the divinity would

Fig. 13.1. Full frontal view of the Temple. From Juan Villalpando and Jerónimo de Prado, *In Ezechielem Explanationes et Apparatus Urbis ac Templi Hierosolymitani Commentariis et Imaginibus Illustratus* (Rome, 1596–1605). Courtesy of the Herzog August Bibliothek, Wolfenbüttel, 17 Theol. 2°.

Fig. 13.2. Eastern face of the sanctuary. From Juan Villalpando and Jerónimo de Prado, *In Ezechielem Explanationes* (Rome, 1596–1605). Courtesy of the Herzog August Bibliothek, Wolfenbüttel, 17 Theol. 2°.

Fig. 13.3. Substructure of the Temple. From Juan Villalpando and Jerónimo de Prado, *In Ezechielem Explanationes* (Rome, 1596–1605). Courtesy of the Herzog August Bibliothek, Wolfenbüttel, 17 Theol. 2°.

hardly design a perfect Temple twice, Villalpando firmly equated the historical Temple described in 2 Chronicles with the one set by the "hand of the Lord" into the mind of his prophet (Ezek. 40–45). Since "the only style capable of perfection was the classical one," Villalpando also amalgamated the temple design of the Roman architect Vitruvius with Ezekiel's descriptions of courts, gates, and vestibules to arrive at a monumental vision of the Temple.[9] The result resembled, perhaps unsurprisingly, a Renaissance palace whose designs were sublimely symmetrical, replete with fluted columns, decorated friezes, and so on. Some forty-five archways marched across the width of the east face, and the Temple walls reached three full stories (figs. 13.1 and 13.2).[10] This was an enormous place. In fact, it was so large that it dwarfed the top of Mount Moriah, and its size thus forced Villalpando to invent a vast substructure that extended the Temple far beyond the mountaintop. The result was a monumental feat of building that ranged—as the Huguenot Hebraist Louis Cappel later calculated with amazement—some "900 feet high" from basement to roof and launched the temple out into an "empty Democritical space" (fig. 13.3).[11]

Villalpando's fantasy of the Temple was unusual not just for its vast proportions but also for its disregard of traditional Christian commentary, which long distinguished between Ezekiel's vision and the actual Temple of Solomon. In the common reading, Ezekiel's description of "the place of my throne and the place of the soles of my feet, where I will dwell in the midst of the people of Israel for ever" (43:7)—as God's emissary called it—portrayed Christianity triumphant, not an antiquated Jewish building. Most other early modern scholars, like Cappel or (more importantly) Benito Arias Montano, the librarian to Philip II, sharply distinguished the actual from the mystical Temple. Unlike ancient Christian tradition, however, they were strictly attentive to the historical, and particularly Jewish, sources for the Temple of Solomon. When Montano reconstructed the Temple and Tabernacle for the 1572 Antwerp Polyglot—using images later reprinted in his 1593 *Antiquitatum Iudaicarum*—he made special reference to the Mishnah and more specifically to Middoth, which described the Temple and its surroundings in some detail.[12] This emphasis produced images radically different from Villalpando's. Gone was the geometrical perfection of Villalpando's ground plan and in its place was a spare building—the porch was lofty but undecorated, the sanctuary was pierced only by small windows, and its walls were left unadorned by classical motifs (fig. 13.4). This modest building was counterpoised, however, by Montano's detailed sketches of the Temple in use (fig. 13.5). On the outside was what he called the *Mons domus*, an area for animals and commerce; then came

Fig. 13.4. Side view and diagram of the sanctuary of the Temple. From Benito Arias Montano, *Antiquitatum Iudaicarum Libri ix* (Leiden, 1593). Courtesy of the Herzog August Bibliothek, Wolfenbüttel, 58.4 Hist.

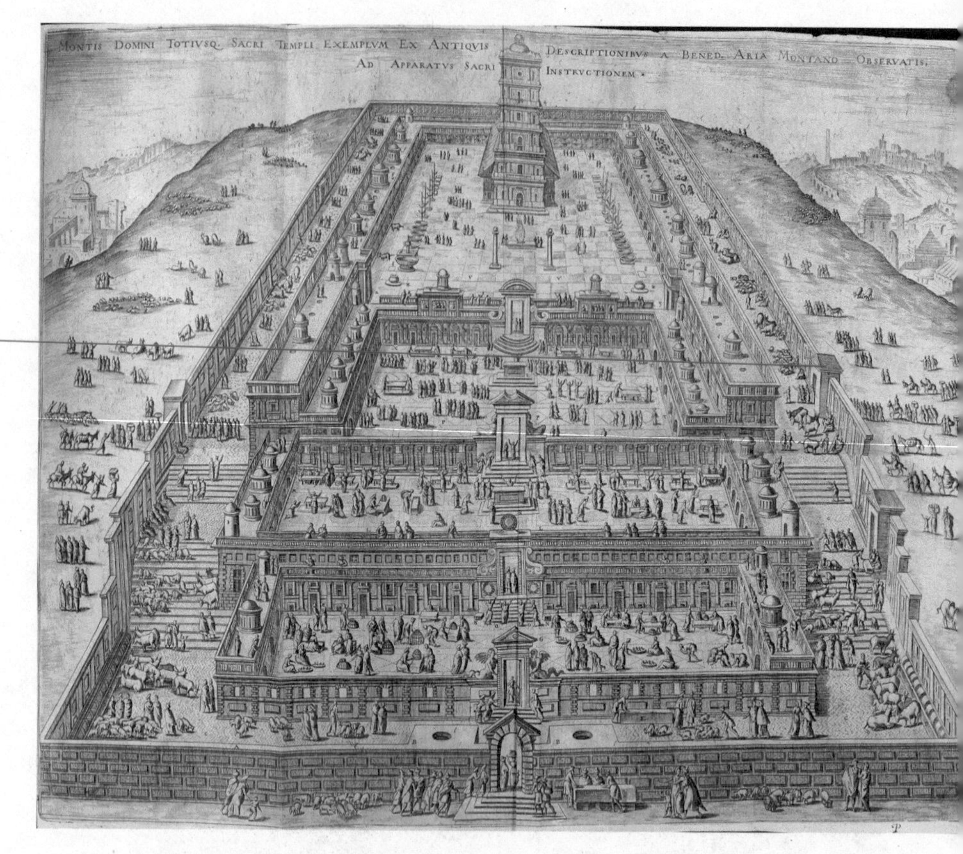

Fig. 13.5. The Temple grounds. From Benito Arias Montano, *Antiquitatum Iudaicarum Libri ix* (Leiden, 1593). Courtesy of the Herzog August Bibliothek, Wolfenbüttel, 58.4 Hist.

the "atrium of the profane," where trading and exchange were permitted; the "atrium of the women," with its "throne of the pontifex"; the "atrium of the Israelites," with the "throne of the king"; the interior section open only to the Levites; and finally the enclosed sanctuaries themselves. And in an even starker contrast with Villalpando, Montano offered his readers exquisitely detailed studies of the ritual objects of Jewish worship. From the garb of the high priest, rendered faithfully with his ephod, breastplate, and pomegranates "round about the skirts of the robe" (Exod. 39.26), to the holocaust altar with its burning sacrifices, the cherubim on the mercy seat, and the seven-branched golden candelabra, the biblical accounts of Temple and Tabernacle were given the most literal visual form (figs. 13.6 and 13.7).

Fig. 13.6. The high priest and his vestments. From Benito Arias Montano, *Antiquitatum Iudaicarum Libri ix* (Leiden, 1593). Courtesy of the Herzog August Bibliothek, Wolfenbüttel, 58.4 Hist.

Fig. 13.7. The Tabernacle, the altar, the candelabra, and the shewbread. From Benito Arias Montano, *Antiquitatum Iudaicarum Libri ix* (Leiden, 1593). Courtesy of the Herzog August Bibliothek, Wolfenbüttel, 58.4 Hist.

Villalpando had little respect for Montano's Temple vision. He thought that Montano's images—"partly because they do not conform precisely to the sacred oracles and, in truth . . . partly also because they are far from the reasons and laws of art"—were not particularly useful for understanding the Temple. How, he asked, could one really "restore the Temple of Solomon without consulting Ezekiel, ignorant of Vitruvius, and disregarding the name of Euclid"?[13] But despite his gorgeous pictures—so attractive that they found their way into the 1653 London Polyglot, accompanied by the criticisms of Louis Cappel—it was less the classicism of the Jesuit than the historicism of the librarian that underwrote the significance of Solomon's Temple in early modern England. Although Villalpando offered a compelling visualization of the Temple, it was nonetheless suspended in a "Democritical space," suspended, that is, above and outside the performance of real religious ritual. Vitruvius was an architect of genius, no doubt, but the place of ancient Jewish religion was grounded, not in Roman design, but rather in historically specific buildings, structures, and rituals. In the end, then, Montano's scholarly interest in the specific place of ancient Jewish religion trumped the speculative musings of the aesthete.

To see the Protestant interest in place, we need look no further than the century's most elaborate explanation of the Temple, the Puritan divine Samuel Lee's 1659 *Orbis Miraculum.*[14] From the outset, Lee laid his scholarly allegiances on the table. Ezekiel, he insisted, described only "a Typical Figure of the spiritual Church."[15] Although "the most learned and laborious Temple student, that ever proceeded into publick light," Villalpando simply could not serve as a "firm ground and Basis for the erection of the true Figure of *Solomon's* Temple."[16] Lee made sure to reproduce Montano's Temple diagram and his images of the Temple high priest, the table of the shewbread, and so on. Though its splendid and opulent construction confirmed it as the "most pleasant . . . strongest . . . most Magnificent" ever to exist in the ancient world, the Temple signified far more than good design.[17] Rather, for Lee, the Temple was an immense representational knot, whose strings needed careful unraveling. The only sure tool for loosing this knot was Christ, whose coming revealed the "Sacred Mysteries lodged within these Palaces" and enabled a panoptic understanding of the Temple and its mysterious meanings.[18]

Penetrating these inner mysteries demanded a precise sense of the place of the Temple. On the one hand, the Temple had to be located in a global space. "[T]he exact distance . . . of this famous City *Jerusalem*" was crucial, argued Lee, "that we may know whereabouts in the World our Discourse lies."[19] After some complex number crunching, Lee determined that his Temple lay exactly "2717 miles, 7 furlongs, 30 poles, and 1/2" from the city of London.[20] From this global space, Lee then moved inward. Beginning at the "Porch" of the Temple, he slowly unwrapped the mysterious center of the place, describing first the inner sanctuary, and then the Holy of Holies, that part of the Temple reserved only for the high priest and, even for him, only on the Day of Atonement. It was this place that had so fascinated nearly all commentators, beginning with the ancient Romans and continuing strongly throughout the early modern period.[21] From the interior, Lee then pulled back, unfolding the courts that surrounded the Temple building and assigning to each their proper ritual objects and holy men.

Only after this tour of the Temple's sacred geography did Lee reveal the true purpose of the *Orbis Miraculum,* which was to weave every aspect of "this Sacred pile" into a synthetic Christological prophecy.[22] The final, ponderous chapter offered a painfully detailed exegesis of what Lee called the "Jewish Gospel, or the streaming forth of the glorious beames of Christ incarnate, whilst he walked under the Vail of *Moses.*"[23] Thus, the sanctuary in general typified "Church Militant on Earth," while the Holy of Holies signified "the place and state of Saints in Glory," and its veil the "vaile of *Christ's Flesh,*" and so on.[24] To integrate all of the Temple's objects, people, and places, however, Lee had to generate what he called "a compendious Map of the then *Terra Incognita,* or the unknown Land of the Gospell," so that the conformity of the entire ancient Temple with "Heavens Geographicall Table" might be perfectly revealed.[25] A perfect typology demanded a perfect geography, for the typological reading was itself structured around the distribution of place within the Temple. In order to produce the correspondences between Jewish antiquity and the Christian land of the gospel, all of their points had to be precisely correlated. For similar reasons, Thomas Fuller relied not on the "strange and uncouth" diagrams of Villalpando but on the sober diagrams of Montano in his 1650 *Pisgah—sight of Palestine.*[26] Fuller's quest for a concrete geography and his attentive analysis of the places of the Holy Land were matched by his careful discussion of the interior spaces of the Temple, right down to the weight of the nails used to build the Holy of Holies. The placement of objects and the attributes of the place of Jewish worship were, for him, essential to any true geography of the land of Christ's birth.[27]

The abiding Protestant interest in the topography of the "Jewish Gospel" was only partly driven by the claims of typology. More significantly, the Protestant curiosity about the Temple grew out of serious theological dilemmas confronting moderate Protestant confessions as they battled both Catholics and their more radical brethren over the course of the seventeenth century. If "Christian Hebraism in early modern Europe was a step-child of theology," as Stephen Burnett has argued, then the theologies of ceremony and sacrament were among the most highly charged and, ultimately, intellectually productive areas.[28] Orthodoxy was not the only concern on the table: ortho*praxis* was just as fraught, especially in England, where mainstream Protestants were caught between the Catholic embrace of ritual and the enthusiastic efforts to jettison ritual *tout court.* In the words of Michel de Certeau, the seventeenth-century loss of religious homogeneity meant that "the believer ha[d] to be distinguished from the unbeliever . . . through his practices."[29] Practices, not essential theological beliefs, marked the line between truth and heresy. After all, Catholics (even savages, according to missionaries) believed in the immortal soul, the devil, and the resurrection. This problem of orthopraxis thus became a bedrock issue for all of the English commentators on the Temple. "So tender and jealous is the Lord of Hosts of every point of his instituted Worship," Lee wrote, that "it is dangerous dallying with his divine Majesty, in using any traditionall rites, or ceremoniall inventions of man in his solemn service."[30] What differentiated Jew and Christian, after all, was their relationship to the so-called ceremonial law, those prescriptions offered to Moses in Exodus 21–34. "The *Ceremonial* or *Ecclesiastical* Law," wrote the Calvinist John Edwards in 1699, "is no other than the Precepts given by God to the *Jews* concerning External Rites belonging to Religion and the Worship of God."[31] If, as controversialist and nonjuror William Sherlock argued in 1688, Christ "has taken away all that was merely External in Religion" and has "laid aside all such External Rites . . . thought to be in themselves Acts of Religion," then ceremonial law had no binding force on Christians.[32]

The greatest density of ceremonies in the scriptures was found precisely in the ancient Jewish Temple. Indeed, the Temple was, for ancient Judaism, the very foundation of ceremonial legitimacy. Sacrifice, for example, was inseparable from its Solomonic location. The Temple, for Bishop Samuel Parker, was built solely to prevent Jews from "sacrificing upon High Places, which were Egyptian Altars built in the form of High Towers, that they might make nearer approaches to the *Sun* in their Devotions."[33] And so when the Anglican divine William Owtram attempted to define the nature of sacrifice in his 1677 *De Sacrificiis Libri Duo,* he did so with explicit reference to its ceremonial center. Why did God take up residence in the Tabernacle and Temple? "In order that you might understand that the sacrifices done *in that place,* which God chose for his people to worship him," Owtram declared, "were in that manner clearly sacred things."[34] Only the place of performance consecrated ritual for the Jews. The question that Protestant antiquarians faced was, however, whether *Christian* ritual and religion also needed their own place. Or, better put, through investigation of the place of Judaism, Christian scholars reflected on and, eventually, came to terms with the place of Christian religion. In doing so, they shifted from local and confessionally specific definitions of the sacred to more broadly defined ones. And in this way, they helped to make religion itself into a category of scientific study.

The Temple and the Place of Christianity

When Benito Montano summarized the rationale for the Temple, he used the language of Deuteronomy:

> Destroy all the places where the gentiles worshiped. . . . Scatter their altars and shatter their statues. Burn their sacred groves with fire and break up their idols: purge all of their names from those places. Do not do this to your Lord God, but at the place, which the Lord God selected for all of your tribes that he might put his name there and dwell in it, come together and offer in that place the holocaust and your victims . . . and your offerings and gifts.[35]

Building the Temple was, for Montano's God, a destructive act, intended to crush all vestiges of Gentile worship not just among the Jews, but among the *ethnoi* themselves. But it was also a constructive one: erecting the Temple created a place for legitimate sacrifice and offering. Like the Tabernacle before it, the *only* rationale for the Temple was to circumscribe the place of true religion: although Jews and Gentiles might perform the same sacrificial rituals at similar altars, Jewish ritual was made legal by its performance in the proper place of the Temple. The Temple thus served both to centralize religious worship and to define the difference between proper and improper human worship. In this sentiment, Montano echoed Josephus, who had argued that the scent of sacrifice was "a sign of God's being present and dwelling—according to human belief—in the place which had been newly built and consecrated to Him."[36] "According to human belief" (*kat' anthropinen doxan*) was the key qualifier, since Josephus acknowledged that, for a God to whom the "vault of heaven and all its host is but a small habitation," the Temple was merely a human convenience.[37] Thus, Solomon reminded God: "I have built this temple to Thy name so that from it we may, when sacrificing and seeking good omens, send up our prayers in to the air to Thee, and may ever *be persuaded* that Thou art present and not far removed."[38] Solomon (and through him Josephus) admitted that religion needed a place of religion, not for any divine necessity, but as an accommodation to human needs. As the Jewish philosopher Maimonides later put it, "man, according to his nature, is not capable of abandoning suddenly all to which he was accustomed," and so God, through a "divine ruse," instituted the Temple, the Tabernacle, and the ceremonial laws as a means of weaning ancient Jews from the pagan practices of their former Egyptian overlords.[39]

It was always clear to both Jewish and Christian commentators that the Jewish religion was connected in very specific ways to the idea of place. The exodus of Jews out of Egypt was, after all, bound tightly to the idea of a promised land whose capital would be the divinely ordained sacred center of the Jewish religion. The traveler George Sandys, in his 1610 journey, described Jerusalem in full awareness of this sacred geography: "This City, once sacred and glorious, elected by God for his seat, and seated in the midst of Nations; like a Diadem crowning the head of the mountains; the theater of mysteries and miracles."[40] In terms of both pilgrimage and prophecy, the return to Jerusalem had profound significance for Jews and Christians alike: the convergence on Jerusalem in the 1660s of the Jewish followers of Sabbatai Zevi was an emphatic reminder to all of the priority of place in Judaism.[41] At the very center of this sacred place was the Temple, a spring from which holiness flowed to spread its effects across all of Palestine. Early modern scholars recognized this gradation of holiness. John Lightfoot's 1649 *Temple Service as It Stood in the Days of our Saviour,* for example, following Maimonides, laid out in great detail the "different holiness" attending different parts of the Holy Land. Beginning with the idea that "the land of Israel was more holy than other lands," Lightfoot established eleven circles of priority moving inward from Jerusalem to the "mountain of the Temple," to the "Court of the Women," and to the "Court of the Priests."[42] This crescendo of holiness ended inside the Holy of Holies, where lay the physical center of Jewish spirituality.

But for Protestants in general, this notion of *loci sacri* was theologically vexed. All Protestants agreed that God had appointed a specific place of worship for the ancient Jews. Nearly all were equally convinced—at least in theological terms—that God had appointed no such place for Christians. Generally speaking, Protestant theology sublimated its sacred places off the actual earth and put them into the minds of believers. As the Presbyterian John Weemse wrote, "there is now no appointed place for the worship of God, nor ever shall bee."[43] "GOD Dwells not in their TEMPLES made with HANDS," exclaimed the Quaker George Fox in 1682, but rather in "the Saints that are lead by the Holy-Ghost, [who] are the Temples of God and his Holy Ghost."[44] There were, of course, radical strains in Protestantism that fully identified with the Jews and their theology of place: the Puritan praise of the "holy commonwealth" of New England, for example, invested the stones of Massachusetts with the numinous glow of Jerusalem, while later American Methodists described their rural camps as "specimen[s] of heaven on earth."[45] And yet when speaking theologically, even the less sober Protestants (like Fox) still rejected any essential connection between Christianity and the place of its performance. The 1647 Westminster Confession—the creedal document produced by the Puritans during the English Revolution—solemnly declared, after all, that no part of religious worship is "either tied unto, or made more acceptable by any place in which it is performed."[46]

This theological argument could be expressed in historical terms as well. If "our Saviour has taken all Distinction of Places under the Gospel," the Presbyterian Joseph Hill argued, he did this right away, such that "the Primitive Christians neither built, nor had any publick separate Places, appropriate for the publick Worship of God, for above Two hundred Years" after Christ.[47] The advent of Christ cut true religion free of all the world's places. For a missionary religion that prided itself on its commitment to Jew and Gentile alike, this was only one of many dislocations from the realm of the real into that of the imaginary, from "external" to "internal" forms of worship. In the early modern context, however, Protestant radicals used it to justify their own unorthodox sites of worship. The tradition of the "house-church" common in Dutch, English, and German Protestantism contained "religious dissent within spaces demarcated as private," as Benjamin Kaplan has argued, but it also affirmed the irrelevance of specifically demarcated religious places.[48] House-churches were sacralized, of course, but they did not need external social sanction to become so, merely the practice of true religion. Indeed, church could be held virtually anywhere, as the later phenomenon of field preachers made clear. Catholic churches and cathedrals—as symbolic centers of religious power—were obviously the contrast many Protestants had in mind when they emphasized, with William Sherlock, that "*God's* Presence should no longer be confined to any one place . . . consequently the Holiness of places is lost."[49]

But over the course of the seventeenth century, this position was tempered by the fires of religious conflict, and Christianity—like all religions—

began to occupy its own place in the world. Even before the English Revolution, Archbishop Laud and his Anglican followers started reacting against the radical Calvinist antipathy for "external," church structures. They stressed the integral holiness of churches; they rebuilt church structures fallen into disrepair; and they replaced communion tables with altars, the "greatest place of God's residence upon earth," Laud commented in 1637.[50] They wistfully looked back, with Henry Spelman, to a time when "the world burned so with that sacred fire of devotion . . . that every man desired to sanctifie his hand, in the building of Churches."[51] In Ramie Targoff's terms, these reforms were part of a wider effort to "safeguard against the natural weakness of human devotion" through formal techniques of worship.[52] As John Donne commented in 1622: "*I believe in the Holy Ghost,* but do not find him, if I seek him only in private prayer, but *in Ecclesia,* when I go to meet him in the *Church* . . . a shower is poured out upon me."[53] What some have termed "English Arminianism," a movement away from Calvinist orthodoxy that began in the early 1600s and then resurged after 1660, "emphasized the liturgical and, indeed, physical structures of the church as themselves holy," according to Peter Lake. A renewed emphasis on the "visible church" and the virtues of "public worship" pushed the actual space of worship to the forefront of English theological controversy.[54] After the revolution, with the establishment of an "Anglican monopoly of worship," this argument for sacred place resonated across a spectrum of commentators as mainstream Protestants rejected their radical brethren, as a philosophical criticism of religious zealotry and a concomitant language of tolerance developed in Protestant countries, and as absolute states refined their controls over the religious life of their citizens. All of these factors combined to reshape the notion of public worship: on the one hand, rights to private worship were extended; on the other, the value of a distinct and orderly space of public worship was realized. The result was a shift foreign to the ostensible spirit of Protestantism, as churchmen, scholars, and philosophers argued that even Christianity needed a specific place for its performance.

Already in the middle of the century, the lines of this position could be discerned. "There can scarce be a person found so stupid and senselesse," Samuel Lee argued in his *Orbis Miraculum,* "as to think that there was any real inherent holinesse in those beautiful stones of King *Solomon's* building." Rather, Lee argued, we should "understand the holinesse to be ascribed to it . . . under the notion of consecration; as things or persons which are separated from profane or civil use and dedicated to God."[55] Like Josephus's Solomon, then, Lee read the Temple, not as a site of real divine presence, but rather as a place that conformed to the weakness of human beings in their relationship to the divine. Consecration reserved a space for the sacred that was not "real," even in Temple times, but was rather an accommodation to the human condition. In this simple sense—of consecration and the separation of sacred and profane—even a Puritan like Lee had to admit that Christians too had their sacred places, spaces necessary, not because Christ demanded them, but because human beings need them. As Lee wrote: "it is not unlawful even in these times (but very commendable and useful) to erect material Temples, wherein the solemnities of Gospel-Ordinances may be celebrated, and the congregations of his faithful people may more commodiously meet together."[56] Churches were, in other words, the very images of this ancient Temple: places for congregations where the sacred might be divorced from the profane.

Lee's typological method of establishing analogies between Judaism and Christianity would have held little interest for someone like Thomas Hobbes. And yet it was reasoning by analogy that also structured Hobbes's own theory of sacred place. Making the Protagorean *homo mensura* into a foundation of his philosophical anthropology—"men measure, not onely other men, but all other things, by themselves"—Hobbes argued that all men are forced to reason analogically and thus see the world and its God through the distorting lens of their own imagination.[57] Despite these human limitations on orthopraxis, Hobbes believed that churches were not only necessary but also sanctioned by "the authority of them that are our Soveraign Pastors," ultimately the king, in conformance with human weakness.[58]

> To worship God, in some peculiar Place, or turning a mans face towards an Image, or determinate Place, is not to worship, or honor the Place, or Image; but to acknowledge it Holy, that is to say, to acknowledge the Image, or Place to be set apart from common use. . . . But to worship God, as inanimating, or inhabiting, such Image, or place; that is to say, an infinite substance in a finite place, is Idolatry.[59]

Although Gentiles were especially prone to imagine "an infinite substance" contained in a "finite place"—a prerequisite for an idolatrous relationship to divine places—Hobbes's anthropology was ecumenical in its import. On the one hand, all worship was threatened by the specter of idolatry, given the nature of humanity's relation to the world. On the other, all worship needed to happen somewhere, in conscious avowal of the difference between the place of religious ritual and that of civil profane society. This was a pragmatic concession, in Hobbes's mind, to the human interest in religion. It also served to *confine* religion to specific places overseen by the sovereign.

Hobbes's pragmatism echoed the tones of Richard Hooker, who already in 1597 had praised churches not because of their innate virtue but because of the virtues they instilled in worshipers. Though Job might worship "on the dunghill," the "majestie and holines of the place, where God is worshipped, hath *in regard of us* great vertue force and efficacie."[60] For this reason, the church should take priority over all other places. In Spelman's words, churches are endowed with "a certain kind of *Sanctification:* and are not therefore to bee abused to secular and base implyoyments."[61] And these sentiments were replayed with great enthusiasm after 1660, when Hooker was remade into the "patron saint of Anglicanism," and the strict Calvinist eschewal of religious space fell into disfavor.[62] Thus, despite his explicit belief that the "Holiness of places is lost," William Sherlock still acknowledged that "Christian Churches" are "holy things" and that "being employed in the Worship of God, they ought to be separated from common uses." Although Sherlock reversed the direction of consecration, arguing that "whatever Holiness there is in Christian Churches and Oratories, they are sanctified by the Worship that is performed there, not the Worship sanctified by them," he still embraced the holiness of place that Calvinist theology tended to deny.[63] And his reasoning was telling: in the end, such places were necessary, not because of an absolute divine imperative, but as a concession to a Hobbesian human frailty. Holy places and other external rites are, he wrote, "necessary to the decent and orderly performance of Religious Worship."[64]

By the end of the seventeenth century, antiquarians like John Spencer took a similar argument and extended it into a generalized theory of sacred place. In his 1685 *De Legibus Hebraeorum Ritualibus,* Spencer, the master of Corpus Christi College at Cambridge, dedicated himself to the search for parallels between ancient Jewish and Egyptian ritual and religion.[65] His task was "to demonstrate the Egyptian origin of the ritual laws of the Hebrews," and, like Maimonides, he wanted to show that the ritual laws handed down by God to Moses in Exodus were designed to restrain Jews "addicted" to Egyptian rites.[66] Rather than "impos'd upon them by God as a *Punishment,*" the ritual and ceremonial laws were a course of therapy for the Jews, one accommodated to their physical and spiritual limits.[67] In the sacrifice of the Passover lamb, for example, "God wanted to vilify" an animal that "meant the most to the Egyptians," making it sacred (like the Egyptian god Amun's ram), but then killing it.[68] God's rituals handed down in Exodus were, in other words, imitations of and replacements for human inventions. As such, they suited perfectly the "primaeval simplicity" of both the Egyptians and the Jews.[69] God's introduction of the ritual laws was thus in conformity with the nature of humans, not with his own desires. The same logic applied to ancient temples, which, Spencer argued, were a product of an ancient simplicity that required housing its gods in a physical structure. It was only because of the "feebleness of the Hebrews" that God allowed his people to build their own places of worship: the Temple and the Tabernacle.[70] These structures were not, in other words, "ordered by God's direction"; rather, they were erected in response to the natural tendency of the Israelites, "unaided" by God.[71] Like the ancient Gentiles, the Jews, too, became anxious and uneasy unless their God "was stuck to some certain place in order that they might, when afflicted by things, betake themselves to it."[72]

The significance of Spencer lay less in his investigation of the origins of Jewish rites—this was common enough in seventeenth-century scholarship—than in his discovery that origins mattered less than functions. The Jews got their veneration for sacred places not just from the Egyptians but also (and more significantly) from their own post-Egyptian human nature, which could not be satisfied without putting God somewhere in the world. Although Spencer describes the worship of the Jewish patriarchs as "simple, free and (one might almost say) Evangelical," he explicitly acknowledges that virtually no one, even in the Christian world, has achieved this simplicity of mind.[73] Just as Jews after their enslavement retained a human nature that needed sacred places, so too did Christians. "The fathers of the Christian church," declared Spencer, "in order to bring the Jews to the Christian side, adopted their mores and institutions. . . . the temples of the Christians, divided into the atrium, nave, and sacristy, were formed on the example of the Jewish temple."[74] Just as the Jews retained traces of their Egyptian experience, so too do Christians retain traces of their Jewish roots, and it is this continuity over time that justifies the retention of even those institutions deemed unnecessary in the pure light of rationality or divine law. Religion needs a place because *people* need places to perform its rituals.

It was precisely this sense of general human needs that structured a new antiquarian scholarship on the history of Christianity, pioneered in England by the Anglican Joseph Bingham. His monumental *Origines Ecclesiasticae: or, the Antiquities of the Christian Church* (1711) insisted that early Christian churches were not only common but even necessary for the performance of the Christian religion: "we may easily judge, what

Opinion the Ancients had of the Sacredness of Churches. . . . they would as soon deliver up their Bibles to be burnt by the Heathen, as their Churches to be Prophaned by Heretical Assemblies. . . . As to their Publick Behaviour in the Church, it was generally such as expressed great Reverence for it, as the Sanctuary of God, and the Place of his immediate Presence."[75] Bingham's argument was startling. To claim that the earliest and, for Protestants, purest Christians venerated their churches as much as their Bibles and regarded them as God's own home evidenced a sharp turn away from Protestant tradition and toward an ideal of religion as an anthropological category. If some Protestants held that "under the Gospell the sin of sacrilege cannot be committed" because the sacred resides solely inside human beings, Bingham emphatically insisted that, like all ancient peoples, Christians too believed that the sacred was attached to specific places, and thus it was "the highest Prophanation and Sacrilege, to divert any thing to any other use, which was given to God's Service."[76] Like the Jews, like the Egyptians, indeed like all people, Christians, too, need a place for their religion.

The Temple and the Human Sciences

In the decades following the Restoration, and at the end of a century stained by the blood of countless religious conflicts, religion was given its place by scholars, theologians, and commentators on Temple and Tabernacle. These ancient Jewish places of worship provided researchers with a fund of common topoi and historical examples: even Isaac Newton dedicated many manuscript pages to a close analysis of the Temple and its grounds.[77] More significantly, the systematic elaboration of place—in the Bible, in Jewish commentary, in Christian typology and history—gave these structures, in turn, an analytic density that provided the foundation for a rich analysis of religious place. On the one hand, early modern scholars found in the Temple an exemplar of the very general human need to house the sacred in an earthly vessel. On the other hand, they found in the Temple a means for mapping and measuring the sacred in human terms, for quantifying and organizing in abstract ways the relationship between human beings and the holy. Knowledge was *made* at the intersection of polemic theology, religious tradition, and the artifacts and history of the ancient world.

Given the huge early investment in these sacred sites of Judaism, it may not be such a surprise to discover that comparative religion as it emerged as an institutionalized academic discipline in the late nineteenth century similarly seized on the idea of sacred place and conceptualized it with reference to the Temple and Tabernacle of the ancient Jews. If "place is the beginning of our existence, just as a father," as the thirteenth-century scholastic Roger Bacon argued, then the Temple has been one of the fathers of the idea of a "holy place."[78] Making the Temple of Solomon his leading example, the Semitist William Robertson Smith declared in the famous ninth edition of the *Encyclopaedia Britannica* (1875–89): "The temple is an institution common to religions of natural growth . . . the dwelling house of the deity," an extension of the general "conception of a holy place, separated from profane uses."[79] In his *Lectures on the Religion of the Semites* (1889), Smith generalized about holy places, learning from the religion of the Jews and other Semitic tribes that "it is hardly possible to attach a definite sense" of the "holiness of the gods. . . . apart from the holiness of their physical surroundings." More emphatically: "holy persons things and times . . . all presuppose the existence of holy places."[80] Although not given to the abstractions of his contemporary, James Frazer also sensed that the Temple, at the very least, served to condense crucial aspects of the sacrality of place. His *Folklore in the Old Testament* seized, among other things, on the "Keepers of the Threshold" to offer a primitive version of what Arnold van Gennep and Victor Turner would later analyze as a "rite of passage," a sense of the uncanny that haunts the boundaries between one place and another.[81] Like the Syrians, the Chinese, the Fijians, the West Africans, and the Persians, the ancient Jews stationed these men "at the entrance of the sacred edifice to prevent all who entered from treading on the threshold."[82] Contemporary researchers have made this connection between Temple and the idea of a sacred place even more explicit: Mircea Eliade has used "Palestine, Jerusalem, and the Temple" as definitive exemplars of the ideal of a sacred "Center of the World"; Jonathan Z. Smith rooted his argument that sacred texts and objects "are sacred solely because they are used in a sacred place" in an analysis of Ezekiel's Temple vision; and most elaborately, Francis Schmidt has recently offered a full excavation of the "symbolic system" of the Temple and its organization of the spaces of the sacred and profane in terms both anthropological and sociological.[83]

For modern researchers, the Temple and Tabernacle have clearly offered rich materials for developing a comparative and anthropological notion of sacred place. It is just as clear that this ideal of sacred place has become a crucial piece of the modern social scientific definition of religion: it was and is one of those overarching categories—like "ritual," "time," "the sacred," and so on—that have made possible the comparison of religions despite vast differences in theology, social organization, and history.[84] In essence, these categories have given researchers for the past century the tools to dissolve what Jan Assmann has called the "Mosaic distinction," the distinction between religious truth and error that provides so much of the characteristic rhetoric in the autodescription of religion.[85] As Assmann's language implies, this distinction has been particularly crucial to the Western monotheistic religions, whose very essence depends on the unity and indivisibility of their God. The ability of comparative religion to put Judaism and then Christianity onto the same footing as Syrian idol worship testifies to the extreme shift in attitudes toward religion that this new discipline demanded and ultimately to the power of its social scientific categories, for better or worse, to supersede theological commitments.

It was not in the late nineteenth century, however, but in the late seventeenth that this Mosaic distinction began to dissolve in the gentle acids of Christian antiquarianism. This larger transformation has never been fully explored—it is the big project of which this essay is a small part—but the development of a Christian, and more specifically Protestant, idea of "sacred place" was one key step in its unfolding, as it gave to scholars and researchers one of those building blocks in the conceptual architecture of the modern analysis of religion.[86] Insofar as Christianity was, over the course of the seventeenth century, given just as specific a place in the world as any religion, and insofar as its uniqueness diminished in the face of the anthropological similarities between all religions, religion itself could be abstracted from particular theological commitments. The incomparable truths of Christianity might still stand (indeed did stand for nearly all involved), and at the same time, the analogies between it and the religions of the world grew ever stronger. The Mosaic distinction began its slow disappearance.

It is the great irony, of course, that this disappearance should be precipitated, not by the growth of secularism or science, but by the pressures of theological controversy. Indeed, the very entanglement of Temple and Tabernacle in theological and liturgical polemics was what made them into such powerful magnets for creative scholarly inquiry. Orthodoxy and orthopraxis were intimately linked here, making the questions of worship, ceremony, and ritual deeply divisive. Although an essay of this length can hardly do justice to the fierce religious climate that surrounded these antiquarian and theological investigations, it was exactly this climate that forced scholars to renovate fundamental beliefs both within Protestantism specifically and within Christianity more generally. Caught in between a Catholic veneration of holy places and a radical Calvinist evacuation of church sanctity, moderate Anglican scholars found themselves, in the name of religious apologetics, propelled into wholly new domains of biblical study and innovative strategies of biblical interpretation. Like much intellectual labor in the period, the Christian study of Judaism was energized, in Peter Harrison's terms, "by the desire to score points from theological adversaries."[87] But its confessional origins make it all the more interesting, suggesting that theological polemics, as much as the new science or the New World, was responsible for key developments in the human sciences. The enabling role of theology in the creation of a space for the secular is a paradoxical feature of this history, but one utterly in accordance with the temper of the period. As the case of the Temple made clear, the anthropology of religion was produced, not in spite of confessional commitments, but precisely because of them.

CHAPTER FOURTEEN

The Fiscal Logic of Enlightened German Science

ANDRÉ WAKEFIELD

Local fiscal reforms and strategies largely determined the structure of academic knowledge in eighteenth-century Germany. That is my strong claim here. The proposition implies a corollary: to understand the German universities and academies of the eighteenth century, and the sciences they produced, we need to look beyond them to the mines, manufactories, forests, and fields of the waning Holy Roman Empire. In other words, we ought to try looking at eighteenth-century academic institutions from the standpoint of those who founded them, funded them, and organized them; we ought to regard these institutions, that is, as part of a larger fiscal system dedicated to the promotion of state and economy.

Universities and academies founded during the eighteenth century fused the management of state with the management of science (*Wissenschaft*).[1] The University of Göttingen, for example, mobilized knowledge in the service of the Hanoverian government and its fiscal interests. Gerlach Adolph von Münchhausen, first curator of the Georg-August-Universität and minister of the Hanoverian finances, made sure that the new university followed the government's fiscal mission. Farther east, in the Saxon silver town of Freiberg, another high fiscal official, Friedrich Anton von Heynitz, founded the world's first mining academy. The main purpose of these two prominent "scientific" institutions, as understood by those who established and administered them, was largely the same: to enlist academic knowledge in the generation of state revenue. This essay explores the implications of that fiscal logic by comparing the everyday administration of knowledge in Göttingen and Freiberg.

I have chosen Göttingen and Freiberg not only because they represent new models for the creation of knowledge in eighteenth-century Germany but also because they have rarely been considered worth comparing.

Göttingen, probably the most prominent university of the German Enlightenment, has attracted much attention because of the famous professors it housed, fed, and clothed.[2] It has also played a central part in the drama of "the German university," assuming a supporting role somewhere between Wittenberg and Berlin. Freiberg, on the other hand, has become *both* a canonical site for German Romanticism *and* a staging ground for the creation of a German technocratic elite.[3] Needless to say, these narratives generally focus on the great literary and scientific personalities who passed through town. This essay, in contrast, will compare the administrative philosophies and practices that rendered these two places so much alike. In each case, the dreams and demands of powerful fiscal officials gave rise to new kinds of institutions. These new institutions, in turn, generated new types of knowledge.

University

In his 1768 work on German universities, the Göttingen professor Johann David Michaelis explained that the real purpose of a university had little to do with science or scholarship. Could one possibly believe, he asked, that a "mere love of the sciences" would move great lords and their ministers to fund such expensive institutions? Of course not. States founded and supported universities for fiscal reasons (*Cameralnutzen*) with an expectation of profit (*Vortheil*). By attracting wealthy students from at home and abroad, successful universities would draw foreign wealth while keeping domestic money at home. At flourishing universities—those with many wealthy and few poor students—one could expect the average student to spend 300 thalers each year. Multiply that by one thousand "polite" (i.e., propertied) students, Michaelis calculated, and a successful university might add three tons of gold annually to the territory's circulating coin.[4]

Though it may seem crass, Michaelis's account illustrates the kind of administrative philosophy that guided the University of Göttingen through its first several decades. Under the watchful eye of Minister Münchhausen, Göttingen rapidly rose to prominence, and by the second half of the eighteenth century, contemporaries came to regard it as the most successful and fashionable university in central Europe. Naturally enough, later historians tended to locate the cause of Göttingen's success in its famous professors and students.[5] In doing so, however, they neglected what these famous students and professors themselves insisted: namely, that Göttingen was, more than anything, shaped by the everyday administrative efforts of its first curator, Minister Münchhausen.[6]

From Münchhausen's point of view, universities resembled other state institutions like mines and manufactories. Officials in Hanover—and Münchhausen was the most important of them—regarded Göttingen's famous university as only one part of a larger productive system. Within that system, the university came to share a single goal with other state institutions. In a word, it was expected to serve the interests of state and fisc.

When the University of Göttingen first opened its doors in 1734, Münchhausen understood better than anyone that its success depended on the prosperity, appearance, and reputation of the town.[7] It was not enough to have famous professors. Other things—suitable apartments, walking paths, coffeehouses, pleasure gardens, nice streets, good tailors—were needed to attract wealthy and elegant students. In a backwater town like Göttingen, however, that would not be easy. The royal government, which could do only so much to spruce up the town, needed to ensure that Göttingen and its inhabitants enjoyed some modicum of prosperity. Otherwise, the signs of poverty—abandoned lots, dirty streets, shoddy buildings, beggars—might scare off the elegant students that the university was hoping to attract.[8] Münchhausen and his fellow privy councilors had to approach the university as one part in the great machine of state. They had to be aware of how decisions about individual institutions, such as universities and manufactories, might affect the general welfare.

The decision to establish a new university in Göttingen created immediate problems. The royal government hoped that an infusion of money from wealthy students could make the town flourish. But who would convince them to come in the first place? Since Göttingen's reputation was not good, one would have to advertise the place. The authorities, therefore, commissioned and encouraged a series of works that sang the praises of the idyllic little university town on the Leine River.[9] These quasi-official tributes to Göttingen praised its "excellent location, healthy air, good water, and other advantages,"[10] stressing that the "utmost care" had been taken to ensure the presence of good police.[11] In 1756, Münchhausen, perhaps convinced that the town needed yet another makeover, asked the famous cameralist Johann von Justi—at that time Göttingen's chief police commissioner—to advertise the town.[12]

As Münchhausen strove to burnish Göttingen's image, however, a few troublemakers wrote less flattering things. One of them was the young Danish student Johann Georg Bärens. Unlike the writers who had been recruited by authorities in Hanover, Bärens did not sugarcoat life in the upstart university town. "The weather in Göttingen," he complained, "is not the best." It was too hot in the summer, too cold in the winter, and seemed to rain all the time. The constant rain made for muddy, slimy walking paths. Moreover, there were no decent gardens, and the town was filled with "desolate abandoned lots." Before the arrival of the university in 1734, things had reportedly been even worse. The town had been "indescribably dirty" and "smoky," since half of the town's houses had no chimneys. People would expel the smoke through their attic windows after it had "thoroughly seasoned both them and their famous sausages."[13]

Nor did Bärens have any special affection for the locals. "The inhabitants," he wrote, "are basically a coarse, rude, unfriendly lot who cannot, even with the greatest effort, be cured of their uncouth manners [*Sitten*]." Local burghers were not only lazy and selfish but obstinate too. "They have no understanding of commerce and don't want to learn about it; anyone who has seen Bremen or Frankfurt is considered a well-traveled merchant." Especially remarkable was their "immense hatred of outsiders." New arrivals had often found it difficult to buy food, since the locals would sooner "give it to the pigs than sell it to outsiders." Nor did the town's inhabitants have any notion about "good order or police." The Göttingen authorities—judges, mayors, syndics, secretaries, and town councilors—were not much better.[14]

Despite his criticisms of Göttingen and his distaste for the local inhabitants, Bärens found two things there that he did like: the university and the local *Camelott* manufactory.[15] Both institutions had risen to prominence under the watchful eye of Kammerpräsident Münchhausen, but the manufactory also owed its success to the efforts of Johann Heinrich Grätzel, who had emigrated to Göttingen from Saxony (*Kursachsen*). Grätzel, wrote Bärens, had relied on "his industriousness, his skill, his understanding, and perhaps also his luck" to build a manufacturing empire in Göttingen. He had invested an immense sum, over 80,000 thalers, in buildings, and his "factory [*Fabrique*]" provided work for at least 500 people.[16] When George II visited Göttingen in 1748, he not only toured the university but also met with Grätzel and visited his manufactory.[17]

By 1750 the university and Grätzel's manufactory constituted the twin pillars of Göttingen's prosperity. Münchhausen, who had patiently promoted both institutions, understood that their continued growth and vigor would be necessary to anchor the welfare of the entire region. University and manufactory became recognized as the twin engines of territorial prosperity. Bärens, for example, even claimed that the university had been founded *in order to* revive the town.[18] In any case, there can be no doubt that Münchhausen was deeply concerned about the complex interaction between town and university. The sweep of his administrative control meant that the curator had to make decisions on an astounding array of issues every day. He might resolve disputes between competing manufacturers in the morning and turn his attention to cranky professors in the afternoon.

In fact, Bärens's report articulated Hanover's concerns about Göttingen much better than any of the fawning, quasi-official tributes to the town did. Behind closed doors, the privy council had long worried about what to do with Göttingen. During the 1720s, for example, commerce and manufactures seemed completely stagnant, and the town still bore many scars from the Thirty Years' War. In 1724, concerned about the situation in Göttingen, the privy council dispatched an agent to observe things and to offer advice about possible improvements.[19]

Hanover's agent described Göttingen as a fallen town. Once the center of woolens production in Lower Saxony, Göttingen and its cloth makers had been ruined by a combination of many things: the dissolution of the Hanseatic League, the religious troubles, the Thirty Years' War, the two sieges and bombardment of the town, and its eventual conquest.[20] Nor were there many options for reviving Göttingen's productive life. It was not located along a navigable river or a major road. It had no substantial intercourse with neighboring towns. There were no notable merchants or "capitalists" living in the town. In short, Göttingen's sustenance and prosperity depended entirely on its own agriculture, brewing, and manufactures.[21] Ultimately, Hanover's agent proposed that the government concentrate on building up Göttingen's linen and woolen manufactures as the only viable way to make the town prosper.

In the opinion of Hanover's fiscal officials—especially Münchhausen—this was the best possible way to ensure the success of town and university. Put differently, the support of local manufactures *was* an academic policy, and, conversely, university policy *was* mercantile policy. The success of the university depended on manufactories like Grätzel's, and these manufactories in turn could benefit from the money of wealthy foreign students lured to Göttingen. For state officials like Münchhausen, then, university and manufactory were linked symbiotically.

This explicit connection between university and manufactory meant that large-scale production of linens and woolens soon became a model for the academic manufacture of knowledge. Consider, for example, how the Göttingen student Friedrich Böll described Münchhausen's administration: "You, Mister Curator, are the factory director; the teachers at universities [*Akademien*] are the apprentices; the young people who attend them, and their parents and guardians, are the customers; the sciences taught at those universities are the wares; your king is the lord and owner of his scientific factories [*wissenschaftliche Fabricken*]."[22]

Between 1734 and 1770, Münchhausen worked to create a productive system in Göttingen that included both town and university. What made Göttingen dramatically new and strikingly different from all existing universities was not so much its commitment to neohumanism or its emphasis on law or even its demotion of the theology faculty. Rather, Göttingen's novelty lay in its status as an academic factory. This changed the very nature and function of the sciences produced there.

Mining Academy

The same period that saw the success of Göttingen under Münchhausen also witnessed the establishment of the world's first mining academies, all of them in central Europe.[23] The first and most famous of these academies, founded in the little Saxon mining town of Freiberg, opened its doors in 1765. Historians of the German university rarely look to Freiberg or other "technical academies" for useful comparisons, since they are presumed to be things of a different kind—the one a medieval institution, with its peculiar statutes and organization, dedicated to knowledge production and dissemination; the other a technical school dedicated to applied science. But, as William Clark has shown, German universities underwent dramatic change during the eighteenth century as new administrative regimes encroached on the traditional prerogatives of the faculty.[24] From this perspective, the specialized academy of the eighteenth century may have a lot to tell us about the dreams and ambitions of university administrators in places like Halle and Göttingen. In Freiberg, for example, the new mining academy offered the possibility of an administrative utopia, sheltered from the traditional rights and privileges of recalcitrant university professors. Here, in a place run explicitly by and for the state finances, professors would not interfere with the needs of state building. In short, if Göttingen and Halle constituted the great reform universities of eighteenth-century Germany, then the specialized academies founded during this century—the Collegium Carolinum in Braunschweig, the Cameral Academy in Lautern, the Bergakademie in Freiberg—represented what knowledge *could be* if it was freed from the strictures, statutes, and corporate rights of the German university. That at least is how many high fiscal officials, those practical cameralists of the German Enlightenment, viewed the issue.[25]

Friedrich Anton von Heynitz, the chief commissar of Saxony's mines and founder of the Freiberg Bergakademie, was one of these cameralist officials. As a high-level fiscal official who established and administered a prominent academic institution, Heynitz was much like Münchhausen. And yet our historiography conveys a different impression, separating Göttingen, which often foreshadows the German research university of the nineteenth century, entirely from the mining academies, which appear as training grounds for the technocratic elite of the nineteenth century.[26] It may be more useful, however, to view Göttingen and Freiberg, university and technical academy, as things of the same kind. In each place, reforming officials like Münchhausen and Heynitz sought to craft academic institutions that could train a new kind of state official, one capable of handling the increasingly complex fiscal responsibilities of governance. If Münchhausen established the University of Göttingen to pump up the flagging economy of a backwater town, then Heynitz saw the Bergakademie as a way to lure foreign investment and wealthy students to the silver mines of the Erz Mountains. But the new mining academy was also intended to train and professionalize those already meant to govern: the local Saxon nobility. In other words, Heynitz wanted to prepare the local elite to direct the state and its finances. This, more than the education of "mining engineers," constituted the original purpose of the mining academy.

Well-known cameralist writers were among the most prominent and persistent advocates for the establishment of mining academies even before Freiberg was founded.[27] Johann von Justi, who was busy arresting beggars and plugging the good life in Göttingen, took time out to promote the benefits of mining academies. "That the mining sciences prosper," he wrote, "is not unimportant, and one must therefore provide good instruction in both universities and in special mining academies." He felt that the German lands, given their leading role in the mining sciences, should be the first to establish such institutions.[28] Daniel Gottfried Schreber, another prolific cameralist writer, also began hatching plans for a mining academy and an "academy of oeconomic sciences" in the early 1760s.[29] He envisioned an academy for cameralists, separate from the university, and with five professors of the following subjects: (1) cameral science and economy, (2) mathematics and physics, (3) natural history, (4) mineralogy and chemistry, and (5) manufactures, factories, and commerce. When Schreber was appointed as a professor at the University of Leipzig in 1764, he gave up his plans for a cameral and mining academy. But the plan was not lost completely, for in the following year Schreber's good friend, Friedrich Anton von Heynitz, established the Bergakademie in Freiberg.

Heynitz came to Saxony in 1763, lured by the promise of a position on the Kammer- und Berggemach, Saxony's highest administrative body for directing mines and mining. Devastated by decades of war and mismanagement, Saxony was in the midst of a fiscal crisis when Heynitz arrived in Dresden. But Prince Friedrich Christian and a close circle of advisers, led by Thomas von Fritsch, had already begun to remake Saxony's administration. Among the first issues to be addressed by Fritsch and his fellow commissioners was the improvement of Saxony's mines.

"Mining," wrote Fritsch, "is undeniably one of the most important, if not the single most important, pillar of this land's welfare; its repair and maintenance, therefore, deserve the most exact reflection and the most thorough consideration." Fritsch urged the preparation of a comprehensive tabular balance sheet that would allow for systematic comparison of all income and expenditure related to mining. Such an overview, he argued, would demonstrate "how important mining is for the land, and how necessary it is to keep a diligent and watchful eye on the same." Fritsch complained, moreover, that the mines had suffered from bad administration. Foreign investors had lost faith in Saxony's mines. Trust had to be reestablished through a mining administration marked by the "strict oversight of the sovereign." "We lose this trust," he explained, "if we appoint bad or dishonest officials."[30]

Considerations like these prompted the new elector, Friedrich Christian, to add a powerful new voice to the Kammer- und Berggemach in Dresden at the end of 1763. Possibly due to Fritsch's urging,[31] the elector appointed Heynitz, an experienced senior mining official from Brunswick-Wolfenbüttel, as fourth mining councilor in the Berggemach.[32] Heynitz took up his new post in February of 1764. Unfortunately, the elector died suddenly during the following week, leaving his brother, Prince Xaver, as regent until the young heir, Frederick August, came of age to rule. Xaver, however, soon fell out with Fritsch and his allies, and Heynitz, who had hoped for real power in the Saxon government, found himself relegated to an advisory position, with little direct access to the regent or the privy council.[33]

In search of greater influence, Heynitz sent Cabinet Minister Einsiedel a memorandum on the proposed reorganization of Saxony's mining administration on 4 April 1765.[34] He wanted more personal control over the electorate's mines and suggested that members of the Oberbergamt in Freiberg, especially Oberberghauptmann von Oppel, be given a voice in Dresden's Berggemach. Their participation would, in his opinion, be an improvement over the useless *"Medicos"* and *"Chymicos"* in Dresden who directed the central mining administration.[35] Thanks largely to this memorandum, Heynitz was appointed "general commissar of mines" (*Generalbergkommissar*) in June 1765.

The new office did not give Heynitz complete control over Saxony's mining administration. Rather, many of his plans and projects remained subject to the approval of the Berggemach in Dresden.[36] The new position did, however, give Heynitz considerable power over the Oberbergamt in Freiberg, placing him above even the *Oberberghauptmann* there. He thus turned his attention to the Oberbergamt, still animated by the dreams that had originally brought him to Saxony. It was at about this time, in the summer of 1765, that he seems to have embarked on a new approach. If he could not shape Saxony's mining policy from above, in Dresden, then he would reform it from within by taking control over the regional appointment and education of Saxony's mining officials. He would, that is, create a generation of officials in his own image. This strategy, which would eventually lead Heynitz to establish a mining academy in Freiberg, began to take shape in the autumn of 1765.

On 3 September 1765, Heynitz sent a confidential memo to Count Einsiedel.[37] He expressed concern about the poor condition of the Oberbergamt. More particularly, he discussed the poor quality of the mining officials who worked there and the "lack both of those who are now usefully employed and of those who can be recruited to direct affairs." Heynitz felt that the situation in Freiberg was chaotic and unacceptable. Since no one had a view of the whole, the state's mining "household" was in complete disarray. Heynitz proposed to remedy the situation through wholesale reorganization of the mining administration, which, he argued, should be arranged according to the four natural divisions in the great mining household: (1) mining proper, (2) stamping and separation, (3) smelting and assaying, and (4) accounting matters and acquisition of necessary materials (e.g., gunpowder and wood). This form of organization would, in turn, allow officials to specialize. Each "talented subject" could devote himself to one or another branch of the mining household.[38] Heynitz urged Einsiedel to issue a direct "instruction" to the Berggemach about such a reorganization. The new arrangement, he argued, would help to curb abuses and encourage industriousness, allowing for more effective oversight, since each official would be responsible for a discrete aspect of mining.[39]

Heynitz then turned to a specific enumeration and critique of the mining officials in Saxon service. Berghauptmann von Ponikau, at sixty-three, had "little life left in him." Mining Councilor von Wiehmannshausen was not only old, at almost sixty, but had been hampered by a "gouty foot" (*Podagricus*) for many years. Commissions-Rath Meybach was no better. He was also some sixty years old and, with a smattering of knowledge in "speculative chemistry and hydraulics," was quite worthless for the tough work of direction in a collegium. Mining Councilor Pabst von Ohain showed more promise. He had the requisite knowledge, insight, vigor, and zeal. Unfortunately, complained Heynitz, Ohain did not "seem wholly free of the passions, shows too much politics, never follows the truly straight path, does not allow himself to be led, and shows even less evidence of being able to lead others." All of this led Heynitz to the conclusion that there was a "real shortage of capable people to fill posts as *Berghauptleuten* and mining councilors in Freyberg."[40] Heynitz then proposed a solution to the problem:

> The mining district in Freiberg has established a scholarship fund, from which sons of the state's mining servants [*Bergbedienten*] can get money to learn subterranean surveying and assaying as training for subaltern positions. This fund is of great use, and I have already proposed many times in the Berggemach . . . that people who apply themselves to mechanics and other similar sciences should get something from this fund. But because this fund is not adequate for the *kind of people* who want to learn the mining sciences *in order to direct the budget* [*Haushalt*]—for which reason there are considerably more funds in Hungary, Austria, Bohemia, Sweden, and the Harz—and without which it is not easy to guarantee that anyone will take on this always costly profession [*métier*], I see it as my duty to point out that His Royal Highness might see fit to increase the fund with a contribution from the treasury.[41]

Heynitz was already preparing the way for a mining academy. The existing scholarship fund, which had been in place since 1702, no longer seemed adequate to him. It had been designed to provide narrow technical training for subaltern officials—that is, training for the wrong type of official. His proposal, on the other hand, sought to provide funds for educating a completely different kind of mining official. It aimed, that is, at cultivating cameralists who could oversee, control, and direct the mines.

Heynitz proposed a period of training to last three years. The first two years would be spent in one of Saxony's mining towns, probably Freiberg, with a scholarship of 200 thalers per year. During this time, the candidate would study under the direct supervision of the *Oberberghauptmann.* In the third year, the scholarship would increase to 400 thalers, and the candidate would begin touring mines outside Saxony. In certain respects, the proposal resembled the structure of the 1702 scholarship fund, which had provided state support for aspiring young assayers and subterranean geometers to learn a trade from skilled subaltern officials. But Heynitz's plan was significantly different, for it aimed at cultivating high-level officials for service in central bureaus like the Oberbergamt, or even the Berggemach.

Heynitz not only planned to train a new generation of mining officials but aimed also to weed out "useless" officials from Saxon service. He had, for that purpose, begun to prepare an overview, in tabular form, of salaries and other income for all of Saxony's mining officials. He also examined their orders, promising "that many official posts (*Bedienungen*) can be eliminated or combined." In other cases it was simply a matter of replacing old, tired, corrupt, and useless officials with a new generation of better ones.[42] The success of the mines, he believed, depended on cultivating the right kind of mining official. In fact, Heynitz later claimed that the health of Saxony's mines rested completely on God's blessing and on the "diligence, insight, seriousness, application, liveliness, and integrity of the land's mining and smelting officials."[43] He attributed the decline of mines in the Harz and the Hungarian Carpathians, for example, to the absence of good officials there. A small investment in the education of Saxony's mining officials, therefore, would directly benefit the sovereign treasury. "This proposal," he promised, "will soon yield a rich profit [*sich rentiren*], and Your Excellency is already personally acquainted with the importance of Electoral Saxony's mines."[44]

Only two months later, Prince Xaver and the elector's widow, Maria Antonia, visited Freiberg. Heynitz, hoping they would support his plans for a mining academy, put on a show.[45] He had the mine shafts artificially illuminated and the miners' tools restored. He arranged demonstrations of ore stamping and separation. He ordered two officials, Christlieb Ehregott Gellert and Friedrich Wilhelm Charpentier, to perform chemical experiments.[46] And, perhaps most important of all, Heynitz arranged for a dramatic miners' parade to follow the evening meal. Xaver, who had a weakness for military processions, authorized Heynitz to write up a concrete proposal for the mining academy on the spot. Heynitz submitted his plan the very next day.[47]

Heynitz's plan for the new academy did not merely extend the purposes of the existing scholarship fund.[48] Rather, the new academy, as he envisioned it, would prepare young members of the nobility for careers in the upper echelons of Saxony's mining administration. Whereas the existing scholarship fund had been established to support the acquisition of technical skills, especially assaying and subterranean surveying, the new academy would provide broader training in natural history and natural philosophy. Cadets (i.e., the students at the mining academy) were also expected to have legal training, and the plan provided for university study in jurisprudence, financed by the sovereign. Moreover, Heynitz strongly believed in the value of touring Germany's mines, and his plan thus set aside almost half of the total budget for travel costs. He designed his new academy specifically to educate those officials who would staff Saxony's high fiscal bureaus—especially the Berggemach and Oberbergamt—well into the future. Provided with noble titles, legal training, and well-placed connections from their extensive travel, the cadets were being groomed for positions in the upper levels of Saxony's administration. Unlike their predecessors, who had used the scholarship fund simply to learn specific skills, the Bergakademie had more ambitious goals. It would produce good cameralists to direct Saxony's mines.

The Bergakademie, as Heynitz had conceived it, posed a challenge to the old universities. The task of educating officials for state service, whether in law, medicine, or theology, had traditionally been the exclusive province of university education. With its new mandate, the mining academy now began to train state officials of its own.[49] Moreover, Freiberg's increasingly systematic instruction in mineralogy and chemistry offered an alternative to university education, which typically treated these subjects as auxiliary sciences for the medical faculty.[50] But the liberation of these sciences from the medical faculty signaled at the same time their subordination to Saxony's Oberbergamt. The mining academy, that is, gave the state bureaus direct control over certain kinds of knowledge and bypassed the troublesome universities, with their special academic privileges and quasi-autonomous faculties. But the Bergakademie, with its explicit subordination of scientific knowledge to the fiscal interests of the state, was much more like some universities than others; more specifically, its administrative structure and culture resembled Münchhausen's Göttingen more than the older universities in Heidelberg, Vienna, or Leipzig. By harnessing knowledge directly to the needs of the state treasury, Heynitz had perfected Münchhausen's Göttingen model.

Some four decades after the Freiberg Bergakademie first opened its doors, Abraham Gottlob Werner, the famous Freiberg mineralogist, submitted a report to the Saxon authorities in which he explained the importance of the Bergakademie.[51] Prominent men, including the Prussian ministers Baron Stein and Count Reden, had heard lectures in Freiberg. Mining officials from across Europe and the Americas had studied there. But Werner also justified the mining academy on other grounds. It had, he argued, brought money to Freiberg by attracting wealthy foreigners. Like Justi and Heynitz and Münchhausen before him, Werner was a good cameralist who saw in the Bergakademie a many-sided source of sovereign income. He understood that Saxony's silver came not only from the mines of the Erz Mountains but also from the eager hands of wealthy foreign students.

Conclusion

The new model institutions of eighteenth-century Germany shared common goals and purposes. Within each territory, high-level officials like Münchhausen and Heynitz worked to bind institutions, whether manufactories or universities or prisons, into a single productive system. The point was to increase fiscal yields. Cameralists, the German Enlightenment's prophets of prosperity, sought to design the perfect mix of institutions. For them, no one institution could be viewed in isolation from the others or from the system as a whole. Like mines or manufactories, then, universities and academies had to generate silver, and they might do it in a number of ways. It was no accident, for example, that Hanover established its new university in a poor, dirty little town that had almost nothing going for it before 1734. Göttingen aimed to attract wealthy (preferably foreign) students; these wealthy students would draw skilled tradesmen; and skilled tradesmen would establish a basis for manufactures. It was a crazy idea: the place had to be advertised, promoted, and essentially misrepresented. Göttingen often appears as a nascent model for the German research university of the nineteenth century, but from the standpoint of those who ran the university, especially Curator Münchhausen, Göttingen's much-touted academic research also served as an elaborate advertisement, geared to draw money into the land.

The situation was little different in Freiberg, where officials were open about the need to attract wealthy students from at home and abroad. It is not at all clear, for example, that Abraham Gottlob Werner's famous scientific achievements or Novalis's Romantic science did much to fill Saxony's coffers with native silver. It *is* clear, however, that the professors in Freiberg built a towering reputation for the place and that it was teeming with well-heeled students from all over Europe by the end of the

NOTES

INTRODUCTION

1. Adrien Baillet, *Jugemens des savans sur les principaux ouvrages des auteurs* (Amsterdam, 1725; orig. 1685–86), xi (also cited in Ann Blair, "Annotating and Indexing Natural Philosophy," in *Books and the Sciences in History*, ed. Marina Frasca-Spada and Nick Jardine [Cambridge: Cambridge University Press, 2000], 70).

2. Daniel Brewer, "Lights in Space," *Eighteenth-Century Studies* 37, no. 2 (2004): 171–86, quotation on 172.

3. Nick Jardine and Marina Frasca-Spada, "Books and the Sciences," in *Books and the Sciences in History*, ed. Frasca-Spada and Jardine, 1–10.

4. Adrian Johns, *The Nature of the Book: Print and Knowledge in the Making* (Chicago: University of Chicago Press, 1998). The confluence of authorship, technical manuals, and print has been well studied in the scholarship of Pamela O. Long, especially "Power, Patronage, and the Authorship of Ars: From Mechanical Know-How to Mechanical Knowledge in the Last Scribal Age," *Isis* 88 (1997): 1–41; and *Openness, Secrecy, Authorship: Technical Arts and the Culture of Knowledge from Antiquity to the Renaissance* (Baltimore, MD: Johns Hopkins University Press, 2001).

5. See, e.g., Wolfgang Detel and Claus Zittel, *Wissensideale und Wissenskulturen in der frühen Neuzeit: Ideals and Cultures of Knowledge in Early Modern Europe* (Berlin: Akademie Verlag, 2002); and Elizabeth Spiller, *Science, Reading, and Renaissance Literature: The Art of Making Knowledge, 1580–1670* (Cambridge: Cambridge University Press, 2004).

6. Theophilus, *On Divers Arts*, trans. John G. Hawthorne and Cyril Stanley Smith (New York: Dover Publications, 1979), 189–90.

7. *Rechter Gebrauch der Alchimei/ Mitt vil bisher verborgenen uund lustigen Künsten/ Nit allein den fürwitzigen Alchimisten/ sonder allen kunstbaren Wercklеutten/ in und ausserhalb feurs. Auch sunst aller menglichen inn vil wege zugebrauchen* (1531), p. III verso.

8. For a succinct statement of the nature of these analyses and the revisions they have caused, see Molly Faries, "Reshaping the Field: The Contribution of Technical Studies," in *Early Netherlandish Painting at the Crossroads: A Critical Look at Current Methodologies,* ed. Maryan W. Ainsworth (New Haven, CT: Yale University Press, 2001), 70–105.

9. This theme has been explored in the essays contained in Londa Schiebinger and Claudia Swan, eds., *Colonial Botany: Science, Commerce, and Politics in the Early Modern World* (Philadelphia: University of Pennsylvania Press, 2005).

10. We have in mind, e.g., Helen Watson-Verran and David Turbull's "Science and Other Indigenous Knowledge Systems," in *Handbook of Science and Technology Studies,* ed. Sheila Jasanoff, Gerald E. Marble, James C. Peterson, and Trevor Pinch (London: Sage Publications, 1995), 115–39.

11. A classic exposition of distributed cognition is found in Edwin Hutchins, *Cognition in the Wild* (Cambridge, MA: MIT Press, 1995). See also Yrjö Engeström and David Middleton, eds., *Cognition and Communication at Work* (Cambridge: Cambridge University Press, 1996); and Gavriel Salomon, ed., *Distributed Cognitions: Psychological and Educational Considerations* (Cambridge: Cambridge University Press, 1993). Chandra Mukerji has pioneered this approach in her forthcoming study of the Canal du Midi.

12. Seidel's approach builds on earlier investigations of maker's knowledge, such as those of Michael Baxandall and his students. There have also been recent developments in this area: e.g., Michael W. Cole, *Cellini and the Principles of Sculpture* (Cambridge: Cambridge University Press, 2002); Spike Bucklow, "Paradigms and Pigment Recipes: Vermilion, Synthetic Yellows and the Nature of Egg," *Zeitschrift für Kunsttechnologie und Konservierung* 13 (1999): 140–49; Spike Bucklow, "Paradigms and Pigment Recipes: Natural Ultramarine," *Zeitschrift für Kunsttechnologie und Konservierung* 14 (2000): 5–14; and Spike Bucklow, "Paradigms and Pigment Recipes: Silver and Mercury Blues," *Zeitschrift für Kunsttechnologie und Konservierung* 15 (2001): 25–33. For a discussion of narratives of making, see Malcolm Baker, "Limewood, Chiromancy and Narratives of Making: Writing about the Materials and Processes of Sculpture," *Art History* 21 (1998): 498–530.

13. Recent studies that also focus on local knowledge, practices, and universal claims to "science" include the works of Pamela O. Long cited above as well as her essay "The Contribution of Architectural Writers to a 'Scientific' Outlook in the Fifteenth and Sixteenth Centuries," *Journal of Medieval and Renaissance Studies* 15 (1985): 265–98; Lisbet Koerner, *Linnaeus: Nature and Nation* (Cambridge, MA: Harvard University Press, 1999); William Clark, "On the Ministerial Archive of Academic Acts," *Science in Context* 9, no. 4 (1996): 421–86; Michael Bravo and Sverker Sorlin, eds., *Narrating the Arctic: A Cultural History of Nordic Scientific Practices* (Canton, MA: Science History Publications, 2002); Lorraine Daston and Katharine Park, *Wonders and the Order of Nature* (New York: Zone Books, 1999); Otto Sibum, "Les gestes de la mesure: Joule, les pratiques de la brasserie et la science," *Annales: Histoire, Sciences Sociales* 53, no. 4 (1998): 745–74; and William

Clark, Jan Golinski, and Simon Schaffer, eds., *The Sciences in Enlightened Europe* (Chicago: University of Chicago Press, 1999).

14. For literature on colonial science, see Marie-Cécile Bénassy-Berling, ed., *Nouveau monde et renouveau de l'histoire naturelle*, 3 vols. (Paris: Presses de la Sorbonne Nouvelle, 1986–94); John MacKenzie, ed., *Imperialism and the Natural World* (Manchester: University of Manchester, 1990); N. Jardine, J. A. Secord, and E. C. Spary, eds., *Cultures of Natural History: From Curiosity to Crisis* (Cambridge: Cambridge University Press, 1995); David Miller and Peter Reill, eds., *Visions of Empire: Voyages, Botany, and Representations of Nature* (Cambridge: Cambridge University Press, 1996); Roy MacLeod, ed., *Nature and Empire: Science and the Colonial Enterprise* (Chicago: University of Chicago Press, 2000); and Londa Schiebinger, ed., "Forum: Colonial Science," *Isis, Journal of the History of Science* 96 (2005): 52–87.

15. Practices of reading have received much attention of late, often in confluence with studies of the history of the book. See Roger Chartier, *The Order of Books: Readers, Authors and Libraries in Europe between the Fourteenth and Eighteenth Centuries*, trans. Lydia G. Cochrane (Cambridge: Polity Press, 1994); Roger Chartier, *Forms and Meanings: Texts, Performances, and Audiences from Codex to Computer* (Philadelphia: University of Pennsylvania Press, 1995); Frasca-Spada and Jardine, *Books and the Sciences in History*; Kevin Sharpe, *Reading Revolutions: The Politics of Reading in Early Modern England* (New Haven, CT: Yale University Press, 2000); Jennifer Anderson and Elizabeth Sauer, eds., *Books and Readers in Early Modern England: Material Studies* (Philadelphia: University of Pennsylvania Press, 2002); and Johns, *Nature of the Book*. One of the germinal essays in this area remains Lisa Jardine and Anthony Grafton, "'Studied for Action': How Gabriel Harvey Read His Livy," *Past and Present* 129 (1990): 30–78.

16. On autobiographical writing, see, e.g., Winfried Schulze, ed., *Ego-Dokumente: Annäherung an den Menschen in der Geschichte* (Berlin: Akademie, 1996); and Rudolf Dekker, *Childhood, Memory, and Autobiography in Holland: From the Golden Age to Romanticism* (New York: Palgrave, 1999). On the encounter in the Renaissance between new worlds of knowledge and ancient scholarly texts, see Anthony Grafton, with April Shelford and Nancy Siraisi, *New Worlds, Ancient Texts: The Power of Tradition and the Shock of Discovery* (Cambridge, MA: Harvard University Press, 1992); and see also Anthony Pagden, *European Encounters with the New World: From Renaissance to Romanticism* (New Haven, CT: Yale University Press, 1993).

17. See Robert Darnton, "Readers Respond to Rousseau," in *The Great Cat Massacre and Other Episodes in French Cultural History*, by Robert Darnton (New York: Basic Books, 1984), 215–56.

18. Walter J. Ong, *Ramus, Method, and the Decay of Dialogue: From the Art of Discourse to the Art of Reason* (Cambridge, MA: Harvard University Press, 1983).

19. On "active" reading, see Jardine and Grafton, "'Studied for Action.'"

20. Steven Shapin discusses other aspects of Boyle's literary choices in "Pump and Circumstance: Robert Boyle's Literary Technology," *Social Studies of Science* 14 (1984): 481–520.

21. The body of work on the study of the senses and sensory knowledge in the early modern period is large and growing fast. For an overview of recent literature, see Peter Dear, *Revolutionizing the Sciences: European Knowledge and Its Ambitions, 1500–1700* (Princeton: Princeton University Press, 2001); as well as Dear's *Discipline and Experience: The Mathematical Way in the Scientific Revolution* (Chicago: University of Chicago Press, 1995). Steven Shapin's *The Scientific Revolution* (Chicago: University of Chicago Press, 1996) also contains much that is useful in this regard. One of the best overviews of the place of the senses in the new epistemology of the Scientific Revolution remains R. Hooykaas, "The Rise of Modern Science: When and Why?" *British Journal for the History of Science* 20 (1987): 453–73. These studies deal with the epistemology of empiricism. Studies on the five senses themselves are far less common. Although much of the research in this field pertains to sight, see Bruce R. Smith, *The Acoustic World of Early Modern England: Attending to the O-factor* (Chicago: University of Chicago Press, 1999); and Constance Classen, *The Book of Touch* (Oxford: Berg, 2005). For one preliminary attempt to write a history of the senses, see J. R. R. Christie, "The Paracelsian Body," in *Paracelsus: The Man and His Reputation, His Ideas, and Their Transformation*, ed. Ole Peter Grell (Leiden: Brill, 1998).

22. For another examination of the many meanings that objects could possess at this time, see Martin Kemp, "'Wrought by No Artist's Hand': The Natural, the Artificial, the Exotic, and the Scientific in Some Artifacts from the Renaissance," in *Reframing the Renaissance: Visual Culture in Europe and Latin America, 1450–1650*, ed. Claire Farago (New Haven, CT: Yale University Press, 1995), 117–96.

23. One of the best examinations of the ambivalent status of natural and artificial objects in Dutch still lifes and of the senses through which these objects are perceived can be found in Simon Schama, "Perishable Commodities: Dutch Still-Life Painting and the 'Empire of Things,'" in *Consumption and the World of Goods*, ed. John Brewer and Roy Porter (London and New York: Routledge, 1994), 478–88. The history of the body also shows the fine line between sensual and sensory. See, e.g., Lorraine Daston, "Curiosity in Early Modern Science," *Word and Image* 11, no. 4 (1995): 391–404.

CHAPTER ONE

1. L. T. C. Rolt, *Le canal entre deux mers* (Paris: Euromapping, 1994); Jean-Denis Bergasse, ed., *Le Canal du Midi*, vols. 3–4 (Millau: Maury, 1984); André Maistre, *Le canal des deux-mers: Canal royal du Languedoc, 1666–1810* (Toulouse: Éditions Privat, 1998).

2. Archives du Canal du Midi (ACM), Toulouse, no. 23-16.

3. For the artwork at Naurouze that symbolized its importance, see Louis de Froidour, *Lettre à Monsieur Barrillon Damoncourt, Conseiller du Roy en ses Conseils, Maître des Requestes Ordinaire de son Hostel, Intendant de Iustice, Police et Finances en Picardie Contenant la Relation & la description des Travaux qui se sont en Languedoc, pour la communication des deux mers* (A Toulose: Chez I. Dominique Camusat, 1672), 29. On finding capable workers, see ACM, nos. 21-1 and 23-4. For the role of workers in the ceremonial opening of the gate to the Garonne River, see Matthieu de Mourgues, *Relation de la seconde navigation du Canal Royal* (1683), quoted in Phillippe Delvit, "Un Canal au Midi," in *Canal royal de Languedoc: Le partage des eaux*, ed. Conseil d'Architecture d'Urbanisme et de l'Environnement de la Haute-Garonne (Fonsegrives: CAUE-Éditions Loubatieres, 1992), 204–5.

4. For Pons de la Feuille and his supervision of the second enterprise, see ACM, nos. 22-18, 23-4, 23-47, 24-6, 24-12, 24-35; and Chandra Mukerji, "Cartography, Entrepreneurialism and Power in the Reign of Louis XIV: The Case of the Canal du Midi," in *Merchants and Marvels: Commerce, Science and Art in Early Modern Europe*, ed. Pamela Smith and Paula Findlen (New York: Routledge, 2002), 248–76.

5. Reported in Riquet to Colbert, 30 October 1669, ACM, no. 22-27.

6. Rolt, *Le canal entre deux mers*, 52–99; and Michel Adgé, "L'art de l'hydraulique," in *Canal royal de Languedoc*, ed. Conseil d'Architecture, 185–98.

7. Maistre, *Le canal des deux-mers*, 74–75.

8. On 20 October 1668, Riquet wrote that he was staying on in Perpignan to recruit workers for the canal (ACM, no. 21-18). Again, on 15 February 1673, Riquet said he was getting workers from Bigorre (ACM, no. 30-65). See also Isaure Gratacos, *Femmes pyrénéennes, un statut social exceptionel en Europe* (Toulouse: Éditions Privat, 2003), 105–15.

9. A. Marcet-Juncosa, "L'opposition catalane à P. P. Riquet," in *Le Canal du Midi*, ed. Bergasse, 4:143–50.

10. See Mukerji, "Cartography, Entrepreneurialism and Power in the Reign of Louis XIV."

11. Maistre, *Le canal des deux-mers*, 72–76.

12. See Bertrand Gabolde, "Les ouvriers du chantier," in *Le Canal du Midi*, ed. Bergasse, 3:235–36; ACM, no. 1072. These account records with the high file numbers are in the "uncataloged" boxes at the Archives du Canal du Midi. They are not consistently numbered.

13. ACM, no. 1072-18.

14. ACM, no. 1072-40.

15. Louis de Froidour, *Memoire du pays et des états de Bigorre*, with introduction and notes by Jean Boudette (Paris: H. Champion; Tarbes: Baylac, 1892).

16. Froidour, *Lettre à Monsieur Barrillon Damoncourt*.

17. Froidour, *Memoire du pays et des états de Bigorre*, 20–21 (italics in the original).

18. Ibid., 57–58.

19. Le Roy Ladurie, *Montaillou* (New York: Vintage, 1979), 103–19. For revolts in the mountains, and the role of women in these revolts, which gave them their reputation for violence, see Jean François Soulet, *La vie quotidienne dans les Pyrénées sous l'ancien regime* (Paris: Hachette, 1974), 73–74. For women, children, and agriculture, see Soulet, *La vie quotidienne,* 83–88, 101–4; and Froidour, *Memoire du pays et des états de Bigorre,* 31–32. See also ACM, nos. 21-2, 21-18; and Peter Sahlins, *Forest Rites: The War of the Demoiselles in Nineteenth-Century France* (Cambridge, MA: Harvard University Press, 1994), particularly 40–60.

20. Froidour, *Memoire du pays et des états de Bigorre,* 20. For the patterns of collective management of natural resources in different parts of the Pyrenees, see Soulet, *La vie quotidienne,* 37–56, 71–83; and Olivier de Marliave, *Tresor de la mythologie pyreneene* (Luçon: Éditions Sud Oust, 1996), 33–41.

21. The Pyrenees themselves were female in the origin story often told in these areas. The princess Pyrène was seduced by Hercules but afterward was devoured by beasts. Hercules was upset and built her a mausoleum, which was the Pyrenees mountains. See Marliave, *Tresor de la mythologie pyreneene,* 13. The apparitions of the Virgin take many forms. She is associated with some of the mountain water sources, but she also appears in trees and is associated with the sun. See Marliave, *Tresor de la mythologie pyreneene,* 21, 150, 161, 170. See also Gratacos, *Femmes pyrénéennes,* 131–83, 177–80, which describes how the traditional sightings of fairies were turned into apparitions of the Virgin. See also Sahlins, *Forest Rites,* 40–60.

22. We know that recruitment of canal laborers in 1668 was timed to employ seasonal agricultural laborers after the harvest. See ACM, no. 7-02; and Gratacos, *Femmes pyrénéennes,* 105–8, 146–70.

23. Froidour, *Memoire du pays et des états de Bigorre,* 31–32; Marliave, *Tresor de la mythologie pyreneene,* 57; and Gratacos, *Femmes pyrénéennes,* 139–40, 143–46.

24. Bagnères may have been a Roman spa town, but the canals there today serve other purposes. For some details of Roman canal engineering, see Sextus Julius Frontinus, *The Two Books on the Water Supply of the City of Rome of Sextus Julius Frontinus, Water Commissioner of the City of Rome* A.D.*97* (New York: Longmans, Green, 1913).

25. Weirs in the rivers that helped divert water into Mazères and Ustou also helped produce something like a settling pond for river water before it was diverted into the towns. The velocity of the water was reduced as it was taken out of the main stream, allowing materials held in suspension to precipitate. Although the weirs produced a narrowing channel of calmer waters rather than a widened area for settling silt or cobbles, the effect was similar. Where the intake sluices were set over raised stones, they also held back debris from the river. Given the variety of ways in which these kinds of controls were used in the mountains, laborers who came from these areas would have easily understood both why and how to build such structures for the Canal du Midi.

26. Cf. ibid.

27. Gabolde, "Les ouvriers du chantier," 236.

28. Sahlins, *Forest Rites*, 40–60; and Gratacos, *Femmes pyrénéennes*, 143–84.

29. ACM, no. 12-2; Gabolde, "Les ouvriers du chantier," 236.

30. Rolt, *Le canal entre deux mers*, 30–31; Maistre, *Le canal des deux-mers*, 38–41; Inès Murat, "Les rapports de Colbert et de Riquet: Méfiance pour un homme ou pour un système?" in *Le Canal du Midi*, ed. Bergasse, 3:111–12.

31. ACM, nos. 13-3, 12-2, 931-18.

32. ACM, no. 13-14; *Memoire de l'état des travaux du Canal fait par M. Riquet* (n.d., but from the description of the work, apparently the 1670s). See Gabolde, "Les ouvriers du chantier," 237; and Rolt, *Le canal entre deux mers*, 56–57. Also see the account books in ACM, no. 931.

33. Sieur de la Feuille's comments about the women are reported in a letter from Riquet to Colbert, 30 October 1669, ACM, no. 22-27.

34. ACM, nos. 1072-1, 1072-3.

35. ACM, nos. 1072, 1073.

36. Rolt, *Le canal entre deux mers*, 86; and Anne Blanchard and Michel Adgé, "Les ingénieurs du roy et le Canal de Communication des Mers," in *Le Canal du Midi*, ed. Bergasse, 3:18. La Feuille was not just a spy for Colbert. He was an educated engineer, a *geographe du roi*, but he was hostile to Riquet. Blanchard and Adgé, "Les ingénieurs du roy," 3:183–94, particularly 184–88. La Feuille was sent to the Netherlands and Italy to look at the canal construction there. See Rolt, *Le canal entre deux mers*, 76, 93; ACM, no. 10-70. For his complicated relations with Riquet, see Rolt, *Le canal entre deux mers*, 76, 83–87, 93. After Riquet's death, Père Mourgues gained even greater power and ensured that the project was properly finished by Riquet's son and that the contracts with the state were kept.

37. ACM, no. 31-36; and Murat, "Les rapports de Colbert et de Riquet," 3:108. See Riquet's letter of 27 May 1669, ACM, no. 22-16.

38. Rolt, *Le canal entre deux mers*, 89.

39. For Gabolde's contention that the Montagne Noire was a training ground of sorts, see "Les ouvriers du chantier," 236–37. For work in the Somail region, see ACM, nos. 1071, 1072.

40. See Adgé, "L'art de l'hydraulique," 196–97.

41. Ibid., 197, 199, 209.

42. Ibid., 190; Rolt, *Le canal entre deux mers*, 96; ACM, no. 1098.

43. ACM, no. 1098.

CHAPTER TWO

1. Elisabeth Dhanens, *Het retabel van het Lam Gods in de Sint-Baafskathedral te Gent*, Inventaris van het Kunstpatrimonium van Oostvlaanderen (Ghent: KunstPatrimonium, 1965).

2. Paul Coremans et al., *L'agneau mystique en laboratoire: Examen et traitement*, Les Primitifs flamands, vol. 3, Contributions à l'étude des Primitifs

flamands, vol. 2 (Antwerp: De Sikkel, 1953); Erwin Panofsky, *Early Netherlandish Painting: Its Origins and Character* (Cambridge, MA: Harvard University Press, 1953), 178–204; Lotte Brand Philip, *The Ghent Altarpiece and the Art of Jan van Eyck* (Princeton, NJ: Princeton University Press, 1971); Claus Grimm, *Meister oder Schüler? Berühmte Werke auf dem Prüfstand* (Stuttgart: Belser Verlag, 2002); Till-Holger Borchert, "Introduction: Jan van Eyck's Workshop," in *The Age of Van Eyck: The Mediterranean World and Early Netherlandish Painting,* 1430–1530, ed. Till-Holger Borchert (London: Thames and Hudson, 2002), 14–15.

3. *The Writings of Albrecht Dürer,* trans. and ed. William Martin Conway (New York: Philosophical Library, 1958), 117.

4. Quoted in W. H. James Weale, *Hubert and Jan van Eyck,* Their Life and Work (London and New York: John Lane, the Bodley Head, 1908), cxxiv–cxxv; and Otto Pächt, *Van Eyck and the Founders of Early Netherlandish Painting,* ed. Maria Schmidt-Dengler, trans. David Britt (London: Harvey Miller, 1994), 11–12. Münzer's remarks would not have offered a unique understanding of art's intellectual dimensions by that time. For comments on the remarks of Filippo Villani a century earlier, see Charles Harrison, "The Arena Chapel: Patronage and Authorship," in *Siena, Florence, and Padua: Art, Society, and Religion, 1280–1400,* ed. Diana Norman, 2 vols. (New Haven, CT: Yale University Press, 1995), 2:101.

5. Michael Baxandall, "Bartholomaeus Facius on Painting: A Fifteenth-Century Manuscript of the *De viris illustribus,*" *Journal of the Warburg and Courtauld Institutes* 27 (1964): 90–107, especially 102 for the translations cited in my text and 103 for the Latin; and Michael Baxandall, *Giotto and the Orators: Humanist Observers of Painting in Italy and the Discovery of Pictorial Composition, 1350–1450* (Oxford: Clarendon Press, 1971), 97–101.

6. Diminished precision in the description of this work suggests that Fazio's knowledge of it may have been secondhand. Sergio Paccagnini has drawn my attention to alternative readings of the Latin that modify our notion of certain figures' activity.

7. Walter S. Melion, *Shaping the Netherlandish Canon: Karel van Mander's "Schilder-boeck"* (Chicago: University of Chicago Press, 1991), 84–86.

8. Lisa Monnas, "Silk Textiles in the Paintings of Jan van Eyck," in *Investigating Jan van Eyck,* ed. Susan Foister, Sue Jones, and Delphine Cool (Turnhout: Brepols, 2000), 149, 158. See also Rembrandt Duits, "Figured Riches: The Value of Gold Brocades in Fifteenth-Century Florentine Painting," *Journal of the Warburg and Courtauld Institutes* 62 (1999): 60–92.

9. For the latter, see Pächt, *Van Eyck,* 124–70.

10. See Lorne Campbell, "L'organisation de l'atelier," in *Les Primitifs flamands et leur temps,* by Maryan Wynn Ainsworth et al., under the direction of Brigitte de Patoul and Roger van Schoute (Wesmael: La Renaissance du Livre, 1994), 88–99.

11. Theophilus Presbyter, *"On Diverse Arts": The Treatise of Theophilus,* trans. John G. Hawthorne and Cyril Stanley Smith (Chicago: University of Chicago Press, 1963). The earliest known copy of the work is in Vienna (National Li-

brary 2527) and is dated, on internal evidence, after 1126 (John van Engen, "Theophilus Presbyter and Rupert of Deutz: The Early Manual of Arts and Benedictine Theology," *Viator* 11 [1980]: 147–63).

12. Cennino d'Andrea Cennini, *The Craftsman's Handbook: Il libro dell'arte,* trans. Daniel V. Thompson Jr. (New York: Dover, 1954); hereafter page numbers for this work are given in the text.

13. For the Campanile reliefs, see Diana Norman, "The Art of Knowledge: Two Artistic Schemes in Florence," in *Siena, Florence and Padua,* ed. Norman, 2:218–22.

14. Giorgio Vasari retained this trope in the preface to his celebrated history of art, first published in 1550: *Lives of the Most Eminent Painters, Sculptors and Architects,* trans. Gaston DuC. DeVere, 10 vols. (London: Macmillan, 1912–14).

15. Paul was the first to remark that Luke was a physician, an identification that became popular in the West after it was included in Jacobus da Voragine's *Golden Legend.* See H. Höllander, "Lukasbilder," in *Lexikon der Christlichen Ikonographie* (Rome: Herder, 1971), 3:121–22; and Catherine King, "National Gallery 3902 and the Theme of Luke the Evangelist as Artist and Physician," *Zeitschrift für Kunstgeschichte* 48 (1985): 249–55.

16. The Adam and Eve panels have a more complicated history than do the other panels of the altarpiece. Max Friedländer reported that they had been severed from the altarpiece in 1781, after having been judged "obscene." See his *Early Netherlandish Painting,* vol. 1, *The Van Eycks* and *Petrus Christus,* trans. Heinz Norden (Leiden: A. W. Sijthaff, 1967), 29. Elisabeth Dhanens reported that they were removed from their setting in 1816 and placed in the archives (probably referring to the church archives), from whence they were purchased by the Belgian government in 1861 and put on display in the Musée des Beaux Arts in Brussels. See her *Van Eyck: The Ghent Altarpiece* (New York: Viking Press, 1973), 133–37. The altarpiece was restored to its original form only after the Second World War. For a more detailed account of the polyptych's history, along with a discussion of issues related to its production, see Bernhard Ridderbos, "Objects and Questions," in *Early Netherlandish Painting: Rediscovery, Reception, and Research,* ed. Bernhard Ridderbos, Anne van Buren, and Henk van Veen, trans. Andrew McCormick and Anne van Buren (Los Angeles: Getty Publications, 2005), 4–169, especially 42–59. For Jan's revisions as he worked on the panels, see J. R. J. van Asperen de Boer, "A Scientific Re-examination of the Ghent Altarpiece," *Oud-Holland* 93 (1979): 141–214.

17. For a colored reproduction of the Eden illumination, fol. 25v, see *The "Très riches heures" of Jean, Duke of Berry,* introduction by Jean Longnon and Raymond Cazelles, trans. Victoria Benedict (New York: George Braziller, 1969), pl. 20.

18. Margaret R. Miles, *Carnal Knowing: Female Nakedness and Religious Meaning in the Christian West* (Boston: Beacon Press, 1989), 99 and n. 5, citing H. W. Janson's *History of Art* as her source.

19. Marc[us] van Vaernewick, *Mémoires d'un patricienne gantois du XVIe siècle,* 2 vols. (Ghent: Maison d'Éditions d'Art, 1905–6), 2:9.

20. Karel van Mander, *The Lives of the Illustrious Netherlandish and German Painters, from the First Edition of the "Schilder-boeck," 1603–1604*, ed. Hessel Miedema, 2 vols. (Doornspijk: Davaco, 1994–95), 1:62; and Melion, *Shaping the Netherlandish Canon*, 85, 137–38.

21. Pächt, *Van Eyck*, 163–64.

22. Karen Cherewatuk, St. Olaf's College, asked me about Eve's abdominal pigmentation following a presentation I gave on the Ghent altarpiece on that campus in 1996, stimulating my efforts to account for the presence of the darkened line in a satisfactory manner. I have reflected on our conversation ever since and am pleased to be able to acknowledge it in print. For different approaches to this rendering of Eve, see Madeleine Caviness, "Obscenity and Alterity: Images That Shock and Offend Us/Them, Now/Then?" in *Obscenity: Social Control and Artistic Creation in the European Middle Ages*, ed. Jan M. Ziolkowski, Cultures, Beliefs, and Traditions, vol. 4 (Leiden: E. J. Brill, 1998), 1–17; Jacques Paviot, "Les tableaux de nus profanes de Jan van Eyck," Gazette des Beaux-Arts 135 (2000): 265–82.

23. On the image, see I. G. Dox, J. L. Melloni, and H. H. Sheld, *Melloni's Illustrated Dictionary of Obstetrics and Gynecology* (New York: Parthenon, 2000), 210; and Michael J. O'Dowd and Elliot E. Philipp, *The History of Obstetrics and Gynaecology* (New York: Parthenon, 1994), 59. For evidence of physical contact with patients around the start of the fifteenth century, see Danielle Jacquart, "Theory, Everyday Practice, and Three Fifteenth-Century Physicians," in *La science médicale occidentale entre deux renaissances (XIIe s.–XVe s.)*, Variorum Collected Studies Series, CS567 (Aldershot, UK, and Brookfield, VT: Variorum, 1997), 140–60.

24. Keith Moxey, "Art History's Hegelian Unconscious: Naturalism as Nationalism in the Study of Early Netherlandish Painting," in *The Subjects of Art History: Historical Objects in Contemporary Perspective*, ed. Mark A. Cheetham, Michael Ann Holly, and Keith Moxey (Cambridge: Cambridge University Press, 1998), 33–35.

25. Lorne Campbell understands Jan's remarkable signature on the *Arnolfini Portrait* as an assertion of the painter's skill, an observation with which I fully agree. See his *The Fifteenth Century Netherlandish Schools* (London: National Gallery Publications, 1998), 201.

26. Martin Kemp, "Leonardo da Vinci," in *Biography and Early Art Criticism of Leonardo da Vinci*, ed. Claire Farago (New York: Garland Press, 1999), 40–59; and especially Leo Steinberg, *Leonardo's Incessant Last Supper* (Chicago: University of Chicago Press, 2003). For related observations on the relationship between artistic and divine creativity in the fifteenth century, see Pamela H. Smith, *The Body of the Artisan: Art and Experience in the Scientific Revolution* (Chicago: University of Chicago Press, 2004), 53, 259–63.

27. Walter Cahn, *Masterpieces: Chapters on the History of an Idea*, Princeton Essays on the Arts, vol. 7 (Princeton, NJ: Princeton University Press, 1979), 3–22. For more on pigments and painting practices, see Jonathan J. G. Alexander, *Medi-*

eval Illuminators and Their Methods of Work (New Haven, CT: Yale University Press, 1992), 39–40, 159–60nn43–49; and Smith, *Body of the Artisan,* 114, 140–41.

28. See Kenneth Bé, "Geological Aspects of Jan van Eyck's *Saint Francis Receiving the Stigmata,*" in *Jan van Eyck: Two Paintings of "St. Francis Receiving the Stigmata"* (Philadelphia: Philadelphia Museum of Art, 1997), 88–95.

29. Gervase Rosser, "Crafts, Guilds and the Negotiation of Work in the Medieval Town," *Past and Present* 154 (1997): 30–31.

30. See Minta Collins, *Medieval Herbals: The Illustrative Traditions,* British Library Studies in Medieval Culture (London: British Library; Toronto: University of Toronto Press, 2000), especially chap. 5, 239–83.

31. Uta Neidhardt and Christoph Schülzel, "Jan van Eyck's Dresden Triptych," in *Investigating Jan van Eyck,* ed. Foister, Jones, and Cool, 25–39; Raymond White, "Van Eyck's Technique: The Myth and the Reality, II," in *Investigating Jan van Eyck,* ed. Foister, Jones, and Cool, 101–5; "The Methods and Materials of Northern European Painting, 1400–1550," *National Gallery Technical Bulletin* 18 (1997): 6–55.

32. John E. Murdoch, *Album of Science, Antiquity, and the Middle Ages* (New York: Charles Scribner's Sons, 1984), 203, 305–6; Peter Murray Jones, *Medieval Medicine in Illuminated Manuscripts* (London: British Library, 1998), 43–57, 73–75, figs. 33, 35, 36, 46, 48, 66. For a brief overview of medieval medicine, see Charles H. Talbot, "Medicine," in *Science in the Middle Ages,* ed. David C. Lindberg (Chicago: University of Chicago Press, 1978), 391–428.

33. The saint likewise holds a carafe in a border decoration of a *volvelle,* or rotating calendar, made for surgeons in the late fifteenth century (Murdoch, *Album of Science,* 54–57, fig. 48). For a compelling analysis of an early-sixteenth-century painting's extensive relationship to physical well-being, see Andrée Hayum, *The Isenheim Altarpece: God's Medicine and the Painter's Vision* (Princeton, NJ: Princeton University Press, 1989), 13–52.

34. Pearl Kibre, "Hitherto Unnoted Medical Writings by Dominicus de Ragusia (1424–25 A.D.)," *Bulletin of the History of Medicine* 7 (1939): 990–95; for an illustration of such a chart, see Murdoch, *Album of Science,* 305–6. For comments on the altered color of a woman's urine, see Monica H. Green, ed. and trans., *The "Trotula": A Medieval Compendium of Women's Medicine* (Philadelphia: University of Pennsylvania Press, 2001), 67.

35. A section on cosmetics concludes the *Trotula* as well (Green, *Trotula,* 113–24).

36. Michael Baxandall established this point in connection with Florentine practices; see his *Painting and Experience in Fifteenth-Century Italy: A Primer in the Social History of Pictorial Style* (Oxford: Oxford University Press, 1972).

37. Ashok Roy, "Van Eyck's Technique: The Myth and the Reality, I," in *Investigating Jan van Eyck,* ed. Foister, Jones, and Cool, 97–100; and Till-Holger Borchert and Paul Huvenne, "Van Eyck and the Invention of Oil Painting: Artistic Merits in Their Literary Mirror," in *The Age of Van Eyck,* ed. Borchert, 220–25.

38. Francisco Pacheco, *Arte de la pintura,* ed. Bonaventura Bassegoda I. Hugas (Madrid: Ediciones Catedra, 1990); the work was published posthumously in 1649. The anecdote on Coxie is in van Mander, *Lives,* 1:61.

39. Pächt, *Van Eyck,* 13.

CHAPTER THREE

1. Alan St. Hill Brock, *A History of Fireworks* (London: George G. Harrap, 1949), 196. Other general histories of fireworks include Arthur Lotz, *Das Feuerwerk, seine Geschichte und Bibliographie* (Leipzig: Verlag Karl W. Hiersemann, 1941); George Plimpton, *Fireworks, a History and Celebration* (New York: Doubleday, 1989); George Kohler and Alice Villon-Lechner, eds., *Die schöne Kunst der Verschwendung: Fest und Feuerwerk in der europäischen Geschichte* (Zurich and Munich: Artemis Verlag, 1988). On early modern fireworks, see Kevin Salatino, *Incendiary Art: The Representation of Fireworks in Early Modern Europe* (Santa Monica, CA: Getty Research Institute for the History of Art and Humanities, 1997); O. W. Schaub, "Pleasure Fires: Fireworks in the Court Festivals in Italy, Germany, and Austria during the Baroque" (PhD diss., Kent State University, 1978); Eberhard Fähler, *Feuerwerke des Barock: Studien zum öffentlichen Fest und seiner literarischen Deutung vom 16. bis 18. Jahrhundert* (Stuttgart: Metzler, 1974).

2. Quoted in Edgar F. Smith, *James Cutbush, an American Chemist, 1788–1823* (Philadelphia: Lippincott, 1919), 84; James Cutbush, *A System of Pyrotechny* (Philadelphia, 1825).

3. J. R. Partington, *A History of Greek Fire and Gunpowder* (Cambridge: W. Heffer, 1960).

4. Edgar Zilsel, "The Sociological Roots of Science," *Social Studies of Science* 30 (2000): 935–49; Paolo Rossi, *Philosophy, Technology, and the Arts in the Early Modern Era,* trans. Salvator Attanasio (New York: Harper and Row, 1970); J. A. Bennett, "The Mechanics' Philosophy and the Mechanical Philosophy," *History of Science* (1986), 24:1–28; Simon Schaffer, "Experimenters' Techniques, Dyers' Hands, and the Electric Planetarium," *Isis* 88 (1997): 456–83; Pamela H. Smith, *The Body of the Artisan: Art and Experience in the Scientific Revolution* (Chicago: University of Chicago Press, 2004).

5. Pamela Long, *Openness, Secrecy, Authorship: Technical Arts and the Culture of Knowledge from Antiquity to the Renaissance* (Baltimore, MD: Johns Hopkins University Press, 2001); Pamela Long, "The Openness of Knowledge: An Ideal and Its Context in Sixteenth-Century Writings on Mining and Metallurgy," *Technology and Culture* 32 (1991): 318–55; William Eamon, *Science and the Secrets of Nature* (Princeton, NJ: Princeton University Press, 1994).

6. On scientific controversy, see, e.g., H. M. Collins, ed., "Knowledge and Controversy: Studies of Modern Natural Science," special issue, *Social Studies of Science* 11 (1981): 3–158; H. M. Collins, *Changing Order: Replication and Induction in Scientific Practice,* 2d ed. (Chicago: University of Chicago Press, 1992).

Historical studies developing these ideas include Steven Shapin, "The Politics of Observation: Cerebral Anatomy and Social Interests in the Edinburgh Phrenology Disputes," in *On the Margins of Science: The Social Construction of Rejected Knowledge*, ed. Roy Wallis (Keele, UK: Keele University Press, 1979), 139–78; and Steven Shapin and Simon Schaffer, *Leviathan and the Air-Pump: Hobbes, Boyle and the Experimental Life* (Princeton, NJ: Princeton University Press, 1985).

7. On the geography of knowledge, see David N. Livingstone, *Putting Science in its Place: Geographies of Scientific Knowledge* (Chicago: University of Chicago Press, 2003); James Secord, "Knowledge in Transit," *Isis* 95 (2004): 654–72; Marie-Noelle Bourguet, Christian Licoppe, and H. Otto Sibum, eds., *Instruments, Travel and Science: Itineraries of Precision from the Seventeenth to the Twentieth Century* (London: Routledge, 2002).

8. On fireworks laboratories in London and St. Petersburg, see O. F. G. Hogg, *The Royal Arsenal: Its Background, Origin, and Subsequent History*, 2 vols. (London: Oxford University Press, 1963); V. M. Rodzevich, *Istoricheskoe opisanie Sanktpeterburgskogo Arsenala za 200 let ego sushchestvovaniia: 1712–1912 gg.* (St. Petersburg: Tipo-litografiia S. Peterburgskoi Tiur'my, 1914); M. V. Danilov, "Zapiski M. V. Danilova," ed. Pavel Stroev, *Russkii Arkhiv* 3 (1883): 46–47.

9. Bruno Latour, *Science in Action: How to Follow Scientists and Engineers through Society* (Milton Keynes, UK: Open University Press, 1987), 215–57. Russian tsar Peter the Great, an active participant in the production of fireworks, accumulated numerous pyrotechnic recipes, notes, and books during his reign. See P. M. Luk'ianov, *Istoriia khimicheskikh promyslov . . . Rossii do kontsa XIX veka*, 6 vols. (Moscow: Izdatel'stvo Akademii Nauk, 1948–65), vol. 5 (1961), 82–90.

10. Examples of manuscripts include *Ein Buch zusamen gezogenn aus vielenn probirten Künsten und Erfarung, wie ein Zeughaus sampt aller monition anheimisch werden sol . . .* (Staatsbibliothek, Berlin, Handschriftenabteilung, Ms. Germ. Fol. 487); the Dutch *Beschrijving van kunst vuurwerken zoo als dezelven zich in hunne uitwerkingen vertoonen*, ca. 1780 (Getty Research Institute, Los Angeles, Special Collections, acc. no. 970037); and John Maskall's Woolwich Arsenal manuscript, *Artificiall Fireworks*, 3 vols., 1785 (Getty Research Institute, Los Angeles, Special Collections, acc. no. 920091). For discussion of early manuscripts, see Long, *Openness, Secrecy, Authorship*, 117–22.

11. See *Traicté des feux artificielz de joye & de recreation*, Paris, ca. 1649 (Bibliothèque Nationale, Paris, MSS. Ms. fr. 1247); Phyllis Dearborn Massar, "Stefano della Bella's Illustrations for a Fireworks Treatise," *Master Drawings* 7 (1969): 294–302. On manuscripts and patronage, see Pamela Long, "Power, Patronage, and the Authorship of Ars: From Mechanical Know-How to Mechanical Knowledge in the Last Scribal Age," *Isis* 88 (1997): 1–41.

12. Casimir Siemienowicz, *The Great Art of Artillery*, trans. George Shelvocke (London, 1729), 145–47.

13. Robert Norton, *The Gunner, Shewing the Whole Practise of Artillerie* (London, 1628), dedication, unpaginated.

14. George Shelvocke, preface to Siemienowicz, *Great Art of Artillery,* ii.

15. Mary J. Henninger-Voss, "How the 'New Science' of Cannons Shook Up the Aristotelian Cosmos," *Journal of the History of Ideas* 63 (2002): 371–97; Serafina Cuomo, "Shooting by the Book: Notes on Niccolò Tartaglia's *Nova Scientia,*" *History of Science* 35 (1997): 155–88; Long, *Openness, Secrecy, Authorship,* 195–98.

16. Thomas Stutevill, dedicatory verses in John Babington, *Pyrotechnia or, a Discourse of Artificiall Fire-Works* (London, 1635), unpaginated.

17. Siemienowicz, *Great Art of Artillery,* 147.

18. Ibid., 131.

19. Ken Alder, "French Engineers Become Professionals, or: How Meritocracy Made Knowledge Objective," in *The Sciences in Enlightened Europe,* ed. William Clark, Jan Golinski, and Simon Schaffer (Chicago: University of Chicago Press, 1999), 94–125; Ken Alder, *Engineering the Revolution: Arms and Enlightenment in France, 1763–1815* (Princeton, NJ: Princeton University Press, 1997), 136–46. Antoine Picon makes a similar argument in his *French Architects and Engineers in the Age of Enlightenment* (Cambridge: Cambridge University Press, 1992).

20. H. M. Collins discusses "unrecognized knowledge" and other forms of tacit knowledge in *Gravity's Shadow: The Search for Gravitational Waves* (Chicago: University of Chicago Press, 2004), 608–12. On tacit knowledge, see also Mukerji's essay (this volume).

21. On the consequences of travel for knowledge, see Marie-Noëlle Bourguet, Christian Licoppe, and H. Otto Sibum, eds., *Instruments, Travel and Science: Itineraries of Precision from the Seventeenth to the Twentieth Century* (London: Routledge, 2002).

22. "Sur Torré, artificier du roi," *L'Esprit des Journaux,* Dec. 1780, 191–204; Brock, *History of Fireworks,* 61. On the Ruggieri, see below.

23. Sarah Hengler became famous for her fireworks in the eighteenth century; see Anita McConnell, "Hengler, Sarah (ca. 1765–1845)," in *Oxford Dictionary of National Biography,* ed. H. C. G. Matthew and Brian Harrison, 60 vols. (Oxford: Oxford University Press, 2004), 26:353.

24. On Italian fireworks, see Lucia Cavazzi, *Fochi d'allegrezza a Roma dal Cinquecento all'Ottocento* (Rome: Quasar, 1982); Giulio Ferrari, *Bellezze architettoniche per le feste della Chinea in Roma nei secoli 17. e 18.: Composizioni di palazzi, padiglioni, chioschi, ponti, ecc. per macchine pirotecniche* (Turin: C. Crudo, 1919).

25. Michel Foucault, *Society Must Be Defended, Lectures at the Collège de France, 1975–76,* ed. Mauro Bertani and Alessandro Fontana (New York: Picador, 2003), 179; Long, *Openness, Secrecy, Authorship,* 72–96.

26. In other words, patronage could function as an "institution without walls," as Mario Biagioli has put it. See Mario Biagioli, "Galileo's System of Patronage," *History of Science* 28 (1990): 1–62.

27. Patrick Bracco and E. Lebovici, *Ruggieri, 250 ans de feux d'artifice* (Paris: Denoël, 1988), especially 33–50; Henri de Chennevières, "Les Ruggieri: artificiers,

1730–1885," *Gazette des Beaux-Arts* 36 (1887): 132–40; Claude-Fortuné Ruggieri, *Précis historique sur les fêtes, les spectacles et les réjouissances publiques* (Paris, 1830), 77n1 and passim.

28. Ruggieri, *Précis historique*, 77n1; Bracco and Lebovici, *Ruggieri, 250 ans*, 34.

29. Amédée-François Frézier, *Traité des feux d'artifice pour le spectacle. Nouvelle édition, toute changée & considérablement augmentée*, 2d ed. (Paris, 1747), 439.

30. De Chennevières, "Ruggieri," 137.

31. On Gaetano Ruggieri and Giuseppe Sarti, see *A description of the machine for the fireworks, with all its ornaments, and a detail of the manner in which they are to be exhibited in St. James's Park, Thursday, April 27, 1749, on account of the general peace, signed at Aix La Chapelle, October 7, 1748. Published by order of his Majesty's Board of Ordnance* (London, 1749); Danilov, "Zapiski," 37; Bracco and Lebovici, *Ruggieri, 250 ans*, 37, 50.

32. *Catalogue des livres rares et précieux composant la bibliothèque de M. E.-F.-D. Ruggieri. Prix d'adjudication . . . précédés de la notice et de la table des divisions du catalogue* (Paris, 1873).

33. Roger Armand Weigert, "Les feux d'artifice ordonnées par le bureau de la ville de Paris au XVIIe siècle," *Paris et Ile-de-France—Mémoires* (Fédération des Sociétés Historiques et Archéologiques de Paris et de l'Ile-de-France) 3 (1951): 194–98.

34. Amédée-François Frézier, *Traité des feux d'artifice pour le spectacle*, 1st ed. (Paris, 1706); for the 1747 edition see n. 29 above.

35. Frézier, *Traité . . . Nouvelle édition*, 383.

36. Ibid., 446–47.

37. Ibid., 384–92; see also xi–xxvi.

38. Ibid., 384. In the preface (vi), Frézier suggests that his audience is composed of "men of letters."

39. Jean-Louis de Cahuzac, "Feu d'artifice," in *Encyclopédie, ou dictionnaire raisonné des science, des arts et des métiers*, ed. Denis Diderot and Jean d'Alembert, 17 vols. (Paris, 1751–65), vol. 6 (1756), 639.

40. Ibid., 639, translated in Salatino, *Incendiary Art*, 23.

41. J. C. Perrinet d'Orval, "Feux d'artifice (artificier)," in *Encyclopédie*, ed. Diderot and d'Alembert, 6:640–47; Perrinet d'Orval also wrote the article "Artifice," in ibid., vol. 1 (1751), 740–44.

42. John R. Pannabecker, "Representing Mechanical Arts in Diderot's *Encyclopédie*," *Technology and Culture* 39 (1998): 33–73. See also Alder, *Engineering the Revolution*, 127–46; Cynthia J. Koepp, "The Alphabetical Order: Work in Diderot's *Encyclopédie*," in *Work in France: Representations, Meaning, Organization, and Practice*, ed. Steven Laurence Kaplan and Cynthia J. Koepp (Ithaca, NY: Cornell University Press, 1986), 229–57; Simon Schaffer, "Enlightened Automata," in *Sciences in Enlightened Europe*, ed. Clark, Golinski, Schaffer, 129–35.

43. Perrinet d'Orval, "Feux d'artifice (artificier)," 645.

44. J. C. Perrinet d'Orval, *Essay sur les feux d'artifice* (Paris, 1745), 153.

45. J. C. Perrinet d'Orval, *Manuel de l'artificier. Seconde édition, révué, corrigée & augmentée* (Paris, 1757), 114.

46. Claude-Fortuné Ruggieri's later comments also suggest the technique was not secret: "This ingenious contrivance [of communicating fire] at first astonished the physicists [Physiciens] of the time, who said when it was explained to them that nothing could be more simple and that anyone could have done it at once." Claude-Fortuné Ruggieri, *Elémens de pyrotechnie,* 1st ed. (Paris, 1801), 138.

47. Horace Walpole, quoted in Brock, *History of Fireworks,* 165–66.

48. For disputes, see Danilov, "Zapiski," 37; Brock, *History of Fireworks,* 50. New treatises included M. V. Danilov, *Dovol'noe i iasnoe pokazanie po kotoromu Vsiakoi sam' soboiu liuzhet' prigotovliat' i delat' vsiakie feierverki i raznyia illuminatsii* (Moscow, 1779); Robert Jones, *A new treatise on artificial fireworks* (London, 1765).

49. G. W. Leibniz, *Sämtliche Schriften und Briefe: Dritte Reihe: Mathematischer, naturwissenschaftlicher und technischer Briefwechsel, Vierter Band, Juli 1683–1690* (Berlin: Akademie Verlag, 1995), 410, 413.

50. Pierre Nicolas Le Chéron d'Incarville, "Manière de faire les fleurs dans les feux d'artifice chinois," *Mémoires de mathématique et de physique, présentés à l'Académie Royale des Sciences par divers sçavans, et lus dans ses assemblées* 4 (1763): 66–94. On d'Incarville, see R. Simon, "Voyage de P. N. Le Cheron d'Incarville en Chine sur le 'Jason,'" *Archivum Historicum Societatis Iesu* 40 (1971): 423–36.

51. Louis Petit de Bachaumont, *Mémoires secrets pour servir à l'histoire de la république des lettres en France depuis 1762 jusqu'à nos jours, ou journal d'un observateur,* 31 vols. (Paris, 1760–74), 5:111–12, 134–35, 137–39; Robert M. Isherwood, *Farce and Fantasy: Popular Entertainment in Eighteenth-Century Paris* (Oxford: Oxford University Press, 1986), 138–40; Bracco and Lebovici, *Ruggieri, 250 ans,* 49–50.

52. Ruggieri, *Elémens de pyrotechnie*; an English translation of the third edition is available as Claude-Fortuné Ruggieri, *Principles of Pyrotechny,* trans. Stuart Carlton (Paris, 1821; Buena Vista, CA: MP Associates, 1994). On Chaptal's *Elémens de chimie,* see Bernadette Bensaude-Vincent, "A View of the Chemical Revolution through Contemporary Textbooks: Lavoisier, Fourcroy and Chaptal," *British Journal for the History of Science* 23 (1990): 435–60.

53. Jeff Horn and Margaret C. Jacob, "Jean-Antoine Chaptal and the Cultural Roots of French Industrialization," *Technology and Culture* 39 (1998): 671–98.

54. Ruggieri, *Principles of Pyrotechny,* 32–33, 309, 318; Ruggieri, *Précis historique,* 102, 330–35, 342–45, 349–62, for descriptions of Claude's fireworks performances; *Times* (London), 29 July 1819, 2; Brock, *History of Fireworks,* 155. Many details of Claude's life are given in *Précis pour Michel-Marie Ruggieri* (Paris, 1830).

55. On the Jardin Ruggieri, see Ruggieri, *Précis historique,* 77–79; Gilles-

Antoine-Langlois, *Folies, tivolis, et attractions: Les premiers parcs de loisirs parisiens* (Paris: Délégation à l'Action Artistique de la Ville de Paris, Difusion, Hachette, 1991), 71–73; Isherwood, *Farce and Fantasy,* 56–59.

56. Antoine Lavoisier, "Remarques sur la composition de quelques feux d'artifice colores en bleu et en jaune, dont je remets les recettes cachetées entre les mains de l'académie," 1766 (Archive de l'Académie des Sciences, Institut de France, Paris, ref. 333-A); Antoine Lavoisier, *Elements of Chemistry, in a systematic order containing all the modern discoveries* . . ., trans. Robert Kerr (Philadelphia, 1799), 437.

57. Antoine Lavoisier, "Rapport sur les procédés d'artifices proposés par M. Ruggieri," in *Ouevres de Lavoisier,* ed. Edouard Grimaux, 6 vols. (Paris, 1862–93), 4:417–18.

58. Ruggieri, *Elémens de pyrotechnie,* 1st ed., xii–xiii.

59. Ruggieri, *Principles of Pyrotechny,* 36, where Claude also invokes Chaptal to make this point.

60. Ibid., 33–34.

CHAPTER FOUR

This essay is drawn from my *Plants and Empire: Colonial Bioprospecting in the Atlantic World* (Cambridge, MA: Harvard University Press, 2004).

1. Carl Linnaeus, *Critica Botanica* (Leiden, 1737), preface, no. 213.

2. Frans A. Stafleu, *Linnaeus and the Linnaeans: The Spreading of Their Ideas in Systematic Botany, 1735–1789* (Utrecht: A. Oosthoek's Uitgeversmaatschappy, 1971); John Heller, *Studies in Linnaean Method and Nomenclature* (Frankfurt: Verlag Peter Lang, 1983); Tore Frängsmyr, ed., *Linnaeus: The Man and His Work* (Berkeley and Los Angeles: University of California Press, 1983); G. Perry, "Nomenclatural Stability and the Botanical Code: A Historical Review," in *Improving the Stability of Names: Needs and Options,* ed. D. L. Hawksworth (Königstein: Koeltz, 1991), 79–93; Dirk Stemerding, *Plants, Animals, and Formulae: Natural History in the Light of Latour's "Science in Action" and Foucault's "The Order of Things"* (Enschede: School of Philosophy and Social Sciences, University of Twente, the Netherlands, 1991); Londa Schiebinger, *Nature's Body: Gender in the Making of Modern Science* (Boston: Beacon Press, 1993); Lisbet Koerner, *Linnaeus: Nature and Nation* (Cambridge, MA: Harvard University Press, 1999); Paul Farber, *Finding Order in Nature: The Naturalist Tradition from Linnaeus to E. O. Wilson* (Baltimore, MD: Johns Hopkins University Press, 2000).

3. J. McNeill, "Latin, the Renaissance Lingua Franca, and English, the 20th Century Language of Science: Their Role in Biotaxonomy," *Taxon* 46 (1997): 751–57, especially 755. See also D. J. Mabberley, "Robert Brown of the British Museum: Some Ramifications," in *History in the Service of Systematics,* ed. Alwyne Wheeler and James Price (London: Society for the Bibliography of Natural History, 1981), 101–9, especially 108; and A. J. Cain, "Logic and Memory in Linnaeus's

System of Taxonomy," *Proceedings of the Linnean Society of London* 169 (1958): 144–63.

4. Bruno Latour, *Science in Action* (Cambridge, MA: Harvard University Press, 1987), chap. 6.

5. John Merson, "Bio-prospecting or Bio-piracy: Intellectual Property Rights and Biodiversity in a Colonial and Postcolonial Context," in "Nature and Empire: Science and the Colonial Enterprise," ed. Roy MacLeod, special issue, *Osiris* 15 (2000): 282–96. See also Michael Balick, Elaine Elisabetshy, and Sarah Laird, eds., *Medicinal Resources of the Tropical Forest* (New York: Columbia University Press, 1996).

6. Jorge Cañizares-Esguerra, *How to Write the History of the New World: Histories, Epistemologies, Identities in the Eighteenth-Century Atlantic World* (Stanford, CA: Stanford University Press, 2001), 155–60.

7. Michael Craton and James Walvin, *A Jamaican Plantation: The History of Worthy Park, 1670–1970* (Toronto: University of Toronto Press, 1970), 148–49; Jerome Handler and JoAnn Jacoby, "Slave Names and Naming in Barbados, 1650–1830," *William and Mary Quarterly* 53 (1996): 685–728; Trevor Burnard, "Slave Naming Patterns: Onomastics and the Taxonomy of Race in Eighteenth-Century Jamaica," *Journal of Interdisciplinary History* 31 (2001): 325–46.

8. John Garrigus, "Redrawing the Color Line: Gender and Social Construction of Race in Pre-revolutionary Haiti," *Journal of Caribbean History* 30 (1996): 29–50, especially 38.

9. For another example of multiple naming systems in the history of science, see Scott Montgomery, *The Moon and the Western Imagination* (Tucson: University of Arizona Press, 1999).

10. Olof Swartz reclassified the *Poinciana* under *Caesalpinia* in the 1790s; both names are used today (*Observationes Botanicae quibus Plantae Indiae Occidentalis* [Erlangen, 1791], 165–66).

11. On *Quassia*, see David de Isaac Cohen Nassy, *Essai historique sur la colonie de Surinam* (Paramaribo, 1788), 71–76.

12. S. M. Walters, "The Shaping of Angiosperm Taxonomy," *New Phytologist* 60 (1961): 74–84, quotation on 74.

13. Linnaeus, letter to Baeck, cited in Jean-Paul Nicolas, "Adanson, the Man," in *Adanson: The Bicentennial of Michel Adanson's "Familles des plantes,"* ed. George Lawrence (Pittsburgh: Hunt Botanical Library, 1963), 1–121, quotation on 51.

14. Alexander von Humboldt (and Aimé Bonpland), *Personal Narrative of Travels to the Equinoctial Regions of the New Continent, during the Years 1700–1804*, trans. Helen Williams, 7 vols. (London, 1821), 5:208.

15. Pierre Barrère, *Nouvelle relation de la France equinoxiale* (Paris, 1743), 39.

16. Jorge Cañizares Esguerra, "Spanish America: From Baroque to Modern Colonial Science," in *Science in the Eighteenth Century*, ed. Roy Porter, vol. 4 of the *Cambridge History of Science* (Cambridge: Cambridge University Press, 2003), 729.

17. Antonio Lafuente and Nuria Valverde, "Linnaean Botany and Spanish Imperial Biopolitics," in *Colonial Botany: Science, Commerce, and Politics in the Early Modern World,* ed. Londa Schiebinger and Claudia Swan (Philadelphia: University of Pennsylvania Press, 2005), 134–47.

18. Michel Foucault, *The Order of Things: An Archaeology of Human Sciences* (New York: Vintage, 1973), 63–67. Foucault states that "the sign is the pure and simple connection between what signifies and what is signified (a connection that may be arbitrary or not . . .)" (67). On this point, see Staffan Müller-Wille, *Botanik und weltweiter Handel: Zur Begründung eines natürlichen Systems der Pflanzen durch Carl von Linné (1707–1778)* (Berlin: Verlag für Wissenschaft und Bildung, 2000), chap. 5. See also Gordon McOuat, "Species, Rules and Meaning: The Politics of Language and the Ends of Definitions in 19th Century Natural History," *Studies in the History and Philosophy of Science* 27 (1996): 473–519.

19. B. D. Jackson, "The New Index of Plant Names," *Journal of Botany* 25 (1887): 66–71, 150–51, quotation on 68.

20. W. T. Stearn, "The Background of Linnaeus's Contributions to the Nomenclature and Methods of Systematic Biology," *Systematic Zoology* 8 (1959): 4–22, especially 5. See also Edward Lee Greene, *Landmarks of Botanical History* (Washington: Smithsonian, 1909).

21. Linnaeus, *Critica Botanica,* no. 256.

22. See, e.g., Christian Mentzelius, *Index Nominum Plantarum Multilinguis* (Berlin, 1682).

23. Jerry Stannard, "Botanical Nomenclature in Gersdorff's *Feldtbüch der Wundartzney,*" in *Science, Medicine, and Society in the Renaissance,* ed. Allen Debus (New York: Science History Publication, 1972), 87–103.

24. John Gerard, *The Herbal or General History of Plants* (London, 1633), chap. 310, p. 845.

25. Robert McVaugh, *Botanical Results of the Sessé and Mociño Expedition, 1787–1830* (Pittsburgh: Hunt Institute for Botanical Documentation, 2000), 19. See also Francisco Hernández, *Historia natural y moral des las Indias* (Barcelona, 1590); Simon Varey, ed., *The Mexican Treasury: The Writings of Dr. Francisco Hernández* (Stanford, CA: Stanford University Press, 2000).

26. Charles de Rochefort, *Histoire naturelle et morale des Iles Antilles de l'Amerique* (Rotterdam, 1665), 104–6.

27. Charles Plumier, *Description des plantes de l'Amerique* (Paris, 1693).

28. Hendrik Adriaan van Reede tot Drakenstein, *Hortus Indicus Malabaricus,* 12 vols. (Amsterdam, 1678–93).

29. Pierre Barrère, *Essai sur l'histoire naturelle de la France equinoxiale* (Paris, 1741).

30. Jean-Baptiste Pouppée-Desportes, *Historie des maladies de S. Domingue* (Paris, 1770).

31. Linnaeus to Haller, cited in Linnaeus, *Critica Botanica,* vii; see also no. 229.

32. Ibid., no. 238.

33. Ibid.

34. For Linnaeus's list of "names commemorating distinguished botanists," see ibid., at 11. Linnaeus's knowledge of several insect species came exclusively from her work.

35. Olof Swartz, *Flora Indiae Occidentalis,* 2 vols. (London, 1797–1806), s.v. "*Meriania purpurea.*"

36. Londa Schiebinger, *The Mind Has No Sex? Women in the Origins of Modern Science* (Cambridge, MA: Harvard University Press, 1989).

37. Linnaeus, *Critica Botanica,* nos. 218, 229, 238.

38. Paul Hermann, *Horti Academici Lugduno-Batavi Catalogus* (Leiden, 1687), 192.

39. The name Poinci or Poincillade seems to have been given by du Tertre and modified and brought into systematic botany by Tournefort, as *Poinciana.* [Denis Joncquet], *Hortus Regius* (Paris, 1666), 3; Joseph Pitton de Tournefort, *Elemens de botanique* (Paris, 1694), 1:491–92, 3: pl. 391. See also Jean-Baptiste du Tertre, *Histoire generale des Ant-isles* (Paris, 1667–71), 1:125–26.

40. Carl Linnaeus, *Hortus Cliffortianus* (Amsterdam, 1737), 158.

41. Mauricio Nieto Olarte, "Remedies for the Empire: The Eighteenth Century Spanish Botanical Expeditions to the New World" (PhD diss., Department of History of Science and Technology, Imperial College, London, 1993), 116–18.

42. Carl Linnaeus, *Amoenitates Academicae* (Leiden, 1749–90), s.v. "Carl Magnus Blom, *Lignum Quassiae,* 1763"; Nassy, *Essai historique,* 160–62.

43. Charles-Marie de La Condamine, "Sur l'arbe de quinquina" (28 May 1737), in *Histoire mémoires de l'Académie Royale des Sciences* (Amsterdam, 1706–55), 319–46. See also A. W. Haggis, "Fundamental Errors in the Early History of Cinchona," *Bulletin of the History of Medicine* 10 (1941): 417–59; and Jaime Jaramillo-Arango, *The Conquest of Malaria* (London: Heinemann, 1950).

44. Thomas Martyn, cited in David Allen, *The Naturalist in Britain: A Social History* (Harmondsworth, Middlesex: Penguin Books, 1978), 39.

45. Michel Adanson, *Familles des plantes,* 2 vols. (Paris, 1763), 1:iii–cii.

46. Georges-Louis Leclerc, comte de Buffon, *Histoire naturelle* (Paris, 1749), 1:23–25. See Jacques Roger, *Buffon: A Life in Natural History,* trans. Sarah Bonnefoi (Ithaca, NY: Cornell University Press, 1997), 275–78.

47. Buffon, *Histoire naturelle,* 1:8, 13–14, 16–18, 26.

48. Roger, *Buffon,* 275.

49. Ibid., 275–78.

50. Félix Vicq-d'Azyr, *Traité d'anatomie et de physiologie* (Paris, 1786), 1:47–48.

51. Adanson, *Familles des plantes,* 1:xl–xli. Linnaeus's sexual system was widely adopted after 1737 and, until the first decades of the nineteenth century, was generally considered the most convenient system of botanical classification. Linnaeus's system, set out in his *Systema Naturae* (1735), was based on the differ-

ence between the male and female parts of flowers. Linnaeus divided plants into classes based on the number, relative proportions, and position of the male parts, or stamens. These classes were then subdivided into some sixty-five orders based on the number, relative proportions, and positions of the female parts, or pistils. These were further divided into genera (based on the calyx, flower, and other parts of the fruit), species (based on the leaves or some other characteristic of the plant), and varieties. On Linnaeus's sexual system, see Schiebinger, *Nature's Body,* chap. 1.

52. Adanson, *Familles des plantes,* 1:cxlix–clii, clxxiii–clxxiv.

53. Ibid., 1:clxxiii.

54. Ibid., 1:cxlix; 2:318.

55. Ibid., 1:cxxiii; Frans Stafleu, "Adanson and the *Familles des plantes,*" in *Adanson,* ed. Lawrence, 123–259, especially 187.

56. Adanson, *Familles des plantes,* 1:clxxi.

57. Ibid., 1:clxxiii.

58. Nicolas, "Adanson, the Man," 30.

59. Ibid., 57.

60. Cited in Stafleu, "Adanson and the *Familles des plantes,*" 176.

61. Linnaeus, *Amoenitates Academicae,* s.v. "Incrementa botanices, Jac. Bjuur. Wman., 1753."

62. Carl Linnaeus, *Species Plantarum* (1753; London: Ray Society, 1957), Stearn's intro., 11.

63. Adanson was not alone in attempting to reform human language. Constantin-François Chasseboeuf, comte de Volney, like Leibniz before him, wanted to simplify the existing languages by reducing them to a more reasonable, simple grammar and a generally applicable system of writing. See Justin Stagl, *A History of Curiosity: The Theory of Travel, 1550–1800* (Chur, Switzerland: Harwood Academic Publishers, 1995), 288.

64. Nicolas, "Adanson, the Man," 102.

65. Sir William Jones, "The Design of a Treatise on the Plants of India," *Asiatick Researches* 2 (1807): 345–52 (italics in the original), quotations on 346, 348. I thank Roberta Bivins for calling my attention to Jones's work.

66. Whitelaw Ainslie, *Materia Medica of Hindoostan* (Madras, 1813), preface.

67. Lafuente and Valverde, "Linnaean Botany and Spanish Imperial Biopolitics."

68. Richard Drayton, *Nature's Government: Science, Imperial Britain, and the "Improvement" of the World* (New Haven, CT: Yale University Press, 2000), 77; J. F. Michaud, *Biographie universelle* (Graz: Akademische Druck, 1966), s.v. "Aublet."

69. Richard Grove, *Green Imperialism: Colonial Expansion, Tropical Island Edens and the Origins of Environmentalism, 1600–1860* (Cambridge: Cambridge University Press, 1995), 78 (quotation), 89–90.

CHAPTER FIVE

1. Erasmus, *The Colloquies*, trans. Craig R. Thompson (Chicago: University of Chicago Press, 1965), 79–86.

2. John Mandeville, *The Travels* (Harmondsworth: Penguin Books, 1983), 103.

3. H. Pleij, *Dreaming of Cockaigne: Medieval Fantasies of the Perfect Life* (New York: Columbia University Press, 2001), 179–80.

4. K. Niehr, in *Gutenberg Jahrbuch* 76 (2001): 278; H. Kühnel, *Alltag und Fortschritt im Mittelalter* (Vienna: Akademie Verlag, 1986), 5–13; J.-D. Müller, "Curiositas und Erfahrung der Welt im frühen Prosaroman," in *Literatur und Laiendichtung im Spätmittelalter und in der Reformationszeit*, ed. Von L. Grenzmann et al. (Stuttgart: Metzler, 1984), 252. See, in general, B. Smalley, "Ecclesiastical Attitudes to Novelty, c. 1100–c. 1250," in *Church, Society and Politics*, ed. D. Baker (Oxford: Blackwell, 1975), 113–31; E. Gössmann, *Antiqui und Moderni im Mittelalter* (Munich: Ferdinand Schöningh, 1974); H. A. Oberman, *Contra Vanam Curiositatem: Ein Kapitel der Theologie zwischen Seelenheil und Weltall* (Zurich: Theologischer Verlag, 1974), 16–22.

5. Jacques Le Goff, *Sint-Franciscus van Assisi* (Amsterdam: Wereldbibliotheek, 2001), 83, 103, 145; Smalley, "Attitudes," 114.

6. Smalley, "Attitudes," 126; Jan de Weert, *Nieuwe doctrinael of spieghel van sonden*, ed. J. H. Jacobs ('s-Gravenhage: Nijhoff, 1915).

7. Smalley, "Attitudes," 131; Oberman, *Curiositatem*, 10.

8. L. Wierda, " 'Een oetmoedich boeck': Het ideale boek bij de Moderne Devoten," in *Geen povere schoonheid: Laat-middeleeuwse kunst in verband met de Moderne Devotie*, ed. K. Veelenturf (Nijmegen: Valkhof Pers, 2000), 156–58; Oberman, *Curiositatem*.

9. G. Adler, "Die Wiederholung und Nachahmung in der Mehrstimmigkeit," *Vierteljahrsschrift für Musikwissenschaft* 2 (1886): 274; also in art: M. Camille, *Image on the Edge: The Margins of Medieval Art* (London: Reaktion Books, 1992), 36.

10. G. Macropedius, *Aluta (1535)*, ed. J. Bloemendal et al. (Voorthuizen: Florivallis, 1995), 42–43.

11. J. Reynaert et al., *Wat is wijsheid? Lekenethiek in de Middelnederlandse letterkunde* (Amsterdam: Prometheus, 1994), 23.

12. G. van Thienen et al., eds., *Incunabula Printed in the Low Countries: A Census* (Nieuwkoop: De Graaf, 1999), no. 1654 (hereafter *ILC*); E. Neurdenburg, ed., *Van nyeuvont, loosheit ende practike; Hoe sy vrou Lortse verheffen* (Utrecht: A. Oosthoek, 1910).

13. W. L. de Vreese, ed., *Een Spel van Sinne van Charon, de helsche schippere (1551)* (Antwerp: L. Dela Montagne, 1896).

14. H. Pleij, "Nieuws bij Anna Bijns," in *Hoort wonder! Opstellen voor W. P. Gerritsen bij zijn emiritaat*, ed. B. Besamusca et al. (Hilversum: Verloren, 2000), 121–26. On Anna Bijns, see also H. Pleij, "Anna Bijns als pamflettiste? Het refrein over de beide Maartens," *Spiegel der Letteren* 42 (2000): 187–225.

15. For a general impression of medieval and early modern Dutch literature, see Erik Kooper, ed., *Medieval Dutch Literature in Its European Context* (Cambridge: Cambridge University Press, 1994); and especially for the *rederijkers,* see Anne-Laure van Bruaene, "Brotherhood and Sisterhood in the Chambers of Rhetoric in the Southern Low Countries," *Sixteenth Century Journal* 36 (2005): 11–35.

16. H. Pleij, "Carrying Books," in *Medieval Mastery: Book Illumination from Charlemagne to Charles the Bold, 800–1475,* ed. B. Cardon (Louvain: Davidsfonds, 2002), 34–45; H. Pleij, "The Despisers of Rhetoric: Origins and Significance of Attacks on the Art of the Rhetoricians (*Rederijkers*) in the Sixteenth Century," in *Rhetoric—Rhétoriqueurs—Rederijkers,* ed. J. Koopmans et al. (Amsterdam: North-Holland, 1995), 157–74.

17. H. Pleij, *Nederlandse literatuur van de late Middeleeuwen* (Utrecht: HES Publishers, 1990), 158–91. For their strong interest in religious matters, see Gary Waite, *Reformers on Stage: Popular Drama and Religious Propaganda in the Low Countries of Charles V, 1515–1556* (Toronto: University of Toronto Press, 2000).

18. Niehr, in *Gutenberg Jahrbuch,* 278–79.

19. *Kronyk van Vlaenderen, van 580 tot 1467,* 2 vols. (Ghent: Maetschappy der Vlaemsche Bibliophilen, 1839–40), 2:233. In general, see P. Arnade, *Realms of Ritual: Burgundian Ceremony and Civic Life in Late Medieval Ghent* (Ithaca, NY: Cornell University Press, 1996).

20. D. Coigneau, "Matthijs de Castelein 'Excellent poëte moderne,'" *Verslagen en Mededelingen van de Koninklijke Vlaamse Academie voor Taal- en Letterkunde,* 1985, 131.

21. Niehr, in *Gutenberg Jahrbuch,* 278–79; Müller, "Curiositas," 252; L. Arbusow, *Colores rhetorici* (Göttingen: Vandenhoeck and Ruprecht, 1963), 98.

22. H. Pleij and J. Reynaert, "Boekproductie in de overgang van het geschreven naar het gedrukte boek," in *Geschreven en gedrukt: Boekproductie van handschrift naar druk in de overgang van Middeleeuwen naar Moderne Tijd,* ed. H. Pleij and J. Reynaert (Ghent: Academia Press, 2004), 1–17; H. Pleij, "What and How Did Lay Persons Read, or: Did the Laity Actually Read? Literature, Printing and Public in the Low Countries between the Middle Ages and Modern Times," in *Laienlekture und Buchmarkt im späten Mittelalter,* ed. T. Kock and R. Schlusemann (Frankfurt am Main: Peter Lang, 1997), 3–32; Y. Vermeulen, *"Tot profijt en genoegen": Motiveringen voor de produktie van nederlandstalige gedrukte teksten, 1477–1540* (Groningen: Wolters-Noordhoff, 1986), 224–39.

23. *ILC,* no. 1736; F. Loockmans, *LXXI Lustighe Historien oft Nieuwicheden* (Antwerp: H. Verdussen, [ca. 1600]); copy available in University Library, Göttingen; earlier editions from 1577 and 1589. In general, see R. J. Clements and J. Gibaldi, *Anatomy of the Novella* (New York: New York University Press, 1977), 4, 14.

24. *ILC,* no. 420; Vermeulen, *Profijt,* 239; Pleij, *Late Middeleeuwen,* 93.

25. W. Nijhoff and M. E. Kronenberg, *Nederlandsche bibliografie van 1500 tot 1540,* 3 vols. ('s-Gravenhage: Nijhoff, 1923–71), no. 1085 (hereafter NK); Vermeulen, *Profijt,* 239.

26. NK, no. 1667; Vermeulen, *Profijt*, 239.

27. In general, see H. Pleij, "De betekenis van de beginnende drukpers voor de ontwikkeling van de Nederlandse literatuur in Noord en Zuid," *Spektator* 21 (1992): 227–63; Pleij and Reynaert, "Boekproductie."

28. NK, no. 2184; Vermeulen, *Profijt*, 225.

29. NK, no. 0534; Vermeulen, *Profijt*, 225.

30. W. Waterschoot, "De Gentse drukker Joos Lambrecht," *De Zeventiende Eeuw* 8 (1992): 27–32; H. Pleij, "Gent en de stadscultuur in de Nederlanden," in *Carolus—Keizer Karel V, 1500–1558*, ed. H. Soly et al. (Ghent: Snoeck-Ducaju and Zoon, 1999), 125–26.

31. Coigneau, "Castelein."

32. In general, see E. L. Eisenstein, *The Printing Press as an Agent of Change* (Cambridge: Cambridge University Press, 1980), 3–43; M. Giesecke, *Der Buchdruck in der frühen Neuzeit* (Frankfurt am Main: Suhrkamp, 1994); H. Pleij, "De onvoltooide Middeleeuwen: Over de drukpers en het andere gezicht van de Middelnederlandse literatuur," in *Grote lijnen: Syntheses over Middelnederlandse letterkunde* (Amsterdam: Prometheus, 1995), 137–55; Pleij and Reynaert, "Boekproductie."

33. *ILC*, no. 1859; H. Pleij, "Over betekenis en belang van de leesinstructie in de gedrukte proza-*Reynaert* van 1479," in *Geschreven en gedrukt*, ed. Pleij and Reynaert, 207–32.

34. H. Pleij, "Reynard the Fox: The Triumph of the Individual in a Beast Epic," *The Low Countries: Arts and Society in Flanders and the Netherlands* 3 (1995): 233–40.

35. W. Prevenier, "Court and City Culture in the Low Countries from 1100 to 1530," in *Medieval Dutch Literature in Its European context*, ed. E. Kooper (Cambridge: Cambridge University Press, 1994), 22; H. Pleij, "Restyling 'Wisdom,' Remodeling the Nobility, and Caricaturing the Peasant: Urban Literature in the Late Medieval Low Countries," *Journal of Interdisciplinary History* 32 (2002): 689–704.

36. As discussed in H. Pleij, *De sneeuwpoppen van 1511: Stadscultuur in de late Middeleeuwen* (Amsterdam: Meulenhoff, 1998), 321–56.

37. H. Pleij, "Lekenethiek en burgermoraal," *Queeste* 2 (1995): 170–80; W. Blockmans and P. Hoppenbrouwers, *Eeuwen des onderscheids: Een geschiedenis van middeleeuws Europa* (Amsterdam: Prometheus, 2002), 280–85, 349, 442.

CHAPTER SIX

1. Stuart Sherman, *Telling Time: Clocks, Diaries and English Diurnal Form, 1660–1785* (Chicago: University of Chicago Press, 1997). The link between time management and autobiographical writing is also a subject of investigation of the NWO project Controlling Time and Shaping the Self, directed by Dr. Arianne Baggerman, which forms the background for this essay. See www.fhk.eur.nl/onderzoek/egodocumenten (accessed 28 December 2006).

2. The grandfather of the Huygens brothers kept a journal in which he recorded

his children's development. See Arthur Eyffinger, ed., *Huygens herdacht: Catalogus bij de tentoonstelling in de Koninklijke Bibliotheek ter gelegenheid van de 300ste sterfdag van Constantijn Huygens* (The Hague: Koninklijke Bibliotheek, 1987), 79–165. See on this text Rudolf Dekker, *Childhood, Memory and Autobiography in Holland from the Golden Age to Romanticism* (London: Macmillan, 1999). His wife, Susanna Hoefnagels, wrote diary-like letters during the year 1624 to her son Constantijn Sr., who was staying in London at the time. See also Jacob Smit, *De grootmeester van woord- en snarenspel: Het leven van Constantijn Huygens* (The Hague: Martinus Nijhoff, 1980), 19.

3. J. C. G. Boot, ed., "Korte biographische aanteekingen van Constantijn Huygens," *Verslagen en Mededeelingen der Koninklijke Akademie van Wetenschappen Afdeeling Letterkunde* 2e R III (1873): 344–56. See also J. H. W. Unger, "Dagboek van Constantijn Huygens: Voor de eerste maal naar het afschrift van diens kleinzoon uitgegeven," *Oud-Holland* 3 (1885): 1–87.

4. René van Stipriaan, *Het volle leven: Nederlandse literatuur en cultuur ten tijde van de Republiek (circa 1550–1800)* (Amsterdam: Prometheus, 2002), 57.

5. J. A. Worp, ed., "Fragment eener autobiografie van Constantijn Huygens," *Bijdragen en Mededelingen van het Historisch Genootschap* 18 (1897): 1–122; A. H. Kan, ed., *De jeugd van Constantijn Huygens door hem zelf beschreven* (Rotterdam, 1946); C. L. Heesakkers, ed., *Mijn jeugd* (Amsterdam, 1987). A new translation by Frans Blom is available: *Mijn leven verteld aan mijn kinderen*, 2 vols. (Amsterdam: Prometheus/Bert Bakker, 2003).

6. *Journalen van Constantijn Huygens, den zoon Derde Deel* (Utrecht: Kemink en zoon, 1888).

7. *Journaal van Constantijn Huygens, den zoon, gedurende de veldtochten der jaren 1673, 1675, 1676, 1677 en 1678* (Utrecht: Kemink en zoon, 1881).

8. Christiaan Huygens, *Oeuvres complètes*, 22 vols. (The Hague: Martinus Nijhoff, 1888–1950).

9. A. G. H. Bachrach and R. G. Collmer, eds., *Lodewijck Huygens, the English Journal, 1651–1652* (Leiden, 1982). There is also an unpublished text "Reisjournaal van de ambassade naar Spanje," 1660, Repertorium 50 (Koninklijke Bibliotheek, The Hague, Manuscript KA LVII).

In the next generation, the Huygens family declined. Christiaan did not marry and the only son of Constantijn Jr. died young. Lodewijk and Susanna, however, married into families that developed their own traditions of autobiographical writing around the same time and carried these on well into the nineteenth century. These were the families Doubleth and Teding van Berkhout. See George Rataller Doubleth (1600–1655), "Journael van de Mechelse reyse, Dec. 1653" (University Library Leiden, Manuscript Ltk 858); Robert Fruin et al., eds., "Een Hollander op de kermis te Antwerpen in 1654," *Bijdragen voor Vaderlandsche Geschiedenis en Oudheidkunde*, 2e S., 6 (1868): 314–35. On this manuscript, see R. Fruin, *Verspreide geschriften* (The Hague, 1901), 4:195–244. See also Pieter Teding van Berkhout (1643–1713), "Journal contenant mes occupations depuis le 1. de janvier 1669 jusqu'au 15.iesme du moijs de juijllet 1669" (Koninklijke Bibliotheek Manuscript 129 D 16). On this manuscript, see Jeroen Blaak, *Geletterde levens: Dagelijks lezen en schrijven in de vroegmoderne tijd in Nederland 1624–1770* (Hilversum: Verloren, 2004).

10. For more background, see Jonathan Israel, *The Dutch Republic: Its Rise, Greatness and Fall, 1477–1806* (Oxford: Clarendon Press, 1998); and Simon Schama, *The Embarrassment of Riches: An Interpretation of Dutch Culture in the Golden Age* (New York: Alfred A. Knopf, 1987).

11. J. A. Worp, ed., *De gedichten van Constantijn Huygens*, 9 vols. (Groningen: Wolters, 1911–17), 8:348, 24.

12. E. de Jongh, *Portretten van echt en trouw: Huwelijk en gezin in de Nederlandse kunst van de zeventiende eeuw* (Zwolle: Waanders, 1986), 238.

13. Jean Puget de la Serre, *Secretaris d'à la mode* (Amsterdam: Jacob Benjamyn, 1652). See also Ann Jensen Adams, "The Paintings of Thomas de Keyser (1596/7–1667): A Study of Portraiture in Seventeenth-Century Amsterdam" (PhD diss., Harvard University, 1985), 4 vols. in 2 (Ann Arbor, MI: University Microfilms International, 1986); Ann Jensen Adams, "Disciplining the Hand, Disciplining the Heart: Letter-Writing, Paintings, and Practices in Seventeenth-Century Holland," in *Love Letters: Dutch Genre Paintings in the Age of Vermeer*, exhibition catalog, Bruce Museum, Greenwich, CT, and National Gallery of Ireland, Dublin (London: Frances Lincoln, 2003), 63–78; see also the contribution by Peter C. Sutton.

14. J. F. Heijbroek et al., *Met Huygens op reis* (Zutphen: Terra, 1983).

15. See Cornelis D. Andriessen's entry on Christiaan Huygens in *Encyclopedia of the Enlightenment* (Oxford: Oxford University Press, 2003).

16. Huygens, *Oeuvres complètes*, 18:114–23. See also *De Zeventiende Eeuw* 12, no. 1 (1996), an issue devoted to Christiaan Huygens. For his father's poem on this accomplishment, see Worp, *De gedichten van Constantijn Huygens*, 7:33.

17. Huygens, *Oeuvres complètes*, 17:199–237. See also C. A. Davids, *Zeewezen en wetenschap: De wetenschap en de ontwikeling van de navigatietechniek in Nederland tussen 1585 en 1815* (Amsterdam: De Bataafsche Leeuw, [1987]).

18. J. H. Kluiver, "De ontwikkeling van de vormgeving van het Nederlandse uurwerk als gevolg van Huygens' uitvinding van het slingeruurwerk in 1657," *De Zeventiende Eeuw* 12 (1996): 141–51.

19. On egodocuments, see Rudolf Dekker, ed., *Egodocuments and History: Autobiographical Writing in Its Social Context since the Middle Ages* (Hilversum: Verloren, 2002).

20. Rudolf Dekker, "Sexuality, Elites, and Court Life in the Late Seventeenth Century: The Diaries of Constantijn Huygens, Jr.," *Eighteenth-Century Life* 23 (1999): 94–110.

21. Rudolf Dekker, *Uit de schaduw in 't grote licht: Kinderen in egodocumenten van de gouden eeuw tot de romantiek* (Amsterdam: Verloren, 1995), 69–73; Rudolf Dekker, "Upstairs en downstairs: Meiden en knechts in het dagboek van

Constantijn Huygens Jr.," *Mededelingen van de Stichting Jacob Campo Weyerman* 25 (2002): 78–89.

22. D. J. Roorda, "Constantijn Huijgens de zoon en zijn ambt," in *Rond prins en patriciaat,* by D. J. Roorda (Weesp: Fibula-Van Dishoeck, 1984), 94–118.

23. Rudolf Dekker, "De rafelrand van het zeventiende-eeuwse hofleven in het dagboek van Constantijn Huygens de zoon: Magie en toverij," *Mededelingen van de Stichting Jacob Campo Weyerman* 23 (2000): 94–102.

24. Koninklijke Bibliotheek, The Hague, Manuscript KA LIII a, b, c, d, e, f, g. Text editions: see nn. 6–7 above. See also J. H. Hora Siccama, *Aanteekeningen en verbeteringen op het in 1906 door het Historisch Genootschap uitgegeven Register op de journalen van Constantijn Huygens den zoon* (Amsterdam: Johannes Müller, 1915). Cf. J. H. Hora Siccama, "Het journaal van Constantijn Huygens," *De Gids* 42, no. 2 (1878): 1–56. A more complete description can be found in R. Lindeman et al., *Egodocumenten van Noord-Nederlanders uit de zestiende tot begin negentiende eeuw: Een chronologische lijst* (Haarlem: Stichting Egodocument; Rotterdam: Erasmus Universiteit, 1993), no. 136.

25. Koninklijke Bibliotheek, The Hague, Manuscript KA LIV, nos. 1, 2, 3, 4, 5, 6, 7. Text edition: *Journaal van Constantijn Huygens, den zoon,* 2 vols. (Utrecht: Kemink en zoon, 1876–77). Several passages have been omitted from this printing; these have been included in F. Boersma, "Het ongelukkige lot van een dagboekschrijver," *Groniek* 101 (1988): 29–51; and Lindeman et al., *Egodocumenten,* no. 158.

26. On this subject, see Arianne Baggerman and Rudolf Dekker, "Otto's Watch: Time, Pedagogy and Diary Keeping in the Late 18th Century," in *Seen and Heard: The Place of the Child in Early Modern Europe, 1550–1800,* ed. Andrea Immel and Michael Witmore (New York and London: Routledge, 2006), 277–305.

27. P. Gerbenzon, ed., *Het aantekeningenboek van Dirck Jansz,* reedited by Ingrid van der Vlis and Rudolf Dekker (Hilversum: Verloren, 1993).

28. This word is not included in the *Woordenboek der Nederlandsche Taal.*

29. The first mention, according to *Woordenboek der Nederlandsche Taal,* was in 1728: *Groot plakaat boek,* 6:164b.

30. D. J. Roorda, "De loopbaan van Willem Meester," in *Rond prins en patriciaat.*

31. Huygens, *Oeuvres complètes,* 7:401, 408–16, 425, 430, 436–38, 464, 465, 474, 480.

32. Jeroen Salman, *Populair drukwerk in de Gouden Eeuw: De almanak als handelswaar en lectuur* (Zutphen: Walburg Pers, 1999).

33. Davids, *Zeewezen,* 150–54.

34. Svetlana Alpers, *The Art of Describing: Dutch Art in the Seventeenth Century* (Chicago: University of Chicago Press, 1983); and see also Celeste Brusati, *Artifice and Illusion: The Art and Writing of Samuel van Hoogstraten* (Chicago: University of Chicago Press, 1995).

CHAPTER SEVEN

1. J. F. Martinet, *Katechismus der natuur,* 5th ed., 4 vols. (Amsterdam: J. Allart, 1782–89), 4:490. The first edition was published in Amsterdam in 1777–79; all references in this essay are to the fifth edition.

2. Ibid., 1:12.

3. A. Baggerman and R. Dekker, eds., *Het dagboek van Otto van Eck (1791–1797)* (Hilversum: Verloren, 1998), 80 (17 November 1792). In what follows, references will cite only the date of the diary entry.

4. R. R. Palmer and J. Colton, *A History of the Modern World* (New York: Alfred A. Knopf, 1965).

5. Aileen Fyfe wrote on children's reading of scientific literature around 1800 from this perspective: A. Fyfe, "Young Readers and the Sciences," in *Books and the Sciences in History,* ed. M. Frasca-Spada and N. Jardine (Cambridge: Cambridge University Press, 2000), 276–90. For another example of this kind of analysis, see A. Baggerman, *Een drukkend gewicht: Leven en werk van de zeventiende-eeuwse veelschrijver Simon de Vries* (Amsterdam and Atlanta: Rodopi, 1993), 202–26.

6. J. de Kruif, *Liefhebbers en gewoontelezers: Leescultuur in Den Haag in de achttiende eeuw* (Zutphen: De Walburg Pers, 1999).

7. H. Brouwer, *Lezen en schrijven in de provincie: De boeken van Zwolse boekverkopers, 1777–1849* (Leiden: Primavera Pers, 1995); J. J. Kloek and W. W. Mijnhardt, "In andermans boeken is het duister lezen: Reconstructie van de vroeg negentiende-eeuwse leescultuur in Middelburg op basis van een boekhandelsadministratie," *Forum der Letteren* 29 (1988): 15–29.

8. M. de Certeau, *The Practice of Everyday Life* (Berkeley and Los Angeles: University of California Press, 1984), xii–xiii.

9. C. Ginzburg, *The Cheese and the Worms: The Cosmos of a Sixteenth-Century Miller* (London: Routledge and Kegan Paul, 1980).

10. R. Darnton, "Readers Respond to Rousseau," in *The Great Cat Massacre and Other Episodes in French Cultural History,* by R. Darnton (New York: Basic Books and Allen Lane, 1984).

11. J. Brewer, "Cultural Consumption in Eighteenth-Century England: The View of the Reader," in *Frühe Neuzeit—Frühe Moderne? Forschungen zur Vielschichtigkeit von Übergangsprozessen,* ed. R. Vierhaus et al., Veröffentlichungen des Max-Planck-Instituts für Geschichte (Göttingen: Vandenhoeck und Ruprecht, 1992), 366–91. A more recent and different version of this article is J. Brewer, "Reconstructing the Reader: Prescriptions, Texts and Strategies in Anna Larpent's Reading," in *The Practice and Representation of Reading in England,* ed. J. Raven, H. Small, and N. Tadmor (Cambridge: Cambridge University Press, 1996), 226–46. J. Blaak is finishing his PhD thesis on daily reading in early modern Netherlands at the Erasmus University Rotterdam. See also D. D. Hall, "The Uses of Literacy in New England, 1600–1850," in *Printing and Society in Early America,* ed. W. L.

Joyce et al. (Worcester: American Antiquarian Society, 1983), 1–47; A. Baggerman, "Otto van Eck en de anderen: Sporen van jonge lezers in schriftelijke bronnen," in *Tot volle waschdom: Nieuwe hoofdstukken voor de geschiedenis van de kinder- en jeugdliteratuur,* ed. B. Dongelmans et al. (The Hague: Biblion Uitgeverij, 2000), 211–25; A. Baggerman, "The Cultural Universe of a Dutch Child: Otto van Eck and His Literature," *Eighteenth-Century Studies* 31 (1997): 129–34; A. Baggerman, "Lezen tot de laatste snik: Otto van Eck en zijn dagelijkse literatuur (1780–1798)," in *Jaarboek voor Nederlandse boekgeschiedenis,* ed. H. Brouwer et al. (Leiden: Nederlandse Boekhistorische Vereniging, 1994), 1:57–89.

12. C. Ginzburg, "Just One Witness," in *Probing the Limits of Representation: Nazism and the "Final Solution,"* ed. S. Friedlander (Cambridge, MA: Harvard University Press, 1992), 82–96.

13. Baggerman, "Lezen tot de laatste snik."

14. For an excellent Dutch study in this field based on the reactions of real but professional readers, see D. L. van Werven, "Dutch Readings of George Eliot, 1856–1885" (PhD diss., Utrecht, 2001).

15. I coordinate this project at the Erasmus University Rotterdam. See www.fhk.eur.nl/onderzoek/egodocumenten.

16. Ginzburg, "Just One Witness," 85.

17. Martinet, *Katechismus der natuur,* 1:*8v.

18. J. Swammerdam, *Bybel der Nature* (Amsterdam, 1737), 357. Cited in J. Bots, *Tussen Descartes en Darwin: Geloof en natuurwetenschappen in de 18e eeuw in Nederland* (Assen: Van Gorcum, 1972), 141.

19. On this subject, see Bots, *Tussen Descartes en Darwin*; A. Lovejoy, *The Great Chain of Being: A Study of the History of an Idea* (Cambridge, MA: Harvard University Press, 1936). Cf. M. Leathers Kuntz et al., eds., *Jacob's Ladder and the Tree of Life* (New York: Lang, 1987).

20. Bots, *Tussen Descartes en Darwin,* 105.

21. Ibid., 146.

22. N. A. Pluche, *Schouwtoneel der natuur of samenspraaken over de bysonderheden der natuurlyke historie* (1st ed., Paris, 1732; 1st ed., Amsterdam, 1737–49).

23. Martinet, *Katechismus der natuur,* 1:357–62.

24. See also E. Koolhaas-Grosveld, "Van de tuin naar de wildernis: Over de waardering voor de natuur en het landschap in Nederland in de achttiende eeuw," in *Langs velden en wegen: De verbeelding van het landschap in de 18de en 19de eeuw,* ed. W. Loos et al. (Blaricum: V+K Publishing, 1997), 47–70.

25. F. de Haas and B. Paasman, *J. F. Martinet en de achttiende eeuw: In ijver en onverzadelijke lust om te leeren* (Zutphen: De Walburg Pers, 1987), 62–64.

26. Ibid., 62–63.

27. Martinet's nephew Joannes Kuyper is supposed to have been the model for the pupil De V. See ibid., 67. For more on this connection, see Parenteel van Thimothee Martinet, at http://members.tripod.com/~geneologie/parmart.htm.

28. Martinet, *Katechismus der Natuur,* 4:483.

29. Ibid., 1:ix.

30. The first edition of the first volume ran to 1,200 copies (de Haas and Paasman, *J. F. Martinet en de achttiende eeuw,* 64). The earliest imitations were even published within a year of the appearance of the last volume of *Katechismus der natuur:* J. De Vries, *Natuurkundige en ophelderende aanmerkingen over J. F. Martinet's "Katechismus der natuur"* (Amsterdam, 1779; 2d printing, 1780; 4th printing, 1791); Taco Brans, *Catechismus der natuur ten gebruike van kinderen* (Amsterdam, 1779).

31. See, among others, Bots, *Tussen Descartes en Darwin,* 72–81.

32. M. van Dijk, "Spel- en speelcultuur in de negentiende eeuw," *Volkskundig Bulletin* 9 (1983): 53–82.

33. H. Brouwer, *Lezen en schrijven in de provincie: De boeken van Zwolse boekverkopers, 1777–1849* (Leiden: Primavera Pers, 1995), 230.

34. Ibid., 199.

35. Ibid., 16–17, 92, 94, 140, 171, 178, 182.

36. Ibid., 199.

37. M. van der Bijl, *Leeslust baart kunde: 200 jaar leesgezelschap in Alkmaar* (Alkmaar: n.p., 1993), 26; H. van Goinga, *Alom te bekomen: Veranderingen in de boekdistributie in de Republiek, 1720–1800* (Amsterdam: De Buitenkant, 1999), 274.

38. J. van Lennep, *Nederland in den goeden ouden tijd zijnde het dagboek van hunne reis te voet, per trekschuit en per diligence van Jacob van Lennep en zijn vriend Dirk van Hogendorp door de Noord-Nederlandse proventien in den jare 1823* (Utrecht: De Haan, 1942), 44.

39. Nationaal Archief II, Stukken Nieuwenhuis, N. Domela en D. N. Nyegaard 18e en 20e eeuw, inv. 151.

40. Q. M. R. VerHuell, *Levensherinneringen, 1787–1812,* ed. L. Turksma (Westervoort: Van Gruting, 1996), 19.

41. Ibid., 16.

42. Ibid., 17.

43. Ibid., 19–20.

44. Ibid., 19. On collecting mosses, see Martinet, *Katechismus der natuur,* 3:374.

45. VerHuell, *Levensherinneringen,* x–xi.

46. Universiteitsbibliotheek Amsterdam, coll. hss. XVII E 6 A. I am grateful to Jeroen Blaak for pointing out this passage to me.

47. 16 November 1792.

48. This appears, for instance, from notes in the accounts book of De Ruit, the family's country house (Rijksarchief Gelderland, Familiearchief Van Eck, inv. 330).

49. For more on this transformation of the garden, see A. Baggerman and R. Dekker, *De horizon van Otto van Eck: Kind tussen twee werelden* (Amsterdam: Wereldbibliotheek, 2005).

50. 2 December 1792.

51. 17 November 1792; Martinet, *Katechismus der natuur,* 1:14–15.

52. 17 November 1792.

53. 16 November 1792.

54. [J. B. Basedow], *Manuel élémentaire d'éducation* (Berlin, etc., 1774).

55. Martinet, *Katechismus der natuur,* 4:211. See also ibid., 36–41.

56. 17 November 1792.

57. Otto read Samuel Richardson's *Little Grandisson* and *Little Clarissa* in Dutch in the translated and revised editions by Margaretha Geertruida de Cambon-van der Werken: *De kleine Grandisson* (The Hague, 1782) and *De kleine Klarissa* (The Hague: J. F. Jacobs de Age, 1790).

58. On Otto's reception of the other literature he read, see A. Baggerman, "Lezen tot de laatste snik," 75–80; Baggerman, "Cultural Universe of a Dutch Child."

59. 16 November 1792.

60. Raymonde Patmos, ed., *Willem van den Hull: Autobiografie (1778–1854)* (Hilversum: Verloren, 1996), 58.

61. Universiteitsbibliotheek Amsterdam, Coll. hss.: Reveil, D 25c. I am grateful to Ellen Grabowsky for drawing my attention to this autobiography.

62. K. van Berkel, *Citaten uit het boek der natuur: Opstellen over Nederlandse wetenschapsgeschiedenis* (Amsterdam: Bert Bakker, 1998), 265–95; Charlotte Zoë Walker, ed., *Sharp Eyes: John Burroughs and American Nature Writing* (New York: Syracuse University Press, 2000).

63. John Burroughs, *The Light of Day* (Boston: Houghton-Mifflin, 1900), 187, 190.

64. John Burroughs, *Boy and Man* (New York: Doubleday, 1920).

65. As cited in Ralph H. Lutts, "John Burroughs and the Honey Bee: Bridging Science and Emotion in Environmental Writing," in *Sharp Eyes*, ed. Walker, 169.

CHAPTER EIGHT

1. Christopher Haigh, *English Reformations: Religion, Politics, and Society under the Tudors* (repr., Oxford: Oxford University Press, 1995), 280.

2. This school of thought was anticipated by John Bossy, whose *The English Catholic Community* (New York: Oxford University Press, 1976) provided a useful corrective to the overly Protestant-friendly approach of an earlier scholarship. Of the works that followed (after some time had passed), Eamon Duffy's *The Stripping of the Altars* (New Haven, CT: Yale University Press, 1992) remains the best-known and most comprehensive work on the late medieval popularity and lingering attractions of Catholicism in early modern England. Duffy inspired a revisionist trend that downplayed Protestantism's attractions, sometimes to the point of misrepresentation. Ian Green's broad-brush treatment of religious print culture, *Print and Protestantism in Early Modern England* (New York: Oxford University Press, 2000), supplies the same view from a bibliographic and

encyclopedic angle. It should be noted here that such scholarly forces are beginning to give ground: Ethan Shagan's *Popular Politics and the English Reformation* (Cambridge: Cambridge University Press, 2003), for example, makes an innovative and persuasive argument for the popular appeal of Protestantism.

3. This claim originated with Patrick Collinson's "Protestant Culture and Cultural Revolution," in his *The Birthpangs of Protestant England* (New York: St. Martin's Press, 1988); reprinted in Margo Todd, ed., *Reformation to Revolution* (London: Routledge, 1995), 33–52 (further references to this work will be to this reprint of the essay).

4. Catherine Randall, *Building Codes: The Aesthetics of Calvinism in Early Modern Europe* (Philadelphia: University of Pennsylvania Press, 1999), 1–8. The work of Huston Diehl also centers on issues of Protestantism and aesthetics: "'Infinite Space': Representation and Reformation in *Measure for Measure*," *Shakespeare Quarterly* 49, no. 4 (Winter 1998): 393–410; "Into the Maze of Self: The Protestant Transformation of the Image of the Labyrinth," *Journal of Medieval and Renaissance Studies* 16, no. 2 (1996): 281–301.

5. I have written on this topic elsewhere: see Lori Anne Ferrell, "Grasping the Truth: Calvinist Pedagogy in Early Modern England," in *Truth: Interdisciplinary Dialogues in a Pluralist Age,* ed. C. Helmer and K. De Troyer (Leuven, Belgium: Peeters, 2003), 139–45.

6. Lori Anne Ferrell, "Transfiguring Theology: William Perkins and Calvinist Aesthetics," in *John Foxe and His World,* ed. C. Highley and J. N. King (Aldershot, UK: Ashgate, 2002), 168, 175.

7. Or, as art historian Martin Kemp might call it, the "mark of truth."

8. Lori Anne Ferrell, "*The Two Gates of Salvation,*" in *The Oxford Collected Works of Thomas Middleton,* ed. Gary Taylor (Oxford: Oxford University Press, forthcoming).

9. Ferrell, "Grasping the Truth," 145–46.

10. Collinson, "Protestant Culture and Cultural Revolution," 44; Ferrell, "Transfiguring Theology," 164–65.

11. Except, perhaps, in the 1940s, when a brief craze for shorthand in the ranks of literary scholars led to a small pamphlet war waged over whether Bright's charactery was to blame for the variants in the folios of *King Lear.* In 1884 and 1910, Matthias Levy published two talks he gave before the London Shorthand Society: "Shakespeare and Shorthand" (1884); "Shakespeare and Timothy Bright" (1910). See also the very boring booklet by G. I. Duthie, *Elizabethan Shorthand and the First Quarto of King Lear* (Oxford, 1949).

12. See Bright's description of the ordeal in his dedicatory epistle to Walsingham: *An Abridgment of the Booke of Actes and Monuments of the Church* (London, 1589).

13. *Statutes of Sir Walter Mildmay, Kt.* (Cambridge: Cambridge University Press, 1983), 90.

14. The only records we have of Bright's career as a parish priest are related to a

protracted tithe dispute he conducted with his parishioners at Methley, in the West Riding of Yorkshire. They complained of his nonresidence, a typical riposte to a parson's demand for more money: W. J. Carlton, *Timothe Bright, Doctor of Physicke: A Memoir of the 'Father of Modern Shorthand'* (London: Elliot Stock, 1911).

15. The source of Bright's ecclesiastical patronage is also unclear, with his biographer claiming he might have had the support of Whitgift and Aylmer, both well-known hammers of the Puritans, or perhaps Burghley and the queen, or maybe Walsingham—a mixed bag. In any event, we now know enough about the vagaries of patronage to recognize that it is not always the most accurate indicator of an individual's personal beliefs or even the public face of his religion—especially in this interesting time, when it was possible to be a Puritan bishop, like Matthew Hutton (who had been with Bright at Trinity), or a moderate Puritan (like many Calvinists in this period).

16. In the 1580s, in addition to the works cited in this paper, Bright wrote a number of treatises on English medicines and health, both in Latin and in English. These include *English Medicines* (1st ed., 1580), *Hygieina* (1st ed., 1582), *Therapeutica* (1st ed., 1583), and *In Physicam Scriboni* (1st ed., 1584). These were subsequently reprinted, some several times. Bright's bibliographer is Geoffrey Keynes, *Dr. Timothie Bright, 1550–1615: A Survey of His Life with a Bibliography of His Writings* (London: Wellcome Medical Library, 1962).

17. Hereafter, page numbers to Bright's *Treatise of Melancholy* will be given in the text.

18. The condition of "melancholy" in the early modern period has for the most part been of more interest to literary scholars than historians—with the important exception of Michael Macdonald's *Mystical Bedlam* (Cambridge: Cambridge University Press, 1981) and *Sleepless Souls* (Oxford: Oxford University Press, 1990). Macdonald writes as a historian of medicine; in the literary critical treatments, Bright invariably takes a back seat to Robert Burton, except in discussions of whether or not Bright's treatise influenced Shakespeare's creation of the character of Hamlet. See, e.g., Lawrence Babb, *The Elizabethan Malady* (East Lansing: Michigan State College Press, 1951), especially 21–72. In her *Voices of Melancholy* (London: Routledge, 1971), 141, Bridget Lyons briefly notes but does not particularly explicate or analyze Bright's concerns with the states of melancholy and despair. Douglas Trevor's *The Poetics of Melancholy in Early Modern England* (Cambridge: Cambridge University Press, 2004), especially 63–86, is an excellent treatment of the Bright-Shakespeare connection.

19. Hereafter, signature and page designations for Bright's *Characterie* will be given in the text.

20. This sample was produced in the hope that Bright would be admitted to the acquaintance of the younger Cecil, for the purpose of becoming Robert's shorthand instructor: Carlton, *Timothe Bright*, 60–68. The letter of introduction can be found in the Burghley papers, *Lansdowne MS* 51:27, Folger Shakespeare Library, Washington, DC.

21. Escalante, sigs. 30–31. The English translation was prefaced by a Protestant-inflected foreword by John Frampton. For Jesuit and Protestant international rivalry in this period and its connection to mathematics, memory systems, and other humanist concerns, see Jonathan D. Spence, *The Memory Palace of Matteo Ricci* (New York: Viking Penguin, 1984), 132–45.

22. More evidence for Bright's preoccupation with the scribal politics of anti-popery might be found in a manuscript that was discovered by Bright's biographer and that demonstrates Bright's interest in cipher alphabets and his desire to curry favor with William Davidson, the privy counselor held responsible for the execution of Mary Queen of Scots. See W. J. Carlson, "An Unrecorded Manuscript by Dr. Timothy Bright," *Notes and Queries* 11 (1964): 463–65.

23. The unwanted attentions of youthful Protestant groupies had to be confronted by the famous pulpit orator Henry Smith, known as "silver-tongued" for his skill in Puritan plain speaking and his ability to render cases of conscience lucidly and compellingly. It seems that they had transcribed his sermon *The Wedding Garment* and published it. Its aggravated author had it reprinted, as he stated in a new preface, "to control false copies of this sermon . . . patched as it seemeth out of some borrowed notes" (Carlton, *Timothe Bright*, 122–24).

24. Eric Carlson, "English Funeral Sermons as Sources," *Albion* 32, no. 4 (Winter 2000): 576.

25. Patrick Collinson, "The Theatre Constructs Puritanism," in *The Theatrical City: Culture, Theatre and Politics in London, 1576–1649*, ed. D. L. Smith, R. Strier, and D. Bevington (Cambridge: Cambridge University Press, 1995), 166–67.

26. Mary Carruthers, *The Book of Memory: A Study of Memory in Medieval Culture* (Cambridge: Cambridge University Press, 1990), 20, 61–71.

27. See, e.g., the "Table of Places," in Thomas Blundeville, *The Art of Logic, Plainly Taught in the English Tongue . . . As Well According to the Doctrine of Aristotle, as of all other Modern and best Accounted Authors thereof* (London, 1599). Blundeville, Bright, and Egerton's transcriber all shared the same printer, J. Windet.

28. See, e.g., *Statutes of Sir Walter Mildmay, Kt.*, 90.

29. The expression is Randall's: *Building Codes*, 197.

CHAPTER NINE

For helpful comments on earlier drafts of this paper, I would like to thank Corinne Harol, Andrew Franta, Lauren Shohet, and Ben Schmidt.

1. The first claim is made by Steven Shapin and Simon Schaffer, *Leviathan and the Air-Pump: Hobbes, Boyle, and the Experimental Life* (Princeton, NJ: Princeton University Press, 1985), 63; the second by Rose-Mary Sargent, *The Diffident Naturalist: Robert Boyle and the Philosophy of Experiment* (Chicago: University of Chicago Press, 1995), 39.

2. *Sceptical Chymist*, in *The Works of Robert Boyle*, ed. Michael Hunter and Edward B. Davis, 14 vols. (London: Pickering and Chatto, 1999–2000), 2:214; all citations of Boyle refer to this edition.

3. William Cornwallis, *Essayes* (1600–1601), ed. Don Cameron Allen (Baltimore: Johns Hopkins University Press, 1946), 74. See chap. 1 of my *Of Essays and Reading in Early Modern Britain* (New York: Palgrave Macmillan, 2006).

4. See Sargent, *Diffident Naturalist*, 14; and Lawrence M. Principe, *The Aspiring Adept: Robert Boyle and His Alchemical Quest* (Princeton, NJ: Princeton University Press, 1998), 12, 23, 220.

5. See Bruno Latour, *We Have Never Been Modern* (Cambridge, MA: Harvard University Press, 1993), especially 69, 47.

6. Quoted in Steven Shapin, *A Social History of Truth: Civility and Science in Seventeenth-Century England* (Chicago: University of Chicago Press, 1994), 140.

7. [Francis Boyle, Viscount Shannon,] *Moral Essays and Discourses Upon Several Subjects* (London, 1690), A2.

8. Owen Felltham, *Resolves: A Duple Century*, 4th ed. (London, 1631), A4.

9. John Hall, *Horae Vacivae, Or, Essays. Some Occasional Considerations* (London, 1646), A3v, 111–12.

10. Grace Gethin, *Misery's Virtues Whetstone. Relique Gethinianae. Or, Some Remains of the Most Ingenious and excellent Lady, the Lady Grace Gethin, Lately Deceased: Being a Collection of Choice Discourses, Pleasant Apothegmes, and Witty Sentences. Written by Her for the most part, by way of Essay, and at spare Hours* (London, 1699), av, a.

11. Robert Boyle, "A Proemial Essay, wherein, With some considerations touching *Experimental Essays* in General Is interwoven such an Introduction to all those written by the Author, as is necessary to be perus'd for the better understanding of them," in *Works*, 2:10–12.

12. Ibid., 14–15.

13. Ibid., 15. Boyle famously declares himself content to be such an "Underbuilder" (20).

14. Shapin and Schaffer, *Leviathan and the Air-Pump*, 77, 69.

15. For Shapin and Schaffer's account of Boyle's writing in terms of pictures, see ibid., 62, 64; for the portrayal of a trustworthy author, see 65, 69.

16. Shapin, "Pump and Circumstance: Robert Boyle's Literary Technology," *Social Studies of Science* 14 (1984): 484.

17. Shapin and Schaffer, *Leviathan and the Air-Pump*, 61.

18. Michael Hunter, *Robert Boyle: Scrupulosity and Science* (Rochester, NY: Boydell, 2000), 9; Principe, *Aspiring Adept*, 107, 109, 111.

19. For romance, see Lawrence M. Principe, "Virtuous Romance and Romance Virtuoso: The Shaping of Robert Boyle's Literary Style," *Journal of the History of Ideas* 56 (1995): 377–97; for his critique of Shapin, see 396–97. For transmutation histories, see Principe, *Aspiring Adept*, 93, 111; and William R. Newman, *Gehennical Fire* (Cambridge, MA: Harvard University Press, 1994), 3–11.

20. Sargent, *Diffident Naturalist*, 211, 127–28; for her discussion of the "Proemial Essay," see 183–89.

21. James Paradis, "Montaigne, Boyle, and the Essay of Experience," in *One Culture: Essays in Science and Literature*, ed. George Levine (Madison: University of Wisconsin Press, 1987), 60, 77, 86.

22. John Florio says of Montaigne's *Essais:* "Why but Essayes are but mens school-themes pieced together; you may as well say, several texts. Al is in the choise & handling" (*The Essayes of Montaigne*, trans. John Florio [London, 1603], A5v). Dudley M. Marchi notes that in the seventeenth century Montaigne was "read as an author of the *exemplum*, as a compiler of the commonplaces of ancient wisdom" (*Montaigne among the Moderns: Receptions of the "Essais"* [Providence, RI: Berghahn, 1994], 23).

23. Boyle, "Proemial Essay," 2:14. Sargent notes this (*Diffident Naturalist*, 300n8).

24. Richard M. Chadbourne, "French Essay," in *Encyclopedia of the Essay*, ed. Tracy Chevalier (Chicago: Fitzroy Dearborn, 1997), 295; in France the essay "was categorized—and often dismissed—as an English literary type" (306).

25. Pierre Nicole, *Moral Essays* (1677), translation of *Essai de morale* (1671), A2–A2v.

26. John Willis, "An Essay of Dr. John Willis, exhibiting his Hypothesis about the Flux and Reflux of the Sea," in Royal Society of London, *Philosophical Transactions*, vols. 1–2 (New York: Johnson Reprint Corporation, 1963), no. 16 (6 August 1666), 266–67, 281.

27. Oldenburg writes in his preface to "An Essay of Dr. John Willis" that "it was thought fit to offer it by Press to the Publick, that other Intelligent Persons also might the more conveniently and at their leisure examine the *Conjecture* (the Author, such is his Modesty, presenting it no otherwise) and thereupon give in their sense, and what Difficulties may occur to them about it, that so it maybe either confirm'd or laid aside accordingly" (264).

28. "And though it be manifest enough, that *Galileo*, as to some particulars, was mistaken in the account which there he gives of it; yet that may be very well allowed, without any blemish to so deserving a person, or prejudice to the *main Hypothesis:* For that Discourse is to be looked upon onley as an *Essay* of the *general Hypothesis;* which as to *particulars* was to be afterwards adjusted, from a good *General History of the Tides*" (ibid., 265).

29. Ibid., 266.

30. Shapin, *Social History*, 309; see 240 for epistemology grounded in decorum.

31. Ibid., 179.

32. Hunter sees Shapin's *Social History* as "dominated by the concept of self-fashioning" (*Robert Boyle*, 11).

33. "[C]onventions and codes of gentlemanly conversation were mobilized as practically effective solutions to problems of scientific evidence, testimony, and assent" (Shapin, *Social History*, 121).

34. Sargent, *Diffident Naturalist*, 44. She argues that contextualist studies "tend to dismiss the epistemic factors embedded within cultural practices" (43; see also 212). Michael Hunter argues for combining contextualism with sensitivity to "the power and complexity of intellectual traditions in their own right" (introduction to *Robert Boyle Reconsidered*, ed. Michael Hunter [New York: Cambridge University Press, 1994], 5).

35. [Thomas Culpeper,] "Of Knowledge and Reading," in *Essayes or Moral Discourses on Several Subjects. Written by a Person of Honor* (London, 1671), 108.

36. For Sargent on social dimensions, see *Diffident Naturalist*, 18; for Shapin on epistemic ones, see *Social History*, 118, 123. Both Harwood and Altgoer note that Boyle does not reject rhetoric; see John T. Harwood, "Science Writing and Writing Science: Boyle and Rhetorical Theory," in *Boyle Reconsidered*, ed. Hunter, 48; and Diane B. Altgoer, *Reckoning Words: Baconian Science and the Construction of Truth in English Renaissance Culture* (Madison, WI: Fairleigh Dickinson University Press, 2000), 132.

37. For Shapin and Schaffer, "The objectivity of the experimental matter of fact was an artifact of certain forms of discourse and certain modes of social solidarity" (*Leviathan and the Air-Pump*, 77–78). Sargent takes issue with this: "'matters of fact' do not compose a rigid category delineated by linguistic structure. Facts are highly confirmed items of knowledge that may refer to singular effects, regular occurrences, or causal processes. The category of the factual, while foundational, is also dynamic. The material included within the informational basis remains open to revision in the light of further discoveries" (*Diffident Naturalist*, 136).

38. Shapin and Schaffer, *Leviathan and the Air-Pump*, 66. For Sargent's emphasis on the philosophical, see *Diffident Naturalist*, 183, 186.

39. Boyle, "Proemial Essay," 2:16.

40. In the preface to *Christian Virtuoso*, Boyle remarks, "I make frequent use of Similitudes, or Comparisons: And therefore I think myself here obliged to acknowledge, once for all, that I did it purposely. . . . Apposite Comparisons do not only give Light, but Strength, to the Passages they belong to, since they are not always bare Pictures and Resemblances, but a kind of Arguments" (*Works*, 11:287); "proper Comparisons do the Imagination almost as much Service, as Microscopes do the Eye; for, *as* this Instrument gives us a distinct view of divers minute Things, which our naked Eyes cannot well discern . . . *so* a skillfully chosen, and well-applied, Comparison much helps the Imagination, by illustrating Things scarce discernible, so as to represent them by Things much more familiar and easy to be apprehended" (11:287–88). See Harwood, "Science Writing," 51.

41. Robert Boyle, *New Experiments and Observations Touching Cold* (1665), in *Works* 4:222.

42. Robert Boyle, *The Christian Virtuoso, First Part* (1690): "I am now and then busied in devising, and putting in practice, Tryals of several sorts, and making Reflections upon them" (*Works*, 11:291).

43. Ibid., 292.

44. Ibid., 306. See Principe on the *Sceptical Chymist*, "vulgar chymists," and systematizers (*Aspiring Adept*, 52, 33, 46–47); and Sargent on Boyle's critique of speculative theorists (*Diffident Naturalist*, 27, 71–75).

45. "[T]he Person I mean here, is such a one, as by attentively looking about him, gathers Experience, not from his own Tryals alone, but from divers other *matters of fact*, which he heedfully observes, though he had no share in the effecting them; and on which he is dispos'd to make such Reflections, as may (unforcedly) be apply'd to confirm and encrease in him the Sentiments of *Natural Religion*, and facilitate his Submission and Adherence to the *Christian Religion*" (Boyle, *Christian Virtuoso*, 11:306).

46. Ibid., 295; Robert Boyle, *Usefulness of Experimental Natural Philosophy* (1663), in *Works*, 3:222; see also 11:299, 3:200–201, 235.

47. Harwood says that reading is Boyle's most important metaphor and discusses his reading the world in both his moral and his natural philosophy ("Science Writing," 50–51); for a discussion of the "two books" in relation to Boyle, see Sargent, *Diffident Naturalist*, 112–15.

48. Robert Boyle, *Occasional Reflections Upon Several Subjects, Whereto is premis'd A Discourse About such kinds of Thoughts* (1665), in *Works*, 5:16–17, 32.

49. Ibid., 52, 26.

50. Reading books "requires rather that a man be docile than ingenious," whereas collecting "Moral and Spiritual Documents out of a Book of Hieroglyphics, or from a Landscape or a Map, is more than every attentive considerer can do, and is that which argues something of Dexterous and Sagacity that is not very ordinary" (ibid., 27). See also the invocation of the classical figure of the bee, an image of transformative gathering (28, 52).

51. Ibid., 52.

52. Ibid., 30.

53. Ibid.; see also 32.

54. See Principe, "Virtuous Romance," 393.

55. Boyle, *Occasional Reflections*, 5:18, 8, 9. Boyle invites his readers "to exercise their Pens in some such way of Writing: Divers of whom will probably be incouraged to venture upon making some such composures, when they find Excuses for divers of those things that are most likely to be thought to Blemish such Essays" (18).

56. Ibid., 5:49–50.

57. Ibid., 10–11. Hunter dates the composition of *Occasional Reflections* to ca. 1647–48 (5:xi) and "Proemial Essay" to 1657 (2:xi).

58. John Uffley, *Wits Fancies: or, Choice Observations and Essayes, Collected out of Divine, Political, Philosophical, Military, and Historical Authors* (London, 1659), A5, A7, A3–A3v.

59. *Discoveries* 731–41, in *The Oxford Authors: Ben Jonson*, ed. Ian Donaldson (New York: Oxford University Press, 1985), 541.

60. [William Master,] *Logoi eukoiroi, Essayes and Observations Theologicall and Morall* (London, 1653), A4–A5, 9.

61. Lady Chudleigh, *Essays Upon Several Subjects In Prose and Verse* (London, 1710), A3v–A4.

62. Ibid., 12.

63. Robert Boyle, "A Physio-Chymical Essay, Containing An Experiment with some Considerations touching the differing Parts and Redintegration of Salt-Petre" ["Essay on Nitre"], section 39 of *Certain Physiological Essays* (1661), in *Works*, 2:112.

64. For one of Boyle's many accounts of his work being unpolished and unfinished, see *Some Considerations About the Reconcileableness of Reason and Religion* (1675), in *Works*, 8:239–40.

65. Boyle, *Occasional Reflections*, 5:20.

66. Boyle, "Proemial Essay," 2:34.

67. Principe, *Aspiring Adept*, 111, 220.

68. Anthony Grafton, *Defenders of the Text* (Cambridge, MA: Harvard University Press, 1991), 4–5.

69. Anthony Grafton, "The New Science and the Traditions of Humanism," in *The Cambridge Companion to Renaissance Humanism*, ed. Jill Kraye (New York: Cambridge University Press, 1996), 212.

70. Oldenberg, "A Preface to the Third Year of these Tracts," *Philosophical Transactions* 1–2, 410.

CHAPTER TEN

1. The literature on early modern Dutch collecting is not, as in the case of Italy, yet integrated with the literature on the history of science and medicine. On Dutch collecting in general, see O. Impey and A. MacGregor, eds., *The Origins of Museums: The Cabinet of Curiosities in Sixteenth- and Seventeenth-Century Europe*, 2d ed. (London: House of Stratus, 2001); and Ellinoor Bergvelt and Renée Kistemaker, eds., *De wereld binnen handbereik: Nederlandse kunst- en rariteitenverzamelingen, 1585–1735*, 2 vols., exhibition catalog (Zwolle: Waanders; Amsterdam: Amsterdams Historisch Museum, 1992). On the relationship between collections and scientific practice in early modern Italy, see principally Giuseppe Olmi, *L'inventario del mondo: Catalogazione della natura e luoghi del sapere nella prima età moderna* (Bologna: Il Mulino, 1992); and Paula Findlen, *Possessing Nature: Museums, Collecting, and Scientific Culture in Early Modern Italy* (Berkeley and Los Angeles: University of California Press, 1994). In general, see Lorraine Daston and Katharine Park, *Wonders and the Order of Nature, 1150–1750* (New York: Zone Books, 1998); and, for a compelling case study of a collection of pictures that belonged to a professor of practical medicine, see Pamela Smith, "Science and Taste: Painting, Passions, and the New Philosophy in Seventeenth-Century Leiden," *Isis* 90 (1999): 421–61.

2. Frits Lugt, *Repértoire des catalogues de ventes publiques intéressant l'art ou la curiosité: Première période vers 1600–1825* (The Hague: M. Nijhoff, 1938), no.

2. E. W. Moes, "De sonderling-heden oft rariteyten ende wtgelesen sinnelickheden van Christiaen Porret," *Leids Jaarboekje* 2 (1905): 93–100, introduces the collection and the collector; other biographical information is found in Henriëtte A. Bosman-Jelgersma, "De lotgevallen van een apothekersleerling in het 17de-eeuwse Leiden," *Leids Jaarboekje* 79 (1987): 62–81. The auction of Porret's collection is mentioned in passing in Bob van den Boogert et al., *Rembrandt's Treasures,* exhibition catalog (Zwolle: Waanders, 1999).

3. Svetlana Alpers, *The Art of Describing: Dutch Art in the Seventeenth Century* (Chicago: University of Chicago Press, 1983).

4. David Freedberg, "Science, Commerce, and Art: Neglected Topics at the Junction of History and Art History," in *Art in History, History in Art: Studies in Seventeenth-Century Dutch Culture,* ed. D. Freedberg and J. de Vries (Santa Monica, CA: Getty Research Institute, 1991), 377–428.

5. Pamela Smith and Paula Findlen, eds., *Merchants and Marvels: Commerce, Science, and Art in Early Modern Europe* (London: Routledge, 2002), 3.

6. Ibid., 17.

7. Harold J. Cook, "The Cutting Edge of a Revolution? Medicine and Natural History near the Shores of the North Sea," in *Renaissance and Revolution: Humanists, Scholars, Craftsmen and Natural Philosophers in Early Modern Europe,* ed. J. V. Field and Frank A. J. L. James (Cambridge: Cambridge University Press, 1993), 47.

8. The classic resource on Paludanus remains F. W. T. Hunger, "Bernardus Paludanus (Berent ten Broecke), 1550–1633: Zijn verzamelingen en zijn werk," in *Itinerario voyage ofte schipvaert van Jan Huygen van Linschoten naer oost ofte Portugaels Indien, 1579–1592,* ed. H. Kern, 8 vols. (The Hague: J. H. van Linschoten Vereniging, 1910–57), vol. 3 (1934), 249–68. See also Bergvelt and Kistemaker, *De wereld binnen handbereik;* Roelof van Gelder, "Paradijsvogels in Enkhuizen: De relatie tussen Van Linschoten en Bernardus Paludanus," and Florike Egmond, "Een mislukte benoeming: Paludanus en de Leidse Universiteit," both in *Souffrir pour parvenir: De wereld van Jan Huygen van Linschoten,* ed. Roelof van Gelder et al. (Haarlem: Uitgeverij Arcadia, 1998), 30–50 and 51–64, respectively.

9. See Findlen, *Possessing Nature,* passim.

10. "Allerhandt schoene und denkwurdige Rariteten und unerhörten Ding auß China, India, America, Africa, Asia, Peru, Aegypten, Malakk, Hispania, Insulis fortunatis [Canary Islands], auß Turquey, Graecia etc." The account was written in November 1593 by Duke Philip Ludwig II of Hanau-Münzenburg and is cited in van Gelder, "Paradijsvogels in Enkhuizen," 36–37.

11. "Thesaurus Orbis, Totius compendium / Arca universi, sacra Naturae penus, Templumque Mundi." Hugo Grotius, *Poemata* (Leiden: Hieronymum de Vogel, 1639), 276.

12. Roelof van Gelder, "Noordnederlandse verzamelingen in de zeventiende

eeuw," in *Verzamelen: Van rariteitenkabinet tot kunstmuseum*, ed. Ellinoor Bergvelt et al. (Heerlen: Gaade Uitgevers, 1993), 125.

13. See, e.g., Thomas DaCosta Kaufmann, "From Mastery of the World to Mastery of Nature: The *Kunstkammer*, Politics, and Science," in *The Mastery of Nature*, by Thomas DaCosta (Princeton, NJ: Princeton University Press, 1993), 174–94, with an excellent review of the literature. For an important corrective, see Harold J. Cook, "Physicians and Natural History," in *Cultures of Natural History*, ed. N. Jardine, J. A. Secord, and E. C. Spary (Cambridge: Cambridge University Press, 1996), 91–105; and Cook, "The Cutting Edge of a Revolution?"

14. Findlen, *Possessing Nature*, 250 and 252ff., with further bibliography.

15. Tommaso Garzoni, *Piazza universale di tutte le professioni del mondo* (Venice, 1585), 155; as cited in Findlen, *Possessing Nature*, 6. In 1588, Ulisse Aldrovandi, who had been teaching *materia medica* for more than thirty years at Bologna with the help of his vast collection of *naturalia* and, since 1568, of the university botanical garden, wrote to Duke Ferdinando I de'Medici that "I teach what plants one should truly choose for medicinal uses to whomever makes use of medicines." As cited in Findlen, *Possessing Nature*, 254.

16. See, inter alia, Karen Meier Reeds, *Botany in Medieval and Renaissance Universities* (New York and London: Garland, 1991), 58ff.; cf. 72.

17. His *album amicorum* survives in the Koninklijke Bibliotheek, The Hague (MS 133 M63); see K. Thomassen, ed., *Alba amicorum: Vijf eeuwen vriendschap op papier gezet; Het album amicorum en het poëziealbum in de Nederland*, exhibition catalog (Maarssen and The Hague: Museum Meermanno-Westreenianum Museum van het Boek, 1990), cat. no. 12, pp. 55–56.

18. He was asked to join the faculty "met alle zijne 'tsamen vergaerde seltsaemheden, zo van cruyden, vruchten, spruytsels, gedierten, schepselen, mineralen, aerden, veninen, gesteenten, marmeren, coralen etc." (with all his collected rarities, such as plants, fruits, cuttings, animals, shells, minerals, earths, poisons, stones, marbles, corals, etc.). P. C. Molhuysen, *Bronnen tot de geschiedenis der Leidsche Universiteit*, 7 vols. (The Hague: M. Nijhoff, 1913–24), 1:180.

19. See Egmond, "Een mislukte benoeming."

20. On Dirck Cluyt, see Henriëtte A. Bosman-Jelgersma, "Clusius en Clutius," *Farmaceutisch Tijdschrift voor België* 58 (1981): 41–45; Henriëtte A. Bosman-Jelgersma, "Dirck Cluyt: De eerste Leidse hortulanus," *Jaarboekje voor Geschiedenis en Oudheidkunde van Leiden en Omstreken* (*Leids Jaarboekje*) 83 (1991): 75–88; Henriëtte A. Bosman-Jelgersma, "Dirck Outgaertszn, Cluyt," *Tijdschrift voor België* 53 (1976): 525–48; Claudia Swan, *The Clutius Botanical Watercolors: Plants and Flowers of the Renaissance* (New York: Harry N. Abrams, 1998); and Claudia Swan, "*Lectura-Imago-Ostensio:* The Role of the *Libri Picturati* A.16–A.30 in Botanical Instruction at the Leiden University," in *Natura-cultura: L'interpretazione del mondo fisico nei testi e nelle immagini*, ed. G. Olmi, L. Tongiorgi-Tomasi, and A. Zanca (Florence: Olschki, 2000), 189–214.

21. Francis Bacon, *Gesta Grayorum,* in *The Works of Francis Bacon, Baron of Verulam, Viscount of St. Alban, and Lord Chancellor of England,* ed. James Spedding, R. L. Ellis, and D. D. Heath, 14 vols. (London: Longman, 1858–74; New York: Garrett Press, 1968), 8:334–35.

22. Erwin Panofsky, *Galileo as a Critic of the Arts* (The Hague: M. Nijhoff, 1954), 18–19.

23. Ibid., 18.

24. On wonder, see Daston and Park, *Wonders and the Order of Nature;* Philip Fisher, *Wonder, the Rainbow, and the Aesthetics of Rare Experiences* (Cambridge, MA: Harvard University Press, 1998); Mary Baine Campbell, *Wonder and Science: Imagining Worlds in Early Modern Europe* (Ithaca, NY: Cornell University Press, 1999).

25. H. D. Schepelern, "Natural Philosophers and Princely Collectors: Worm, Paludanus, and the Gottorp and Copenhagen Collections," in *Origins of Museums,* ed. Impey and MacGregor, 175.

26. Giuseppe Olmi, "Science-Honour-Metaphor: Italian Cabinets of the Sixteenth and Seventeenth Centuries," in *Origins of Museums,* ed. Impey and MacGregor, 2–3.

27. Findlen, *Possessing Nature,* 245; cf. 246.

28. Ibid., 245.

29. Ibid., 246.

30. Daston and Park, *Wonders and the Order of Nature,* 149. See also Findlen, *Possessing Nature.*

31. Daston and Park, *Wonders and the Order of Nature,* 266.

32. On Bernard Palissy, see Ernst Kris, "Der Stil 'rustique': Die Verwendung des Naturabgusses bei Wenzel Jamnitzer und Bernard Palissy," *Jahrbuch der Kunsthistorischen Sammlungen in Wien* 1 (1926): 137–208; Leonard Amico, *Bernard Palissy in Search of Earthly Paradise* (New York: Flammarion, 1996); Bernard Palissy, *Recette véritable* (1563), ed. Frank Lestringant (Paris: Macula, 1996).

33. On the garden, which opened in 1593, see H. Veendorp and L. G. M. Baas Becking, *Hortus Academicus Lugduno Batavus: The Development of the Gardens of Leyden University* (Haarlem: Enschedaiana, 1938); A. J. F. Gogelein et al., eds., *Leidse Universiteit 400, Stichting en Eerste Bloei 1575–ca. 1650,* exhibition catalog (Amsterdam: Rijksmuseum, 1975), 166–84; L. Tjon Sie Fat and E. de Jong, eds., *The Authentic Garden: A Symposium on Gardens* (Leiden: Clusius Stichting, 1991), especially E. de Jong, "Nature and Art: The Leiden Hortus as 'Musaeum,'" 37–60. The construction of the anatomical theater was first proposed in 1591; the first public dissections held in it date to November 1593. See Gogelein et al., *Leidse Universiteit 400,* 100ff.; H. J. Witkam, *Iets over Pieter Paaw en zijn theatrum anatomicum en over het bouwen van de anatomieplaats en de bibliotheek* (Leiden: H. J. Witkam, 1967); and T. H. Lunsingh Scheurleer, "Un amphithéâtre d'anatomie moralisée," in *Leiden University in the Seventeenth Century: An Exchange of*

Learning, ed. T. H. Lunsingh Scheurleer and G. H. Posthumus Meyes (Leiden: Brill, 1975), 217–77.

34. Jan Janszn Orlers, 1614, pp. 143–44.

35. "Inventaris van de Rariteyten opde Anatomie en inde twee gallerijen van des Universiteyts Kruythoff" (ca. 1617), in Erik de Jong, *Natuur en kunst: Nederlandse tuin- en landschapsarchitectuur, 1650–1740* (Amsterdam: THOTH, 1993), 232–34.

36. Pieter Pauw, *Hortvs Pvblicvs Academiæ Lvgdvno-Batavæ. Eivs Ichnographia, Descriptio, Vsus, Addito quas habet stirpium numero, & nominibus* (Leiden, 1601).

37. Hondius must have come to know Porret's collection as a student in Leiden; he enrolled in 1596 and was among the students who benefited from the early years of activity in the Leiden garden and anatomical theater alike. Though he studied botany diligently while at Leiden, he followed in his father's footsteps and entered the ministry. See P. J. Meertens, *Letterkundig leven in Zeeland* (Amsterdam: N. V. Noord-Hollandsche Uitgevers Maatschappij, 1943).

38. Erasmus, preface to *Sancti Hieronymi Opera* (Basel, 1516); *Opus Epistolarum Des. Erasmi Roterodami*, ed. P. Allen (Oxford: Clarendon, 1910), II, 2:56. Otto Brunfels cites the story in *Herbarum Vivae Eicones ad Naturae Imitationem* (Strasbourg, 1530), 1:13.

39. On Calzolari's theriac, see, e.g., Findlen, *Possessing Nature*, 272–77.

40. See Karen Meier Reeds, "Renaissance Humanism and Botany," *Annals of Science* 33 (1976): 519–42; Reeds, *Botany in Medieval and Renaissance Universities*, 24ff. and 74–75 (Montpellier); cf. H. A. Bosman-Jelgersma, "Dodoens en de farmacie," *Handelingen van de Koninklijke Kring van Oudheidkunde Letteren en Kunst van Mechelen* 89 (1986): 137–39; and Vivan Nutton, "Idle Old Trots, Cobblers and Costardmongers: Pieter van Foreest on Quackery," H. A. Bosman-Jelgersma, "Inleiding tot het handschrift tegen kwakzalverij en onbevoegde geneeskunde," and the text of the manuscript against quacks, "Van der Empiriken, lantloeperen ende valscher Medicyns bedorch," in *Pieter van Foreest, de Hollandse Hippocrates*, ed. H. A. Bosman-Jelgersma et al. (Knijnenberg: Pieter van Foreest Stichting, 1996), 245–58, 259–64, and 265–92, respectively.

41. Antonius Musa Brasavolus, *Examen Omnium Simplicium Medicamentorum, Quorum in Officinis Usus Est* (Rome, 1536).

42. Edward Lee Greene, *Landmarks of Botanical History* (Stanford, CA: Stanford University Press, 1983), 2:658–701.

43. From Fernel's *Methodo Medendi*; quoted in Caspar Bauhin, *Catalogus Plantarum circa Basileam Sponte Nascentium . . .* (Basel, 1622), preface to medical students. Cited in Reeds, *Botany in Medieval and Renaissance Universities*, 25–26.

44. "sij meynden dat alsulcken scientie oft kennisse haer niet en betaemde, maar alleen toebehoorde den Apothekers of sommighen anderen ongheleerden . . . die scientie ende kennisse van den cruyden alle medcijns seer nootelijck ende betaemelick es." R. Dodonaeus, *Cruijdeboeck* (Antwerp: Jan van der Loe, 1554),

ccclxxxvii. As cited in Bosman-Jelgersma, "Dodoens en de farmacie," 132. On the production of *dispensatoria* in Antwerp in the sixteenth century, see F. de Nave et al., eds., *Botany in the Low Countries (End of the 15th Century–ca. 1650)*, exhibition catalog (Antwerp: Museum Plantin-Moretus, 1993). Dodoens's sentiments echo those of Gaspare Gabrieli, the first professor of simples at the University of Ferrara. In a lecture given in 1543, Gabrieli wrote: "In my opinion [the lack of interest among physicians in *materia medica*] derives solely from the belief that the part of medicine dealing with knowledge of plants does not concern them. They leave the entire study of this branch [of medicine] to chemists, apothecaries, and wise-women. Thus at present the entire medicine of herbs is in the hands of the unlearned, the foolish, and superstitious wise-women. Not surprisingly, infinite errors occur from this incompetence." Cited in Findlen, *Possessing Nature*, 251.

45. Egmond, "Een mislukte benoeming"; Swan, *Clutius Botanical Watercolors*; and Swan, *"Lectura-Imago-Ostensio."*

46. See above, n. 10.

47. Daston and Park, *Wonders and the Order of Nature*, especially 146–72.

48. Ibid., 137.

49. Petrus Hondius, *Moufe-schans* (Leyden, 1621), prefatory material.

CHAPTER ELEVEN

1. Recent studies, such as Margaret J. Osler, ed., *Rethinking the Scientific Revolution* (Cambridge: Cambridge University Press, 2000); and Stephen Pumfrey, "The Scientific Revolution," in *Companion to Historiography*, ed. Michael Bentley (New York: Routledge, 1997), 293–306, offer new approaches and bibliography.

2. Worm's notes on two theological lectures in Giessen in 1606 are preserved at the Cathedral School in Aarhus, Denmark.

3. H. D. Schepelern, ed., *Breve fra of til Ole Worm*, 3 vols. (Copenhagen: Munksgaard, 1965), 1: no. 3.

4. See Worm's autobiography cited in E. Hovesen, *Lægen Ole Worm* (Aarhus: Aarhus Universitetsforlag, 1987), 45–47, especially 46.

5. Schepelern, *Breve*, 1: no. 8.

6. Worm's autobiography cited in Hovesen, *Worm*, 46.

7. H. D. Schepelern, *Museum Wormianum* (Odense: Wormianum, 1971), 72.

8. Schepelern, *Breve*, 1: no. 16.

9. O. P. Grell, "Caspar Bartholin and the Education of the Pious Physician," in *Medicine and the Reformation*, ed. O. P. Grell and A. Cunningham (London: Routledge, 1993), 78–100, especially 84–85.

10. See Hovesen, *Worm*, 47, 63.

11. For Worm's friendship with de Mayerne, see Schepelern, *Breve*, 1: no. 13 and 2: no. 1040.

12. For Worm's early interest in Rosicrucianism, see the introduction to his *Oratio Inaugura;is de Fratrum R. C. Philosophiam Reformandi Conatu, 10 May*

1619 (Copenhagen, 1619), cited in Schepelern, *Museum Wormianum,* 115–16. For the identification of Johannes Hartmann as Worm's source, see H. Hotson, *Johann Heinrich Alsted, 1588–1638: Between Renaissance, Reformation, and Universal Reform* (Oxford: Oxford University Press, 2000), 99.

13. This is especially interesting in light of recent work arguing that Moritz himself was the author of the pamphlets: Heiner Borggrefe, Thomas Fusenig, and Anne Schunicht-Rawe, eds., *Moritz der Gelehrte: Ein Renaissancefürst in Europa* (Eurasburg: Edition Minerva, 1997).

14. Schepelern, *Breve,* 1: no. 20.

15. Ibid., no. 25.

16. Ibid., no. 38.

17. Ibid., no. 41. Oedipus, king of Thebes, was of course famous for having solved the riddle of the Sphinx.

18. Ibid., no. 48. See also J. Shackelford, "Rosicrucianism, Lutheran Orthodoxy, and the Rejection of Paracelsianism in Early Seventeenth-Century Denmark," *Bulletin of the History of Medicine* 70 (1996): 181–204, especially 194. As will appear from what follows, my conclusion is at odds with Shackelford's. Paracelsianism, in its Severinian form, continued to appeal to leading Danish scholars such as Ole Worm, even after the Rosicrucian debacle.

19. Schepelern, *Breve,* 1: no. 78; see also my "The Acceptable Face of Paracelsianissm," in *Paracelsus: The Man and His Reputation, His Ideas, and Their Transformation,* ed. O. P. Grell (Leiden: Brill, 1998), 245–67, especially 264.

20. Schepelern, *Breve,* 1: no. 142.

21. Ibid., no. 43. For Jean Beguin and Etienne de Clave, see A. Debus, *The French Paracelsians: The Chemical Challenge to Medical and Scientific Tradition in Early Modern France* (Cambridge: Cambridge University Press, 1991), 80–82, 70–71.

22. Schepelern, *Breve,* 1: nos. 131 and 133.

23. For Davidson, see Debus, *French Paracelsians,* 124–25.

24. Schepelern, *Breve,* 2: nos. 729 and 747.

25. For Guy de la Brosse, see Debus, *French Paracelsians,* 82–84.

26. Schepelern, *Breve,* 2: nos. 845 and 862.

27. Ibid., nos. 1378, 1536, and 1694.

28. Ibid., 3: no. 1734.

29. For palingenesis, see Debus, *French Paracelsians,* 161–62. For its links to apocalypticism and the search for the Philosophers' Stone, see F. Secret, "Palingenesis, Alchemy, and Metempsychosis in Renaissance Medicine," *Ambix* 26, pt. 2 (July 1979): 81–92.

30. Schepelern, *Breve,* 3: no. 1572.

31. Ibid., no. 1575. For Estienne de Clave, see Debus, *French Paracelsians,* 71, 80.

32. Schepelern, *Breve,* 3: no. 1578.

33. Ibid., no. 1738.

34. Ibid., no. 1747.

35. Ibid., no. 1740.

36. J. T. Young, *Faith, Medical Alchemy and Natural Philosophy: Johann Moriaen, Reformed Intelligencer, and the Hartlib Circle* (Aldershot, UK: Scolar Press, 1998), 183–216.

37. Schepelern, *Breve,* 1: no. 401.

38. Ibid., no. 396.

39. Ibid., no. 501. On 6 October 1633 Worm returned Svabe's copies of Harvey and Spiegel. For the significance of publishing and the Frankfurt Book Fair in this period, see I. Maclean, *Logic, Signs and Nature in the Renaissance: The Case of Learned Medicine* (Cambridge: Cambridge University Press, 2002), especially chap. 2.

40. Schepelern, *Breve,* 1: no. 413.

41. Ibid., no. 423.

42. See O. Worm, *Controversarium Medicarum Exercitationes* VI, (Copenhagen, 1632); see also Hovesen, *Worm,* 149, 169–74, especially 170–71.

43. Schepelern, *Breve,* 1: no. 520.

44. Ibid., 2: no. 762.

45. Ibid., nos. 768 and 773.

46. Ibid., no. 793.

47. Ibid., nos. 832, 837, and 838.

48. Ibid., no. 837. See also Roger French, "Harvey in Holland: Circulation and the Calvinists," in *The Medical Revolution of the Seventeenth Century,* ed. Roger French and Andrew Wear (Cambridge: Cambridge University Press, 1989), 46–86.

49. For Emilio Parigiano, see Roger French, *William Harvey's Natural Philosophy* (Cambridge: Cambridge University Press, 1994), 228–37. For Worm's letter of 11 May 1640, see Schepelern, *Breve,* 2: no. 844.

50. Schepelern, *Breve,* 2: no. 851.

51. Ibid., no. 1132. For Fortunio Liceti and his ideas about circulation, see French, *Harvey,* 251–54.

52. Schepelern, *Breve,* 2: no. 1141.

53. Ibid., no. 1153.

54. Ibid., no. 1178. For Herman Conring, see French, *Harvey,* 154, 258–63. Having used the Leiden edition of 1646, French seems unaware of the earlier Helmstedt edition.

55. Schepelern, *Breve,* 3: no. 1373 (9 January 1646). Accordingly, the hitherto-prevailing view of Worm as having maintained his opposition to Harvey's discovery of the circulation, as originally argued by E. Gotfredsen in his article "The Reception of Harvey's Doctrine in Denmark," *Journal of the History of Medicine,* no. 12 (1957): 202–8, and accepted by both Schepelern and French, can no longer be maintained. It is also interesting to note that, like Worm, Conring was an admirer of the Danish Paracelsist Petrus Severinus; see Maclean, *Logic, Signs and Nature,* 54–72.

CHAPTER TWELVE

I would like to thank the participants of Pomona College's 2003 Westergaard Workshop "Knowledge and Its Making" for their suggestions and comments on an earlier draft of this paper.

1. For recent discussions, see Paula Findlen, *Possessing Nature: Museums, Collecting, and Scientific Culture in Early Modern Italy* (Berkeley and Los Angeles: University of California Press, 1994); and Pamela Smith and Paula Findlen, eds., *Merchants and Marvels: Commerce, Science, and Art in Early Modern Europe* (New York: Routledge, 2002). For the first important overview, see Julius von Schlosser, *Die Kunst- und Wunderkammern der Spätrenaissance* (Leipzig: Klinkhardt and Biermann, 1908).

2. R. J. W. Evans, *Rudolf II and His World* (Oxford: Oxford University Press, 1973), 176–83, 243–49; Eliška Fučíková, "The Collection of Rudolf II at Prague," in *The Origins of Museums*, ed. Oliver Impey and Arthur MacGregor (Oxford: Clarendon Press, 1985), 47–53; and the thoughtful assessment by Thomas DaCosta Kaufmann in *The Mastery of Nature* (Princeton, NJ: Princeton University Press, 1993), 174–94, 293–312. For seventeenth-century Catholic censorship, see Paula Findlen, "The Museum: Its Classical Etymology and Renaissance Genealogy," *Journal of the History of Collections* 1 (1989): 59–78.

3. H. J. Schroeder, ed., *Canons and Decrees of the Council of Trent* (St. Louis: B. Herder, 1941), Latin text, 483–85; English translation, 215–17.

4. In this first generation of Protestant reformers, Andreas Karlstadt (author of *Von Abtuhung der Bylder*, 1522) and Ulrich Zwingli (various writings from 1521 to 1524) are best known. Lee Wandel's *Voracious Idols and Violent Hands* (Cambridge: Cambridge University Press, 1994) examines the resultant iconoclasm, and Carlos Eire's *War against the Idols* (Cambridge: Cambridge University Press, 1986) discusses later outbreaks.

5. Eire, *War against the Idols*, 197–213, 225–33. Calvin warned against the dangers of relics as early as the 1540s.

6. Hubert Jedin, *Geschichte des Konzils von Trent* (Freiburg: Verlag Herder, 1975), 4/2:165–66.

7. Eire, *War against the Idols*, 279–82; Andrew Pettegree, *Emden and the Dutch Revolt* (Oxford: Clarendon Press, 1992), 115–32. For the Reformation and images more broadly, see Hans Belting, *Likeness and Presence* (Chicago: University of Chicago Press, 1994); David Freedberg, *The Power of Images* (Chicago: University of Chicago Press, 1989); Sergiusz Michalski, *The Reformation and the Visual Arts* (London: Routledge, 1993); and Joseph Koerner, *The Reformation of the Image* (Chicago: University of Chicago Press, 2004).

8. Alan Copus [Nicholas Harpsfield], *Dialogi Sex contra Summi Pontificatus . . .* (Antwerp: Christopher Plantin, 1566), especially 384–87, 780–83.

9. See their relative importance in Johann Baptista Fickler, *Replica wider das ander uberhaufft/ falsch/ . . .* (Munich: Adam Berg, 1592), 49, 102–9.

10. Robert Bellarmine, *Opera Omnia*, ed. Justin Fèvre, 12 vols. (1870–74; repr., Frankfurt a. M.: Minerva, 1965), 1:60–61.

11. James Brodrick, *Robert Bellarmine* (Westminster: Newman Press, 1961), 51–63; Bellarmine, *Opera Omnia*, 3:199–266.

12. Robert W. Richgels provides a statistical analysis of Bellarmine's targets in "The Pattern of Controversy in a Counter-Reformation Classic," *Sixteenth Century Journal* 11 (1980): 3–15.

13. Bellarmine, *Opera Omnia*, 3:190–253, 7:489–503. For pre-Tridentine precursors John of Damascus and Thomas Aquinas, see Wandel, *Voracious Idols*, 39–49.

14. Bellarmine, *Opera Omnia*, 3:236–40.

15. Peter Canisius, *Parvus Catechismus Catholicorum* (Vienna: [Gulielmus Sulienus], 1559), fols. 18v–19r; and, after Trent, Peter Canisius, *Catechismus Catholicus* (Ingolstadt: Wolfgang Eder, 1583), 23–24.

16. For Augsburg, see Christopher S. Wood, "Germany's Blind Renaissance," in *Infinite Boundaries*, ed. Max Reinhart (Kirksville, MO: Sixteenth Century Journal Publishers, 1998), 237–39.

17. Johannes Molanus, *De Picturis et Imaginibus Sacris* (Louvain: Hieronymus Wellaeus, 1570); David Freedberg, "Johannes Molanus on Provocative Paintings," *Journal of the Warburg and Courtauld Institutes* 34 (1971): 229–33; Christian Hecht, *Katholische Bildertheologie im Zeitalter von Gegenreformation und Barock* (Berlin: Gebr. Mann, 1997), 302–12.

18. *Franckfurter ankunfft/ oder verzaichnuß aller Potentaten/ Chur vnd Fürsten/. . ./ Doctorn vnd Geleerten etc. . . . die personlich erschinen vnd gewesen seind* (Augsburg: Ulhart, 1562), fol. Giii r.

19. Samuel Quiccheberg, *Inscriptiones, vel Tituli Theatri Amplissimi* (Munich, 1565), fol. ciii v. For the context of his organizational principles, see Thomas DaCosta Kaufmann, *Court, Cloister, and City* (London: Weidenfeld and Nicolson, 1995), 178–79.

20. Quiccheberg, *Inscriptiones*, fols. aii r–aiv r, ciii v–fiii r.

21. Barbara Gutfleisch and Joachim Menzhausen, " 'How a Kunstkammer Should Be Formed,' " *Journal of the History of Collections* 1 (1989): 8–9.

22. Joachim Menzhausen, "Elector Augustus's *Kunstkammer*," in *The Origins of Museums*, ed. Impey and MacGregor, 69–74.

23. Thomas DaCosta Kaufmann, "From Treasury to Museum: The Collections of the Austrian Habsburgs," in *The Cultures of Collecting*, ed. John Elsner and Roger Cardinal (Cambridge, MA: Harvard University Press, 1994), 137–54; Carina L. Johnson, "Negotiating the Exotic" (PhD diss., University of California, Berkeley, 2000), 191–206.

24. Michael Camille, *The Gothic Idol* (Cambridge: Cambridge University Press, 1989).

25. Bartolomé de Las Casas, *Apologética Historia Sumaria*, 3d ed., ed. Edmundo O'Gorman, 2 vols. (Mexico: Universidad Nacional Autónoma de México,

1967); Sabine MacCormack, *Religion in the Andes* (Princeton, NJ: Princeton University Press, 1991), 212–25.

26. Georges Baudot, *Utopia and History in Mexico* (Niwot: University Press of Colorado, 1995), 490–515.

27. F. J. Sánchez Cantón, ed., *Inventarios reales: Bienes muebles que pertenecieron a Felipe II*, 2 vols. (Madrid: Real Academia de Historia, 1956–59).

28. Ibid., 332–52. The Caribbean term *xeme* was applied to American deities in sixteenth-century Europe.

29. José de Acosta, *Historia natural y moral de las Indias* (1590), in *Obras de P. José de Acosta*, Biblioteca de autores españoles, vol. 73 (Madrid: Atlas, 1954), prologue to books 5–7 and book 5, chaps. 1–6, pp. 139–47. Summaries of Acosta's thought and career can be found in Anthony Pagden, *The Fall of Natural Man* (Cambridge: Cambridge University Press, 1982), 149–200; MacCormack, *Religion*, 264–80; D. A. Brading, *The First America* (Cambridge: Cambridge University Press, 1991), 184–95.

30. Acosta, *Historia*, book 1, chap. 23, pp. 36–37; and book 6, chap. 11, pp. 191–92.

31. José de Acosta, *De Procuranda Indorum Salute* (Cologne: Birkmann, 1596), book 5, chaps. 9–10, pp. 474–77.

32. Valentinus Fricius, *Yndianischer Religionstandt der gantzen newen Welt* (Ingolstadt: Wolfgang Eder, 1587). Fricius dedicated the book, an expansion of Diego de Valadés's work, to Archduke Matthias amid religious dissent in the Netherlands.

33. Acosta, *De Procuranda*, book 5, chap. 9, p. 470. MacCormack explores this theme in *Religion*, 265–66.

34. "Illud magis cogitandum est haereditarium esse impietatis morbum, qui ab ipsis matrum visceribus ingenitus, ipso uberum lacte nutritus, paterno & domestico exemplo confirmatus" (Acosta, *De Procuranda*, book 5, chap. 9, pp. 472–73).

35. Ibid., book 5, chaps. 10–11, pp. 478–83.

36. Acosta, *Historia*, book 5, chap. 2, p. 141.

37. Bellarmine, *Opera Omnia*, 3:202–18.

38. Vincenzo Cartari, with supplement by Lorenzo Pignoria, *Le vere e nove imagini de gli dei delli antichi* [1615] (Padua: Tozzi, 1626). See also Sabine MacCormack, "Limits of Understanding: Perceptions of Greco-Roman and Amerindian Paganism in Early Modern Europe," in *America in European Consciousness, 1493–1750*, ed. Karen Kupperman (Chapel Hill: University of North Carolina Press, 1995), 87–93.

39. Often characterized as an aspect of Neoplatonic *prisca sapientia* or ancient theology, this idea circulated broadly at the beginning of the sixteenth century.

40. Wolfgang Braunfels, "Cuius Regio Eius Ars," in *Wittelsbach und Bayern*, vol. 2:1, *Um Glauben und Reich: Kurfürst Maximilian I* (Munich: Hirmer, 1980), 133–40; Jeffrey Chipps Smith, "The Jesuit Church of St. Michael's in Munich," in *Infinite Boundaries*, ed. Reinhart, 147–70; J. Stockbauer, *Die Kunstbestrebungen*

am Bayerische Hofe unter Herzog Albert V. und seinem Nachfolger Wilhelm V. (Vienna: Braumüller, 1874); and Lorenzo Seelig, "The Munich *Kunstkammer,* 1565–1807," in *The Origins of Museums,* ed. Impey and MacGregor, 76–89. Lorenzo Pignoria cited the "Idoli del Mexico" that he saw when visiting the Wittelsbach collection (Cartari, *Le vere e nove imagini,* 555).

41. *Inventarium oder Beschreibung aller deren Stuckh und sachen frembder und Inhaimischer bekanter und unbekanter seltzamer und verwunderlicher ding . . .* The inventory was compiled by Johann Baptista Fickler and is extant in two copies in the Bayerisches Staatsbibliothek (BStB): cgm 2133 and cgm 2134. I cite from cgm 2134.

42. BStB, cgm 2134 (Tafeln 3–4), fols. 14r–19v.

43. BStB, cgm 2134, fol. 134v, no. 1618; for other examples, see fols. 133r–134v.

44. For princely involvement, see Bayerisches Hauptstaatsarchiv (BStA), KaA, 4855, *Libri Antiquitatum V.*

45. Uwe Müller, "Herzog Wilhelm V. und das Reichsheiltum," *Mitteilungen des Vereins für Geschichte der Stadt Nürnberg* 72 (1985): 117–29. Also BStA, GR Fasz 513, no. 65a.

46. For interest in relics under Albrecht V, see *Der Schatz vom Heiligen Berg Andechs* (Munich: Kloster Andechs, 1967), 74–75.

47. Fickler, *Replica,* fol. *i v.

48. Josef Steinruck, *Johann Baptist Fickler* (Münster: Aschendorff, 1965).

49. Johann Baptista Fickler, *Theologia Juridica, Seu Ius Civile Theologicum . . .* (Dillingen: Sebaldus Mayer, 1575).

50. *Censur oder Urtheil* (Ingolstadt: David Sartorius, 1583); *Anderer und Dritter Theil* (Ingolstadt: David Sartorius, 1585); *Spongia contra Praedicantum* (Ingolstadt: David Sartorius, 1585); *Rettung der Concilien Catholischen Glaubens . . .* (Ingolstadt: Wolfgang Eder, 1590); and *Replica* in 1592.

51. Fickler, *Rettung,* 45, 90–95; Fickler, *Replica,* 124.

52. Evans, *Rudolf II,* 89–91.

53. Rotraud Bauer and Herbert Haupt, eds., "Das Kunstkammerinventar Kaiser Rudolf II, 1607–1611," *Jahrbuch der Kunsthistorischen Sammlungen in Wien,* n.s., 72 (1976): 32–34, 43.

54. See Anthony Grafton, *Joseph Scaliger* (Oxford: Oxford University Press, 1993), vol. 2; Eric Iversen, *The Myth of Egypt and Its Hieroglyphs* (Princeton, NJ: Princeton University Press, 1993); Walter Mignolo, *The Darker Side of the Renaissance* (Ann Arbor: University of Michigan Press, 1995); Gutfleisch and Menzhausen, "'How a Kunstkammer Should Be Formed,'" 15. For the early-seventeenth-century tendency to derive the false gods of Mexico and Japan from Egypt, see MacCormack, "Limits of Understanding," 88.

55. Camille, *Gothic Idol,* 1–9; Avril Henry, *Biblia Pauperum* (Ithaca, NY: Cornell University Press, 1987), 56–59, 133–34.

56. R. J. W. Evans, *The Making of the Habsburg Monarchy, 1550–1700* (Oxford: Oxford University Press, 1979), 32–37; and Evans, *Rudolf II,* 85–92, 196–242.

57. Ingrid Rowland, *The Culture of the High Renaissance* (Cambridge: Cambridge University Press, 1998); Francis Haskell and Nicholas Penny, *Taste and the Antique* (New Haven, CT: Yale University Press, 1981), 14–15.

58. Frederick McGinness, *Right Thinking and Sacred Oratory in Counter-Reformation Rome* (Princeton, NJ: Princeton University Press, 1995), 152, 179–83; Anthony Grafton, "The Ancient City Restored," in *Rome Reborn* (Washington, DC: Library of Congress, 1993), 112–23; Domenico Fontana, *Della trasportatione* (Rome: Domenico Basa, 1590); Eric Iversen, *Obelisks in Exile* (Copenhagen: Gad, 1968), 1:29–44, 51–54, 62–64. I thank Anthony Grafton for the obelisk example.

59. Georg Eder, *Catechismus Catholicus qui Antea Quidem ex Decreto Concilii Tridentini, . . .* (Cologne: Gervinus Calenius, 1569), 305–8; Robert Bellarmine, *Außführliche Erklärung Christlicher Catholischer Lehr* (Augsburg: Christoph Mang, 1607), 141–49. For the catechism's popularity, see Brodrick, *Robert Bellarmine*, 153–55.

60. Bellarmine, *Außführliche Erklärung*, 11, 142

CHAPTER THIRTEEN

1. Sir James Frazer, *The Golden Bough: A Study in Magic and Religion* (New York: Macmillan, 1951), 1–2.

2. Ibid., 633.

3. Mircea Eliade, *The Sacred and the Profane: The Nature of Religion* (New York: Harcourt Brace Jovanovich, 1987), 20–22.

4. Rudolf Otto, *The Idea of the Holy* (Oxford: Oxford University Press, 1950), 126. This translation of Genesis is Otto's.

5. Victor Turner, "Liminality and Communitas," in his *The Ritual Process: Structure and Anti-structure* (Chicago: Aldine, 1969), 98.

6. See, e.g., Hartmann Schedel, *Liber Chronicarum* (Nuremberg, 1493).

7. For an overview, see Helen Rosenau, *Vision of the Temple: The Image of the Temple of Jerusalem in Judaism and Christianity* (London: Oresko Books, 1979), 91.

8. Juan Villalpando and Jerónimo de Prado, *In Ezechielem Explanationes et Apparatus Urbis ac Templi Hierosolymitani Commentariis et Imaginibus Illustratus* (Rome, 1596–1605).

9. Wolfgang Herrmann, "Unknown Designs for the 'Temple of Jerusalem' by Charles Perrault," in *Essays in the History of Architecture Presented to Rudolf Wittkower*, ed. Douglas Fraser, Howard Hibbard, and Milton J. Lewine (London: Phaidon, 1967), 143. On Villalpando's commentary more generally, see R. C. Taylor, "El padre Villalpando (1552–1608) y sus ideas estéticas," *Academia: Anales y Boletín de la Real Academia de Bellas Artes de San Fernando* 1, no. 4 (1952): 411–73; and Jaime Lara, "God's Good Taste: The Jesuit Aesthetics of Juan Bautista Villalpando in the Sixth and Tenth Centuries B.C.E.," in *The Jesuits: Cultures, Sciences, and the Arts (1540–1773)*, ed. John W. O'Malley et al. (Toronto: University of Toronto Press, 1999), 505–21.

10. Villalpando is often connected with the building of the Escorial; see Cornelia von der Osten Sacken, *San Lorenzo el Real de el Escorial: Studien zur Baugeschichte und Ikonologie* (Munich: Mäander Kunstverlag, 1979), 218ff.; Marie Tanner, *The Last Descendant of Aeneas: The Hapsburgs and the Mythic Image of the Emperor* (New Haven, CT: Yale University Press, 1993), 169–70; and René Taylor, "Architecture and Magic: Considerations on the *Idea* of the Escorial," in *Essays in the History of Architecture*, ed. Fraser, Hibbard, and Lewine, 97.

11. Louis Cappel, *Trisagio, sive Templi Hierosolymitani Triplex Delineatio*, in *Critici Sacri*, ed. John Pearson (London, 1660), 9:3746.

12. Benito Arias Montano, *Antiquitatum Iudaicarum Libri ix* (Leiden, 1593). Maurice Simon, trans., "Middoth," in *The Babylonian Talmud*, ed. I. Epstein (London: Soncino Press, 1948), vol. 6. His use of these Jewish sources got Montano in trouble with the papacy: see B. Rekers, *Benito Arias Montano (1527–1598)* (London: Warburg Institute, 1972), 49–50; *Arias Montano y su tiempo* (Mérida: Junta de Extremadura, 1998); *Philip II and the Escorial: Technology and the Representation of Architecture* (Providence, RI: Brown University, 1990), 55; Rosenau, *Vision*, 94.

13. José Corral Jam, "Arquitectura y canon, el proyecto de Villalpando para el Templo de Jerusalem," in Juan Bautista Villalpando, *El tratado de la arquitectura perfecta en la última visión del profeta Ezequiel* (Madrid: Colegio Oficial de Arquitectos de Madrid, 1990), 34.

14. On Lee, see Theodore Hornberger, "Samuel Lee (1625–1691), a Clerical Channel for the Flow of New Ideas to Seventeenth-Century New England," *Osiris* 1 (1936): 341–55.

15. [Samuel Lee], *Orbis Miraculum, or the Temple of Solomon* (London, 1659), b recto.

16. Ibid., a2 verso, b recto.

17. Ibid., a2 verso.

18. Ibid., b2 verso.

19. Ibid., 9.

20. Ibid., 11.

21. On the Temple in the ancient world, see Stephen Weitzman, *Surviving Sacrilege: Cultural Persistence in Jewish Antiquity* (Cambridge, MA: Harvard University Press, 2005), 79–95.

22. Lee, *Orbis*, 15.

23. Ibid., 166.

24. Ibid., 212, 191.

25. Ibid., 166.

26. Thomas Fuller, *Pisgah-sight of Palestine* (London, 1650), 371 (on the nails).

27. Lee, *Orbis*, 188.

28. Stephen G. Burnett, *From Christian Hebraism to Jewish Studies: Johannes Buxtorf (1564–1629) and Hebrew Learning in the Seventeenth Century* (Leiden: E. J. Brill, 1996), 240. On developments in sacramental and ceremonial thought in

England, see Bryan D. Spinks, *Sacraments, Ceremonies, and the Stuart Divines: Sacramental Theology and Liturgy in England and Scotland, 1603–1662* (Aldershot, UK: Ashgate, 2002).

29. Michel de Certeau, "Making History: Problems of Method and Problems of Meaning," in his *The Writing of History*, trans. Tom Conley (New York: Columbia University Press, 1988), 25.

30. Lee, *Orbis*, 95.

31. John Edwards, *A Compleat History or Survey of all the Dispensations and Methods of Religion* (London, 1699), 155. See also Richard Baxter, *Five Disputations of Church-Government and Worship* (London, 1659), especially book 5.

32. William Sherlock, *The Second Part of the Preservative against Popery* (London, 1688), 32.

33. Samuel Parker, *Reasons for Abrogating the Test, Imposed on All Members of Parliament Anno 1678 Octob. 30* (London, 1688), 120.

34. William Owtram, *De Sacrificiis Libri Duo* (London, 1677), 218.

35. Montano, *Antiquitatum Iudaicarum Libri ix*, 86 (Deut. 12:2–6).

36. Josephus, *Jewish Antiquities* (Cambridge, MA: Harvard University Press, 1950), 627 (book 8.102).

37. Ibid., 633 (8.114).

38. Ibid., 629–31 (8.108) (my italics).

39. Maimonides, *Guide of the Perplexed*, ed. Schlomo Pines (Chicago: University of Chicago Press, 1963), 526. On Maimonides in the seventeenth century, see Aaron L. Katchen, *Christian Hebraists and Dutch Rabbis: Seventeenth Century Apologetics and the Study of Maimonides' "Mishneh Torah"* (Cambridge, MA: Harvard University Press, 1984); and Jonathan Elukin, "Maimonides and the Rise and Fall of the Sabians: Explaining Mosaic Laws and the Limits of Scholarship," *Journal of the History of Ideas* 63 (October 2002): 619–37.

40. George Sandys, *Travailes: Containing a History of the Original and Present State of the Turkish Empire*, 6th ed. (London, 1658), 120.

41. See Gerschom Scholem, *Sabbetai Sevi: The Mystical Messiah* (Princeton, NJ: Princeton University Press, 1973).

42. John Lightfoot, *The Temple Service as It Stood in the Days of our Saviour*, in *The Works of the Reverand and Learned John Lightfoot* (London, 1684), 1:897–98. For a modern version, see Seth D. Kunin, *God's Place in the World: Sacred Space and Sacred Place in Judaism* (London: Cassell, 1998), 25.

43. John Weemse, *An Exposition of the Ceremonial Laws of Moses* (London, 1636), 33.

44. George Fox, *An Epistle to All Christians, Jews, and Gentiles* (London, 1682).

45. See Sacvan Bercovitch, *The American Jeremiad* (Madison: University of Wisconsin Press, 1978), 15; Ann Taves, *Fits, Trances, and Visions: Experiencing Religion and Explaining Experience from Wesley to James* (Princeton, NJ: Princeton University Press, 1999), 114. On the more general identification of Calvinists

with the people of Israel, see Christopher Hill, *The English Bible and the Seventeenth-Century Revolution* (London: Penguin, 1993); Charles H. Parker, "French Calvinists as the Children of Israel: An Old Testament Self-Consciousness in Jean Crespin's *Histoire des Martyrs* before the Wars of Religion," *Sixteenth-Century Journal* 24 (Summer 1993): 227–48.

46. "Westminster Confession of Faith," chap. 21, §6, in *The Creeds of Christendom*, ed. Philip Schaff (New York, 1931), 3:648.

47. Joseph Hill, *Dissertation Concerning the Antiquity of Temples* (London, 1696), unpaginated preface.

48. Benjamin J. Kaplan, "Fictions of Privacy: House Chapels and the Spatial Accommodation of Religious Dissent in Early Modern Europe," *American Historical Review* 107 (October 2002): 1036. See also Horton Davies, *Worship and Theology in England* (Princeton, NJ: Princeton University Press, 1975), 2:7.

49. Sherlock, *Preservative*, 35.

50. Julian Davies, *The Caroline Captivity of the Church: Charles I and the Remoulding of Anglicanism, 1625–1641* (Oxford: Oxford University Press, 1992), 73ff.; Nicholas Tyacke, *Anti-Calvinists: The Rise of English Arminianism, c. 1590–1640* (Oxford: Oxford University Press, 1987), 202 (quotation).

51. Henry Spelman, *De Non Temerandis Ecclesiis* (London, 1616), 36.

52. Ramie Targoff, *Common Prayer: The Language of Public Devotion in Early Modern England* (Chicago: University of Chicago Press, 2001), 5.

53. Ibid., 53.

54. Peter Lake, "The Impact of Early Modern Protestantism," *Journal of British Studies* 28 (July 1989): 302.

55. Lee, *Orbis*, 207.

56. Ibid.

57. Hobbes, *Leviathan, or the Matter, Forme, and Power of a Commonwealth Ecclesiastical and Civil* (London, 1651), 4.

58. Ibid., 360.

59. Ibid.

60. Richard Hooker, *Of the Laws of the Ecclesiastical Polity, Book V*, in *The Folger Library Edition of the Works of Richard Hooker*, ed. W. Speed Hill (Cambridge, MA: Harvard University Press, 1977), 2:61.

61. Spelman, *De Non Temerandis Ecclesiis*, 76–77.

62. Peter Lake, *Anglicans and Puritans: Presbyterianism and English Conformist Thought from Whitgift to Hooker* (London: Unwin Hyman, 1988), 229; see also Margot Todd, "The Godly and the Church: New Views of Protestantism in Early Modern Britain," *Journal of British Studies* 28 (October 1989): 423.

63. Sherlock, *Preservative*, 36.

64. Ibid., 32.

65. On Spencer, see Guy Stroumsa, "John Spencer and the Roots of Idolatry," *History of Religions* 41 (August 2001): 1–23; and Jan Assmann, *Moses the Egyp-*

tian: The Memory of Egypt in Western Monotheism (Cambridge, MA: Harvard University Press, 1997).

66. Assmann, *Moses*, 56. See also John Spencer, *De Legibus Hebraeorum Ritualibus* (Cambridge, 1685), 23.

67. Edwards, *Compleat History*, 241.

68. Assmann, *Moses*, 64.

69. Spencer, *De Legibus*, 675.

70. Ibid., 822.

71. Ibid., 825.

72. Ibid., 827.

73. Ibid., 829.

74. Ibid., 542.

75. Joseph Bingham, *Origines Ecclesiasticae: or, the Antiquities of the Christian Church* (London, 1711), 3.2:281, 287.

76. Fuller, *Pisgah-sight*, 403; Bingham, *Origines*, 3.2:277.

77. Isaac Newton, *On Solomon's Temple*, no date (Burndy Library at the Huntington, Babson Collection, MS 434).

78. Jonathan Z. Smith, *To Take Place: Toward Theory in Ritual* (Chicago: University of Chicago Press, 1987), 46.

79. *Encyclopaedia Britannica*, 9th ed. (New York, 1886), s.v. "temple."

80. William Robertson Smith, *Lectures on the Religion of the Semites*, 3d ed. (New York: Macmillan, 1927), 141.

81. See Turner, *Ritual Process*; Arnold van Gennep, *The Rites of Passage* (Chicago: University of Chicago Press, 1960).

82. James Frazer, *Folklore in the Old Testament* (London: Macmillan, 1919), 3:2.

83. Eliade, *Sacred and Profane*, 43; J. Z. Smith, *To Take Place*, 106, 58–59; Francis Schmidt, *How the Temple Thinks: Identity and Social Cohesion in Ancient Judaism* (Sheffield: Sheffield Academic Press, 2001), 85–86. More generally, see Michael Fishbane, "The Sacred Center: The Symbolic Structure of the Bible," in *Texts and Responses: Studies Presented to Nahum M. Glatzer* (Leiden: Brill, 1975), 6–27.

84. For more elaboration, see Jonathan Sheehan, "Sacred and Profane: Idolatry, Antiquarianism, and the Polemics of Distinction in the Seventeenth Century," *Past and Present* 192 (August 2006): 35–66.

85. Assmann, *Moses*, especially 1–6.

86. On religion as "an objective systematic entity," see Wilfred Cantwell Smith, *The Meaning and End of Religion* (New York: Macmillan, 1963), 51. Also see Talal Asad, *Genealogies of Religion: Discipline and Reasons of Power in Christianity and Islam* (Baltimore, MD: Johns Hopkins University Press, 1993). For a history of the term *religio*, see Ernst Feil, *Religio: Die Geschichte eines neuzeitliche Grundbegriffs*, 3 vols. (Göttingen: Vandenhoeck und Ruprecht, 1986).

87. Peter Harrison, *"Religion" and the Religions in the English Enlightenment* (Cambridge: Cambridge University Press, 1990), 146.

CHAPTER FOURTEEN

1. To be clear: I use the term "science" here as synonymous with *Wissenschaft*—that is, a body of systematic knowledge that can be taught.

2. Contemporary accounts include Johann Stephan Pütter, *Versuch einer academischen Gelehrten-Geschichte der Georg-Augustus Universität zu Göttingen*, 2 vols. (Göttingen, 1765 and 1788); and Ernst Brandes, *Ueber den gegenwärtigen Zustand der Universität Göttingen* (Göttingen, 1802). More general histories of the German universities include Christoph Meiners, *Geschichte der Entstehung und Entwicklung der hohen Schulen unsers Erdtheils*, 4 vols. (Göttingen, 1802–5); and Johann David Michaelis, *Raisonnement über die protestantischen Universitäten in Deutschland*, 4 vols. (Frankfurt am Main, 1768–76). Nineteenth- and early-twentieth-century accounts include Emil Franz Rössler, ed., *Die Gründung der Universität Göttingen: Entwürfe, Berichte und Briefe der Zeitgenossen* (Göttingen: Vandenhoeck und Ruprecht, 1855); Friedrich Paulsen, *Das deutsche Bildungswesen in seiner geschichtlichen Entwicklung* (Leipzig: Teubner, 1906); Friedrich Paulsen, *Geschichte des Gelehrten Unterrichts auf den deutschen Schulen und Universitäten*, 2d ed., 2 vols. (Leipzig: Veit, 1896); and Götz von Selle, *Die Georg-August-Universität zu Göttingen, 1737–1937* (Göttingen: Vandenhoeck und Ruprecht, 1937). Some more recent work of note includes R. Steven Turner, "The Prussian Universities and the Research Imperative, 1806 to 1848," (PhD diss., Princeton University, 1973); Charles E. McClelland, *State, Society and University in Germany, 1700–1914* (Cambridge: Cambridge University Press, 1980); and Luigi Marino, *Praeceptores Germaniae: Göttingen, 1770–1820* (Göttingen: Vandenhoeck und Ruprecht, 1995). Göttingen also plays a central role in William Clark's *Academic Charisma and the Origins of the Research University* (Chicago: University of Chicago Press, 2006), which sees the origins of the research university largely in the bureaucratic machinations of eighteenth-century administrators like Münchhausen.

3. For examples of these themes, see Theodore Ziolkowski, *German Romanticism and Its Institutions* (Princeton, NJ: Princeton University Press, 1990); and Donata Brianta, "Education and Training in the Mining Industry, 1750–1860: European Models and the Italian Case," *Annals of Science* 57 (2000): 267–300.

4. Michaelis, *Raisonnement*, 1:1–5.

5. Cf. Selle, *Die Georg-August-Universität*; and Marino, *Praeceptores Germaniae*.

6. Oddly, the only full-length biography of Münchhausen, Walter Buff's *Gerlach Adolph Freiherr von Münchhausen als Gründer der Universität Göttingen* (Göttingen: Kaestner, 1937), did much to reinforce the idea that Münchhausen (and, by extension, the institution he founded) paved the way for the University of Berlin.

7. There was concern from the very beginning about adequate housing, reasonable prices, and suitable amenities. Most of this fell under the domain of "good police." Emil Rössler's published collection of sources, for example, illustrates Hanover's campaign to establish good police in Göttingen. Münchhausen, espe-

cially, dedicated himself to this effort. See Rössler, *Die Gründung der Universität Göttingen,* 66–74, 385–91, 415.

8. Münchhausen even wanted to attract wealthy English students to Göttingen. "My intention," he wrote Privy Secretary Hattorf in London, "is to draw English lords to Göttingen, where they will fare just as well as in Holland, where they already spend their money abundantly." See Selle, *Die Georg-August-Universität,* 49–53.

9. These officially sanctioned advertisements included the three-part *Zeit- und Geschicht-Beschreibung der Stadt Göttingen* (Hanover and Göttingen, 1734); and Johann Christian Claproth's 1748 work, *Der gegenwärtige Zustand der Göttingischen Universität* (Göttingen, 1748). Münchhausen personally directed Claproth, and then Justi, to advertise the wonders of Göttingen. See Johann von Justi to Johann David Michaelis, 13 April 1756 (Niedersächsische Staats- und Universitätsbibliothek Göttingen, Handschriften Abteilung, 2 Cod. MS Michael. 324); Wieland Sachse, *Göttingen im 18. und 19. Jahrhundert: Zur Bevölkerungs- und Sozialstruktur einer deutschen Universitätsstadt* (Göttingen: Vandenhoeck und Ruprecht, 1987), 57–58.

10. *Zeit- und Geschicht-Beschreibung,* 1: preface.

11. See [Claproth], *Der gegenwärtige Zustand der Göttingischen Universität,* 8–9.

12. A 13 April 1756 letter from Justi to Johann David Michaelis (Handschriften Abteilung Göttingen, 2 Cod. MS Michael. 324) indicates that Münchhausen personally directed Claproth, and after him Chief Police Commissioner Justi, to advertise the wonders of Göttingen (fol. 490). Justi suggested to Michaelis that he would begin drafting the work during his upcoming vacation at the baths. I was unable, however, to locate any record of it.

13. Johann Georg Bärens, "Kurtze Nachricht von Göttingen," in "Ein Bericht über Göttingen, Stadt und Universität, aus dem Jahre 1754," ed. Ferdinand Frensdorff, *Jahrbuch des Geschichtvereins für Göttingen und Umgebung* 1 (1908): 55–59.

14. Ibid., 58–60. Between 1732 and 1807 Hanover steadily insinuated itself into local politics by assuming control over the appointments of certain mayors and town councilors. See Heinz Mohnhaupt, *Die Göttinger Ratsverfassung vom 16. bis 19. Jahrhundert* (Göttingen: Vandenhoeck und Ruprecht, 1965), 108–17.

15. *Camelott,* a double-woven woolen fabric, was made with carded wool and a chain of worsted yarn. See Johann von Justi, *Vollständige Abhandlung von den Manufacturen und Fabriken,* 2 vols. (Copenhagen, 1758), 2:50–51.

16. For comparison, consider that annual maintenance costs for the entire university were 16,600 thalers in 1733. See Rössler, *Die Gründung der Universität Göttingen,* 57.

17. Diether Koch, *Das Göttinger Honoratiorentum vom 17. bis zur Mitte des 19. Jahrhunderts* (Göttingen: Vandenhoeck und Ruprecht, 1958), 119.

18. Frensdorff found this assertion so jarring that he made a special point to reject it out of hand. See his introduction to Bärens, "Kurtze Nachricht," 53.

19. Koch, *Das Göttinger Honoratiorentum,* 61n7. The memo is still in the Göttingen City Archive today, but under a different call number than during Koch's time; i.e., Stadtarchiv Göttingen AA, Industrie, Fabrik- und Manufaktursachen, no. 15 (hereafter cited as "1724 Memorandum").

20. 1724 Memorandum, §5. Norbert Winnige's detailed study of Göttingen has demonstrated that the town had begun to experience an economic recovery by 1720. Nevertheless, as the 1724 memo indicates, the town still appeared stagnant and run down to contemporary observers. See Norbert Winnige, *Krise und Aufschwung einer frühneuzeitlichen Stadt: Göttingen, 1648–1756* (Hanover: Hahnsche Buchhandlung, 1996), 406. Göttingen's construction boom, which is a better index of the town's changing appearance, did not really begin in earnest until the early 1730s, during the first years of the university. See Sabine Kastner, "Wohnen und Bauen in Göttingen," in *Studien zur Sozialgeschichte einer Stadt,* ed. Hermann Wellenreuther (Göttingen: Vandenhoeck und Ruprecht, 1988), 182–205; Winnige, *Krise und Aufschwung,* 258.

21. 1724 Memorandum, §1. See Winnige, *Krise und Aufschwung,* 126–33.

22. [Böll], *Das Universitätswesen in Briefen* (n.p., 1782), 4. Quoted in Martin Gierl, "Vom gelehrten Streiter in der 'Trutzburg der Wahrheit' zum höflichen Konkurrenten der Aufklärungsfabrik" (unpublished).

23. For the best recent survey of central European mining academies, see Brianta, "Education and Training in the Mining Industry," 267–83. Still worthwhile are the excellent accounts out of Freiberg: Hans Baumgärtel, "Bergbau und Absolutismus: Der sächsische Bergbau in der zweiten Hälfte des 18. Jahrhunderts und Massnahmen zu seiner Verbesserung nach dem Siebenjährigen Kriege," *Freiberger Forschungshefte* D44 (1963): 1–192; Hans Baumgärtel, "Vom Bergbüchlein zur Bergakademie," *Freiberger Forschungshefte* D50 (1965): 1–169; and Otfried Wagenbreth, ed., *Die Technische Universität Freiberg und ihre Geschichte* (Leipzig and Stuttgart: Deutscher Verlag für Grundstoffindustrie, 1994).

24. Cf. Clark, *Academic Charisma,* 239–96; and William Clark, "On the Ministerial Archive of Academic Acts," *Science in Context* 9 (1996): 421–86.

25. On cameralism as an eighteenth-century profession, see André Wakefield, "Police Chemistry," *Science in Context* 13, no. 2 (2000): 231–67.

26. On Göttingen in this regard, see especially Paulsen, *Das deutsche Bildungswesen;* Selle, *Die Georg-August-Universität;* and Marino, *Praeceptores Germaniae.* For the Bergakademie in Freiberg as the prototypical German technocratic engineering school, see Brianta's extensive overview: "Education and Training in the Mining Industry," 267–300.

27. For more on this connection, see Wolfhard Weber, *Innovationen im Frühindustriellen deutschen Bergbau und Hüttenwesen—Friedrich Anton von Heynitz* (Göttingen: Vandenhoeck und Ruprecht, 1976), 152–54.

28. Johann Heinrich von Justi, *Grundsätze der Policey-Wissenschaft* (Göttingen, 1756), 98.

29. Daniel Gottfried Schreber, "Entwurf von einer zum Nutzen eines Staats zu errichtenden Academie der öconomischen Wissenschaften," *Sammlung verschiedener Schriften* 10 (1763): 417–36; Wilhelm Stieda, *Die Nationalökonomie als Universitätswissenschaft* (Leipzig: Teubner, 1906), 55ff.; Weber, *Heynitz*, 154; Keith Tribe, *Governing Economy: The Reformation of German Economic Discourse, 1750–1840* (Cambridge: Cambridge University Press, 1988), 91–94.

30. Horst Schlechte, ed., *Die Staatsreform in Kursachsen, 1762–1763: Quellen zum Kursächsischen Retablissement nach dem Siebenjährigen Kriege* (Berlin: Rütten und Loening, 1958), 218–20.

31. This is what Horst Schlechte suggests. See ibid., 72. On the other hand, Wolfhard Weber (*Heynitz*, 117) finds no direct evidence for the claim.

32. On Heynitz's appointment and activities in the Saxon civil service, see Weber, *Heynitz*, 116–67; Baumgärtel, "Bergbau und Absolutismus," 67–99; and Schlechte, *Staatsreform*, 72–75.

33. Weber, *Heynitz*, 116–19.

34. Sächsisches Hauptstaatsarchiv Dresden (StADresden), Loc. 1327, 1–7.

35. Ibid., 1–7; Weber, *Heynitz*, 129; Baumgärtel, "Bergbau und Absolutismus," 71.

36. On this see Weber, *Heynitz*, 120–21.

37. StADresden, Loc. 1327, 21–24. For another interpretation of this memo's meaning and significance, see Weber, *Heynitz*, 157.

38. StADresden, Loc. 1327, 21–22.

39. Ibid., 22.

40. Ibid., 22–23.

41. Ibid. (my italics).

42. Ibid., 24.

43. Heynitz to Elector Friedrich August, 27 January 1769 (StADresden, Loc. 36216, 2).

44. StADresden, Loc. 1327, 21–24.

45. Weber, *Heynitz*, 156–57.

46. Both would later become teachers at the mining academy.

47. StADresden, Loc. 514, 1–6.

48. For this view, see Baumgärtel, "Vom Bergbüchlein zur Bergakademie," 142ff. Weber takes a different view (see *Heynitz*, 156).

49. Though many of the Bergakademie's students also attended university, the mining academy began to usurp some of the functions that university education had once provided.

50. See Christoph Meinel, "Reine und angewandte Chemie," *Berichte zur Wissenschaftsgeschichte* 8 (1985): 25–45.

51. Universitätsarchiv der TU Bergakademie, Freiberg, Akte OBA 7917, vol. 1, 231–33.

CONTRIBUTORS

Arianne Baggerman directs the research program "Controlling Time and Shaping the Self: The Rise of Autobiographical Writing since 1750" at Erasmus Universiteit Rotterdam. Her books include *Een drukkend gewicht* (Rodopi, 1993), about a seventeenth-century polygraph, Simon de Vries. Her dissertation, *Een lot uit de loterij* (SDU, 2000), about publishing and book trade in the Netherlands in the eighteenth and nineteenth centuries, will be published in English in 2008 by Brill. With Rudolf Dekker she wrote a study about culture, politics, and society in the late eighteenth century (*Kind van de toekomst,* Wereldbibliotheek, 2005) which will be published in English in 2008 by Brill in the series Egodocuments and History, of which she is an editor. She is writing a book about the links between self-representation, publishing, and commerce in the nineteenth century.

Scott Black is associate professor of English at the University of Utah and author of *Of Essays and Reading in Early Modern Britain* (Palgrave Macmillan, 2006). He is currently working on a project on Heliodorus in English and the art of romance in the age of the novel.

Rudolf Dekker teaches history at Erasmus Universiteit Rotterdam. His books include *Humour and Dutch Culture of the Golden* Age (Palgrave, 2001) and *Ego-documents and History* (Verloren, 2002, editor). With Arianne Baggerman he wrote a study about culture, politics, and society in the late eighteenth century (*Kind van de toekomst,* Wereldbibliotheek, 2005) which will be published in English in 2008 by Brill. He is now writing a history of the Netherlands, 1500–2000, and preparing a study on Dutch and English culture and society in the late seventeenth century based on the diaries of Constantijn Huygens Jr.

Lori Anne Ferrell is professor of early modern history and literature at Claremont Graduate University. She is the author of *Government by Polemic* (Stanford,

1998), as well as the coeditor (with David Cressy) of *Religion and Society in Early Modern England* (2nd ed., Routledge, 2005) and (with Peter McCullough) *The English Sermon Revised* (Manchester, 1999). Her latest work, *The Bible and the People: A Cultural History of Christianity and the English Language Bible,* will be published by Yale in 2008.

Ole Peter Grell is reader in history at the Open University in UK. He has published *Calvinist Exiles in Tudor and Stuart England* (Ashgate, 1996) and *The Four Horsemen of the Apocalypse: Religion, War, Famine, and Death in Reformation Europe* (with A. Cunningham) (Cambridge, 2000–2002). He has recently finished coediting a series of four volumes, Health Care and Poor Relief in Europe 1500–1900, published between 1997 and 2006 (Routledge and Ashgate). His most recent work is a coedited volume, *Medicine and Religion in Enlightenment Europe* (Ashgate, 2007).

Carina L. Johnson is assistant professor of history at Pitzer College. Her current research examines political authority, religious orthodoxy, and the reception of Mexican and Ottoman cultures in the sixteenth-century Habsburg Empire.

Chandra Mukerji is professor of communication and science studies at the University of California, San Diego. She works on the role of the built environment in social life, and is author of *Territorial Ambitions and the Gardens of Versailles* (Cambridge, 1997); *Rethinking Popular Culture,* coedited with Michael Schudson (California, 1991); *A Fragile Power: Science and the State* (Princeton, 1989); and *From Graven Images: Patterns of Modern Materialism* (Columbia, 1983). She is currently writing a book about the Canal du Midi, exploring the role of social intelligence in the formation of the French state.

Herman Pleij is professor of medieval and early modern Dutch literature at the University of Amsterdam. He is the author of *Dreaming of Cockaigne: Medieval Fantasies of the Perfect Life* (Columbia, 2001), and *Colors Demonic and Divine: Shades of Meaning in the Middle Ages and After* (Columbia, 2004). He is currently working on a book about Anna Bijns, an Antwerp writer and schoolteacher during the sixteenth century.

Londa Schiebinger is the John L. Hinds Professor of History of Science and the Barbara D. Finberg Director of the Michelle R. Clayman Institute for Gender Research at Stanford University. Her books include *The Mind Has No Sex? Women in the Origins of Modern Science* (Harvard, 1989); *Nature's Body: Gender in the Making of Modern Science* (Beacon, 1993/Rutgers, 2004); *Has Feminism Changed Science?* (Harvard, 1999); and *Plants and Empire: Colonial Bioprospecting in the Atlantic World* (Harvard, 2004). She has edited *Feminism and the Body* (Oxford, 2000) and *Gendered Innovations in Science and Engineering* (Stanford, 2008), and coedited, with Elizabeth Lunbeck and Angela Creager, *Feminism in Twentieth-Century Science, Technology, and Medicine* (Chicago, 2001); with Claudia Swan, *Colonial Botany: Science, Commerce, and Politics in the Early Modern World* (Pennsylvania, 2005); and with Robert N. Proctor, *Agnotology: The Making and Unmaking of Ignorance* (Stanford, 2008). Schiebinger is the recipient of numerous prizes and awards, including the Alexander von Humboldt Research Prize.

BENJAMIN SCHMIDT is professor of history at the University of Washington, Seattle, and the author of *Innocence Abroad: The Dutch Imagination and the New World, 1570–1670* (Cambridge, 2001), *The Discovery of Guiana by Sir Walter Ralegh* (Bedford/St. Martins, 2007), and coeditor of *Going Dutch: The Dutch Presence in America, 1609–2009* (Brill, 2008). He is completing a book on European geography, exoticism, and globalism around 1700.

LINDA SEIDEL is Hanna Holborn Gray Professor Emerita at the University of Chicago where she taught art history for twenty-seven years. She is the author of three books on French Romanesque sculpture as well as *Jan van Eyck's Arnolfini Portrait: Studies of an Icon* (Cambridge, 1993). Most recently, she edited and provided the introduction to Meyer Schapiro, *Romanesque Architectural Sculpture: The Charles Eliot Norton Lectures* (Chicago, 2006). Her current work reprises longstanding interests in Romanesque art and Roman remains.

JONATHAN SHEEHAN is associate professor of history at the University of California, Berkeley. He is the author of *The Enlightenment Bible: Translation, Scholarship, Culture* (Princeton, 2005), and a number of articles on early modern religious culture that have appeared in *Past & Present*, the *American Historical Review*, and the *Journal of the History of Ideas*.

PAMELA H. SMITH is professor of history at Columbia University and the author of *The Business of Alchemy: Science and Culture in the Holy Roman Empire* (Princeton, 1994) and *The Body of the Artisan: Art and Experience in the Scientific Revolution* (Chicago, 2004). She coedited with Paula Findlen *Merchants and Marvels: Commerce, Science, and Art in Early Modern Europe* (Routledge, 2002). In her present research, she attempts to reconstruct the vernacular knowledge of early modern European metalworkers from a variety of disciplinary perspectives.

CLAUDIA SWAN is associate professor of art history at Northwestern University and the author of *Art, Science, and Witchcraft in Early Modern Holland: Jacques de Gheyn II (1565–1629)* (Cambridge, 2005) and *The Clutius Botanical Watercolors* (Harry N. Abrams, 1998). She is currently at work on a book entitled *The Aesthetics of Possession: Art, Science, and Collecting in Early Modern Holland.*

André Wakefield is assistant professor of history at Pitzer College in Claremont, California. His book, *The Disordered Police State,* and his edited translation (with Claudine Cohen) of G. W. Leibniz's *Protogaea* are both forthcoming from the University of Chicago Press.

Simon Werrett is assistant professor in the Department of History at the University of Washington, Seattle. He recently completed a monograph, "Philosophical Fireworks: Science, Art, and Pyrotechnics in Early Modern Europe," currently under review. He has published on the history of science and spectacle in Europe and Russia, and is presently researching scientific and artistic performances on early-nineteenth-century Russian voyages of exploration.

INDEX